METALS AND ALLOYS DICTIONARY

by

M. MERLUB-SOBEL, PH.D.

Consulting Chemical and Metallurgical Engineer; Director of Research, Virginia-Carolina Chemical Corp., Addressograph Multigraph Corp.; Chemical Engineer, Beryllium Corp. of America; Research Engineer, Kemet Labs. Co.

1944

CHEMICAL PUBLISHING CO., INC.

BROOKLYN N. Y.

Metals and Alloys Dictionary

ISBN: 978-0-8206-0031-4

Chemical Publishing Company:
www.chemical-publishing.com
www.chemicalpublishing.net

First Edition:

© Chemical Publishing Company, Inc. - New York 1944

Second Impression:

Chemical Publishing Company, Inc. - 2011

Printed in the United States of America

PREFACE

Any new book, certainly one dealing with technical matters, ought to submit something akin to a "certificate of necessity." While "necessity" is highly subjective, and capable of many shades of meaning, the plain fact of this case is that no concise dictionary of metallurgical terms, adapted to American needs, has hitherto been available. And so, if the need but be granted, the author may be permitted to hope that this volume, in greater or lesser measure, fills the void.

Not that perfection is expected—the ideal may not even be prayed for in this all-too-imperfect world. Indeed, I realize that this book must inevitably contain at least a normal quota of error—sins of omission and commission both. For such, ignorance and/or carelessness must take full responsibility; in the interest of future editions, I shall be grateful for such corrections as may appear justified to the reader.

No defense can be made, nor is any intended, for positive errors of commission. In a volume of this type, however, omissions may frequently reflect differing opinions as to what ought to be included. Probably no two students of metallurgy would quite agree on the scope of material to be presented. The problem here has been further complicated by a desire to serve, in this volume, lay workers at the bench and behind the torch no less than technologists with university training. As a consequence of this desire, many words and phrases in sciences peripheral to metallurgy have been included; setting a proper course has frequently been most difficult, threatened at once by the Scylla of incompleteness and the Charybdis of undue bulk. In its final form, this dictionary represents one man's effort at intelligent compromise.

By its very nature, a dictionary can present little of new content. So many sources of information have found reflection in this volume that I am prevented from adequately acknowledging them; wherever possible, in fact, I have checked one source against another. Such metallurgical "demi-bibles" as *Metals Handbook, Engineering Alloys* (by Woldman and Dornblatt), and *The Making, Shaping and Treating of Steel* (by Camp and Francis) have served extensively as guides, and the reader is urged to refer to them and their technical congeners for amplification wherever this dictionary proves inadequate. The remarkably inclusive *Engineering Alloys* is particularly recommended for its listing and description of thousands of alloys which have not seemed important enough for description in this dictionary.

There is only one exception to the generalization that nothing new has been presented. In the realm of the rare metals, where some of the published data are fragmentary, and some erroneous, I have drawn directly on my own specialized experience, and have included some amount of data hitherto unblackened by printers' ink.

Generous unselfish help has been my good fortune. In particular, I am deeply in the debt of Mr. Harry B. Pulsifer and Prof. George L. Kehl—neither needs introduction to American workers in metallurgy—for reviewing many of the definitions of this book; many an error has been eliminated by their far-greater-than-mine knowledge. My co-workers of the Virginia-Carolina Chemical Corporation, Dr. James C. Alexander and Mr. Jack J. Press, have similarly been generous with their aid and advice, as well as in reading of proof, and I would like here to express my sincere thanks to them.

The conversion of an outrageously illegible manuscript into neatly typewritten copy, ready for the printer, was the years-long Augean task which Miss Mary Korpita genially undertook; in that highest of heavens reserved for the hewers of wood and the drawers of water, a special seat of glory, please, for a gallant helper. To Miss Gertrude Schutze, a bow for similar assistance. It is a pleasure, also, to acknowledge the clerical help of Mrs. Mildred Maeth.

To the reader, be he professional or novice, I speak the hope that this volume proves of value. If it does, even in small measure, I shall feel amply repaid in the only coin legitimate to the bibliogenetic realm.

M. M.-S.

EXPLANATORY NOTE

DEFINITIONS ARE LISTED on a strictly alphabetical basis. Each entry is treated as a unit, i.e., as though it were one single word. The following sequence is typical:

Air Cooling
Aircraft Metals
Air Drain
Aired Bars
Air Furnace

Connective words, such as "and", "in", etc., are not considered definite parts of an entry and are consequently not treated as part of the alphabetical sequence.

Each entry is in *bold-face* type, and any entry referred to in the course of a definition is similarly in *bold-face* type, frequently followed by "(see)". The latter is to be understood, of course, as indicating that the preceding word or words in *bold-face* type will be found, separately defined, in the proper alphabetical sequence.

Å (or Å.U.). Abbreviation for Angstrom unit, 1/100,000,000 cm, used for submicroscopic measurement (e.g. of wave-length or dimensions of molecules).

Abel's Reagent. Metallographic etching agent, consisting of 10% chromium trioxide (chromic acid) in water, used in the study of steels.

AB Microhardness Test. See **Microhardness.**

Abnormality. See **Abnormal Steel.**

Abnormal Steel. Steel in which the ferrite and cementite form coagulated units instead of normal laminated pearlite. The exact cause of this abnormality, known as "Ferrite Divorcement," is uncertain.

Abnormal Structure. See **Abnormal Steel.**

Abrasion Hardness. Hardness (see) of a material, as judged by resistance to abrasion.

Abrasion Mark. Surface scratch, caused by friction between two sheets of metal, such as will occur when a pile of sheets is improperly bound during shipment.

Abrasion Test. Test, for determining resistance of a metal to abrasive influences, usually conducted so as to duplicate, as nearly as possible, the expected conditions of service.

Abrasive Hardness. See **Abrasion Hardness.**

Absolute Temperature. Any temperature scale, the values of which are based on a zero point at the theoretical absolute minimum temperature, 273.18 degrees below zero Centigrade, or 459.72 degrees below zero Fahrenheit. The absolute Centigrade scale is also known as the Kelvin scale.

Absorption. Ability of a porous solid material to hold relatively large quantities of gases or liquids.

Absorption Spectrum. Spectrum of radiation coming through a selectively absorbing medium.

Abyssinian Gold. Copper-base alloy, with 8–10% zinc, fractional percentages of other elements, and approximately 1% gold, plated or rolled on the surface.

Ac. Chemical symbol for **Actinium** (see).

Accumulator. See **Storage Battery.**

Acheson Graphite. Artificial graphite, made from coke by electric furnace heating.

Acicular. Spine or needle-shaped; crystal whose length is three or more times its width.

Acid. Compound which yields hydrogen ions when dissolved in water or other solvents. Most metals are soluble in stronger acids (those containing a high concentration of hydrogen ions). Noble metals, such as gold or platinum, are soluble only in "aqua regia", a mixture of nitric and hydrochloric acids. In metallurgical technology, materials are considered acidic if they are susceptible to union with basic substances, such as lime and magnesia. Silica, SiO_2, is the most common acidic material of this type.

Acid Bessemer Steel. See **Bessemer Steel.**

Acid Bottom. Bottom of furnace, made of acid-reacting materials such as silica.

Acid Brittleness. Brittleness resulting from pickling steel in acid; hydrogen, formed by the interaction between iron and acid, is partially absorbed by the metal, causing acid brittleness.

Acid Bronze. Copper-base alloy, containing 74–78% copper, 8–10% tin, 2–17% lead, and 0–2% zinc; resistant to mildly acid conditions; used for chemical equipment.

Acid Electric Process. Process of steel manufacture or melting, in which the metal is melted in an electric arc furnace, using an acid hearth and slag.

Acid Flux. Metallurgically acid material (usually some form of silica) used as a **Flux** (see).

Acid-Forming Element. Element, compounds of which tend, in greater or lesser degree, to yield **Anions** (see) in water solution; non-metals are distinctly acid-forming, metalloids less so, and metals are base-forming.

Acid Hearth. See **Open-Hearth Process.**

Acid Lining. Lining of furnace made of acid-reacting materials such as silica.

Acid Open-Hearth Process. See **Open-Hearth Process.**
Acid Open-Hearth Steel. Steel made by the acid **Open-Hearth Process** (see).
Acid Pig (Iron). See **Bessemer Pig (Iron).**
Acid Process. (a) See **Open-Hearth Process.** (b) See **Flux Process.**
Acid Refractory. See **Refractory.**
Acid Slag. Slag in which the ratio of acidic materials (silica and alumina) exceeds that of the basic materials (lime and magnesia).
Acid Steel. Steel made or melted in contact with an acid-reacting furnace lining.
Acieral. Aluminum-base casting alloy, with about 6% copper and fractional percentages of nickel, zinc, and silicon.
Acker Process. Process of electrolyzing sodium metal from fused sodium chloride, using molten lead as cathode; the resulting sodium-lead alloy is decomposed by water to yield pure sodium hydroxide and lead which is recirculated.
Acme (Stainless) Steel. Series of **Stainless Steels** (see) (Acme Steel Co.).
Acorn (Babbitt). Series of tin-base bearing metals, with about 8% antimony and 4% copper (A. W. Cadman Mfg. Co.).
Actinic Rays. Rays affecting photographic film.
Actinium (chemical symbol Ac). Radioactive element, No. 89 of the periodic system; not yet isolated. Theoretically, its metallic properties are similar to those of **Yttrium** (see) and **Lanthanum** (see), and its theoretical valence is likewise three. Radioactivity renders all actinium compounds unstable, causing their constant decomposition into less radioactive substances.
Activated Bath. Cyaniding (see) bath in which alkali cyanide or alkali cyanide-chloride mixtures are used, with activating addition materials present, such as calcium cyanamide, etc.
Active Hydrogen. Hydrogen activated e.g. by adsorption on a metal surface.
AD Aluminum. Copper-base alloy, used in tube form, consisting of approximately 15% zinc, 2% aluminum, and 1% tin (Chase Brass & Copper Co.)
Adamantine. (a) Crystalline **Boron** (see). (b) Low-alloy, heat-treated steel, with about 0.7% chromium and manganese, used for making hard steel balls (Babcock & Wilcox Co.).
Adamite. Low-alloy iron used for wear-resistant objects, such as dies, containing 0.50–1.50% chromium, nickel up to 0.75%, silicon up to 2%, and 1.2–3.0% carbon (Mackintosh-Hemphill Co.).
Adamite Rolls. Rolls made of a mixture of steel, pig iron, and some alloying elements.
Addition Agent. Material added to an electroplating bath to improve operation of the bath, brighten the deposited metal, etc.
Addition Element. Any element added, in relatively small quantity, to an alloy for scavenging or modifying the properties of the alloy.
Admiralty (Alloy). See **Admiralty Brass.**
Admiralty Brass. Copper-base alloy containing approximately 29% zinc and 1% tin, originally developed in Great Britain for condenser tubes; one of the most commonly used nonferrous tube materials, available in rod, wire, and sheet form, it lends itself well to cold-working, except machining, but is generally unsuitable for hot-working.
Admiralty Bronze. Copper-base alloy, containing approximately 10% tin and 2% zinc; used for castings, valves, gears, etc.
Admiralty Gun Metal. See **Gun Metal.**
Admiralty Metal. See **Admiralty Brass.**
Adnic. Copper-base alloy, containing approximately 29% nickel and 1% tin; because of strength and corrosion resistance, used in condenser tubing and for similar purposes (Scovill Mfg. Co.).
Adsorption. Ability of a solid surface to attach to itself firmly gases or liquids with which it comes in contact. Adsorption power varies widely with different types of surfaces.
Advance Metal. Copper-nickel alloy of relatively high, constant resistance, consisting approximately of 55% cop-

per and 45% nickel (Driver-Harris Co.). See also **Constantan.**

Aerocarb (Carburizing Compounds). Series of salt mixtures used for case-hardening and other steel treatment (American Cyanamide & Chemical Corp.).

Aerocase (Case Hardening Compound). Series of case-hardening salt mixtures containing sodium and calcium chlorides and calcium cyanide (American Cyanamide & Chemical Corp.).

Aerolite. (a) Aluminum-base alloy containing 1–12% copper, fractional percentages of magnesium and silicon, and, optionally, a low percentage of manganese or iron. Used in automotive and aircraft industries. (b) Metallic or mineral material falling to earth out of interplanetary space.

Aero Metal. Aluminum-zinc casting metal containing approximately 67% aluminum, 28% zinc, 4% copper, and fractional percentages of iron and silicon (Garford Engineering Co.).

Aerosiderite. Aerolite [see **Aerolite** (b)] composed principally of iron.

Aerugo. Green coating on old copper or copper-base articles, formed by oxidation.

Aeterna. Modified yellow **Brass** (see) of about 40% zinc content; the modifying elements, present in fractional percentages, are tin, lead, and manganese.

Afterblow. Maintaining the "blow" (see **Bessemer Steel**) in the basic Bessemer process, so as to remove phosphorus, after carbon has been eliminated.

After-Flow. Plastic flow in solids, continuing after external forces have ceased to act.

Ag. Chemical symbol for **Silver** (see).

Agate Ware. Enameled iron or steel, particularly as used for household utensils. See **Vitreous Enamel.**

Agathon (Steel). See **U. M. A. Steel.**

Age-Hardening. Spontaneous hardening of some alloys, such as Duralumin and beryllium-copper, occurring at ordinary or relatively low temperatures; presumably caused by the precipitation, in sub-microscopic form, of a component held in supersaturated solution. The trend of age-hardening is greatly modified by previous heat treatment.

Age Hardness. See **Age-Hardening.**

Age Strengthening Range. See **Blue Brittle Range.**

Aggregate. Mixture of physically separable substances.

Aggregates. See **Crystal Aggregates.**

Aging. Spontaneous change in the physical properties of some metals, which occurs on standing, at relatively low temperatures. Frequently, synonymous with the term **Age-Hardening** (see).

Aging Steel. Steel which shows **Aging** (see) characteristics at room temperature or below 150°F, such as increased hardness, strength, and yield point.

Aging Test. Test of physical properties showing the characteristics of a material during **Aging** (see).

Agitator Pickling. Pickling (see) operation in which either the acid solution or the material being treated is maintained in motion, during the operation.

Agraphitic Carbon. See **Combined Carbon.**

Agricultural Ply Steel. Composite consisting of two or more layers of steel of different composition (hard steel on the outside, most commonly, for wear resistance). Generally made by casting one grade of steel on another.

Aich Metal. Yellow brass (about 40% zinc) containing approximately 1–2% iron.

Air-Acetylene Welding. See **Gas Welding.**

Air Blasting. Cleaning surface of metal by air blast, with sand or finely divided steel used as abrasive.

Air Bridge. Bridge, in furnace, admitting heated air to gases passing over it, for improving combustion efficiency.

Air Condenser. Surface **Condenser** (see) utilizing air as a condensing medium.

Air Cooling. Cooling of heated metal, intermediate in rapidity between slow **Furnace Cooling** (see) and **Quenching** (see) in which the metal is permitted to stand in the open air.

Aircraft Metals. Aluminum-base and magnesium-base alloys; less strictly, also alloy steels used for seamless tubing in the fabrication of aircraft.

Air Drain. Passage in **Mold** (see) permitting escape of gases when metal is being cast into it.

Aired Bars. Blister Steel (see) which has been exposed to air during the cooling process, resulting in decarburization of the surface through oxidation, and subsequent formation of scale.

Air Furnace. (a) Furnace, depending exclusively on natural draft. (b) Reverberatory furnace, used for lead smelting. (c) Reverberatory furnace for melting iron in the manufacture of malleable cast iron.

Air Gap. Insulating film of air formed, under certain conditions, between a solidified ingot and its mold, owing to expansion of the mold and contraction of the ingot.

Air Hardening Steel. Alloy steel which may be hardened by cooling in air from a temperature above the critical point. Such steels attain their martensitic structure without going through the quenching process. Additions of chromium, nickel, molybdenum and manganese are effective toward this end.

Air Line Pipe. Steel pipe suitable for airbrake systems, on railroads.

Air Patenting. Process of **Patenting** (see) steel products in which the cooling is accomplished in air.

Airplane Sheet. Steel sheet, frequently of non-corroding composition, in ductile and weldable form, for use in aeronautic construction.

Air Port. Inlet for air in furnace, particularly of the reverberatory type.

Air Quenching. Quenching (see) in air as the cooling medium (either still air or air in blast or jet form).

Air Reduction Process. See **Roast-Reaction Process.**

Air Separation. Separation of powdered or crushed material into fractions of approximately the same size, effected by air currents.

A.I.S.I. Steels. Steels of the American Iron and Steel Institute. Common and alloy steels have been numbered in a system essentially the same as the **S.A.E.** (see), series, to which the A.I.S.I. has added the NE series, representing **National Emergency Steels** (see). The A.I.S.I. system is more elaborate than the S.A.E. in that all numbers are preceded by letters: A represents basic open-hearth alloy steel, B acid Bessemer carbon steel, C basic open-hearth carbon steel, CB either acid Bessemer or basic open-hearth carbon steel, E electric furnace alloy steel.

Aitch Metal. See **Aich Metal.**

Ajax Bronze. Series of copper-base alloys, manufactured by Ajax Metal Co.

Ajax Bull. Bearing metal containing approximately 17% antimony, 7% tin, and 76% lead (Ajax Metal Co.).

Ajax-Northrup Furnace. Type of coreless **Induction Furnace** (see) in which the metal acts as secondary and a water-cooled coil, carrying high frequency current, as primary.

Ajax-Wyatt Furnace. Core-type **Induction Furnace** (see), operating at low frequencies.

Al. Chemical symbol for **Aluminum** (see).

Alabamine. See **Helvetium.**

Albata. See **Albatra.**

Albatra. Copper-base alloy, containing about 20% nickel and 20% zinc, used for hardware and similar products.

Alberti Furnace. Reverberatory furnace for roastng mercury ores, the mercury being condensed in iron tubes and brick chambers.

Albion Metal. Composite (see **Cladding**) made by coating lead with tin and rolling together.

Alclad. Trade name for composite metal consisting of one or more layers of pure aluminum rolled onto a layer of aluminum alloy. The former adds chemical resistance, the latter mechanical strength. Alclad 17S and Alclad 24S are standard forms, the number indicating the type of alloy used as the basic layer (Aluminum Company of America).

Alcoa. Aluminum-base alloys (Aluminum Co. of America). The more important types are: Alcoa 2S—Commercially pure aluminum; 3S—Same with 1.2% manganese; 4S—Same with 1.2% manganese and 1.0% magnesium; 17S—Same with 4% copper, 0.5% manganese, and 0.5% magnesium (see **Duralumin**); 24S—Same, with 4.4% copper, 0.5% manganese, and 1.5% magnesium (see **Duralumin**); 25S—Same with 4.5% copper, 0.8% silicon, and 0.8% manganese; 47S—Same with 12.5% silicon (see **Alpax**); 52S—Same with 2.5% magnesium and 0.25% chromium.

Alcrosil (Steel). Series of low-alloy steels, with 4–6% chromium and normally less than 0.15% carbon (Timken Roller Bearing Co.).

Alcumite. Corrosion-resistant **Aluminum Bronze** (see) with about 8% aluminum, 2% iron, and 1% nickel (Duriron Co.).

Alcunic. Copper-base alloy, with approximately 16–27% zinc, 2% aluminum, 1% nickel and, optionally, 1% tin (Scovill Mfg. Co.).

Alfenide. Modified **Nickel Silver** (see) with about 30% zinc, 10% nickel and 1% iron.

Alfol. Crumpled aluminum foil, used for heat insulation.

Alger Metal. Tin-base bearing metal, with up to 25% antimony, and optionally up to 5% copper.

Algier Metal. See **Alger Metal.**

A_3 Line. The A_3 Point (see **Critical Point**) varies from 1670°F at 0.00% carbon to 1333°F at the eutectoid value (about 0.83% carbon).

A_{cm} Line. Line on temperature-composition diagrams for iron-carbon, delineating decomposition temperatures of austenite into proeutectoid cementite.

Alkali (Alkaline). See **Base.**

Alkali Metals. Metals of group IA of the periodic system, including lithium, sodium, potassium, rubidium, cesium, and virginium (eka-cesium). They react vigorously, sometimes violently, with water, forming highly alkaline hydroxides. They are all very reactive, burning readily in air. See individual elements.

Alkaline-Earth Metals. Reactive metals of group II of the periodic system, including calcium, strontium, barium, and radium. So called because their oxides, alkaline in reaction, were known as "earths" in early chemical terminology. See individual elements.

Alkaline Storage Battery. Electric storage battery in which the positive plates consist of nickel and nickel peroxide, and the negative plates of finely divided iron. A strong solution of potassium hydroxide acts as electrolyte. The normal voltage varies between 1.0 and 1.5 volts per cell.

Allan (Red) Metal. Mechanical mixture (not a true alloy) containing approximately equal parts of copper and lead. Used primarily for bearings (A. Allan & Son).

Allegheny-Ludlum Steel. Series of alloy steels, most commonly of the high-chromium and stainless types, manufactured by Allegheny Steel Co.

Allegheny Metal. Series of corrosion and heat-resistant stainless steels, varying in chromium content between 11% and 30% and containing up to 21% nickel and small optional percentages of manganese, molybdenum, columbium and titanium (Allegheny-Ludlum Steel Corp.).

Allen-O'Hara Furnace. Horizontal double-hearth furnace, used for calcining sulfide ores.

Alligator. See **Squeezer.**

Alligator Shears. Metal shears, with one jaw movable and pinned to the other at the end, commonly used on light gage metal.

All-Mine Pig. Pig iron made entirely by the smelting of ore, without addition of scrap metal.

Allotrimorphic Crystal. See **Idiomorphic Crystal.**

Allotrope. See **Allotropy.**

Allotropy. Appearance of certain elements such as phosphorus, carbon, and iron, in varying forms called allotropes. Evidence of allotropy may be seen in such changes as occur in the atomic lattice arrangements. Allotropic forms of an element differ in practically all physical, as well as many chemical properties. The allotropy of iron, particularly the changes between the alpha body-centered cubic form and the gamma face-centered cubic form, is of fundamental importance in the hardening of steel.

Alloy. Mixture of two or more elements, at least one of which is a metal, forming an apparently homogeneous mass possessing metallic characteristics. Alloying may result in (1) a mere solution of one metal in another, or (2) a chemical combination of the elements present dissolved in the excess of one of the elements. Limited solubility of one metal in another will result in limited alloying, and metals practically insoluble in each other will tend to separate into two layers. Generally, alloying lowers the melting point, decreases electrical conductivity, and increases hardness. The latter is especially true if metal-compound formation takes place.

Alloyage. Alloying precious metals with base metals, for coin manufacture.

Alloy Cast Iron. Cast Iron (see), containing substantial proportions of alloying elements, such as nickel, chromium, molybdenum, etc.

Alloyed Casting. See **Alloy Cast Iron.**

Alloyed Chilled Rolls. Chilled Rolls (see) composed of cast iron, containing appreciable quantities of alloying elements.

Alloyed Grain Rolls. Rolls, for rolling metal, composed of sand-cast iron, containing appreciable quantities of alloying elements.

Alloyed Iron Casting. See **Alloy Cast Iron.**

Alloyed Iron Rolls. Rolls, for rolling metal, composed of cast iron, containing appreciable quantities of alloying elements. See **Rolling Mill.**

Alloyed Rolls. Cast iron, or cast steel rolls, containing appreciable amounts of alloying elements. See **Rolling Mill.**

Alloying. Addition of a metal or **Alloy** (see) to another metal or alloy, the latter usually being in the molten form.

Alloying Element. Element added to a basic metal in order to change its properties. In steel, the term covers elements, other than carbon, which are present in quantities greater than in the standard steels.

Alloy Iron. See **Semi-Steel.**

Alloy Pig Iron. Pig Iron (see), containing one or more alloying elements in appreciable quantity.

Alloy Steel. Steel containing substantial quantities of elements other than carbon and the commonly-accepted quotas of manganese, sulfur, silicon, and phosphorus. Addition of such alloying elements is usually for the purpose of increased hardness, strength, or chemical resistance. The metals most commonly used for forming alloy steels are: nickel, chromium, silicon, manganese, tungsten, molybdenum and vanadium. "Low alloy" steels are usually viewed as those totalling less than 5% of such added constituents.

Alloy Steel Strip. Strip Steel (see) containing one or more alloying elements in appreciable quantity. Typical are stainless steel strip and electric sheet steel.

Alloy System. See **Constitution Diagram.**

Alloy-Treated Steel. Simple steel to which one or more alloying elements have been added in small amounts, to improve general quality, without any significant effect on physical characteristics.

Alloy Wrought Iron. Wrought Iron (see), containing one or more alloying elements in appreciable quantity (most commonly, nickel, up to 3%, and copper up to 1.25%).

Alluvial Gold. Gold found native, in association with water-worn material.

Alluvial Tin. See **Stream Tin.**

Almond Furnace. Furnace in which litharge, from silver refining, is reduced to lead by means of charcoal.

Alnico. Alloy series containing approximately 20% nickel, 12% aluminum, 5% cobalt, 0.4% manganese and silicon combined, 62% iron, and not more than 0.15% carbon. Valuable in making permanent magnets of high coercive force. Normally subjected to heat treatment and stabilization by aging (General Electric Co.).

Alpaca. Nickel-Silver (see), containing approximately 64% copper, 20% zinc, 14% nickel, and 2% silver.

Alpakka. See **Alpaca.**

Alpax. Corrosion-resistant, aluminum-base alloy, containing 10–13.5% silicon; used as a casting alloy. Modifications of this alloy may contain approximately 1.2% of magnesium and manganese.

Alpha Antimony. See **Yellow Antimony.**

Alpha Arsenic. See **Yellow Arsenic.**

Alpha-Beta Brass. Brass containing between 39.0 and 45.5% zinc; a mixture of **Alpha** and **Beta Brasses** (see).

Alpha Brass. Face-centered cubic form of brass, found in all brasses, containing 39% zinc or less, which are solid solutions of zinc in copper.

Alpha-Forming Element. Element which, alloyed with iron, tends to suppress the **Gamma** form of **Iron** (see) by raising the A_3 temperature (see **Critical Point**) and lowering the A_4 temperature, to form the "gamma loop," outside of which the alloy exists in the alpha form. Aluminum, molybdenum, tungsten, silicon, and vanadium are typical among the alpha-forming elements.

Alpha Iron. Allotropic form of iron, stable at 1670°F and below, composed of body-centered cubic crystal structure.

Alpha Particle. Nucleus of helium atom, having a positive charge; emitted by radioactive materials.

Alpha Ray. Stream of alpha particles emitted by a radioactive element.

Alpha Tin. See **Tin Pest.**

Alplate. Aluminum-clad steel made by passing hydrogen-heated steel through molten aluminum (Reynolds Metal Co.). See **Cladding.**

Alpro. Series of aluminum-base alloys, most commonly with copper and nickel, manufactured by Alloys & Products, Inc.

Alrok Process. Process of forming a corrosion-resistant oxide film on the surface of aluminum-base metals by immersion in a hot solution of alkali carbonate and chromate.

Alsifer. Trade name for an alloy, containing approximately 40% iron, 40% silicon, and 20% aluminum. Carbon is absent. Used as a deoxidizer in steel, both the aluminum and silicon being effective reactants (Vanadium Corp. of America).

Alternate Stress. See **Fatigue.**

Alternating Current Arc Welding. **Arc Welding** (see), utilizing alternating current at the arc.

Aludur. Age-hardening aluminum alloy, containing approximately 0.7% silicon, 0.5% magnesium, and 0.5% iron.

Alumac. Series of aluminum-base alloys, with 5–13% silicon and, optionally, up to 4% copper (United States Aluminum Co.).

Alumalun. **Antislip Metal** (see) with aluminum as matrix (American Abrasive Metals Co.).

Alumel. Nickel-base alloy, containing approximately 2.5% manganese, 2% aluminum, and 1% silicon. Stable at temperatures up to 2200°F. Used as electrical resistance alloy and as a component of pyrometric thermocouples of the chromel-alumel type (Hoskins Mfg. Co.).

Alumetized Steel. Aluminum-coated steel, made by spraying the aluminum onto the steel and heating to cause surface alloying.

Alumiliting. See **Anodizing.**

Aluminium. British spelling of **Aluminum** (see). Justified by the fact that the chemical names for most metals end in "ium."

Aluminized Steel. Aluminum-clad mild steel. See **Cladding.**

Aluminothermic Process. See **Thermit.**

Aluminothermic Reduction. See **Thermit.**

Aluminum (chemical symbol Al). Element No. 13 of the periodic system; atomic weight 26.97; silvery white metal of valence 3; melting point 1220°F; boiling point approximately 4118°F; specific gravity 2.70; ductile and malleable; stable against normal atmospheric corrosion, but attacked by both acids and alkalies; cannot be electro-deposited. Although present in many minerals and an important constituent of all clays, aluminum can be prepared today economically only from bauxite, which is first converted into pure aluminum oxide, then dissolved in molten cryolite and electrolyzed. Aluminum is used extensively in articles requiring lightness, corrosion resistance, electrical conductivity, etc. The only important commercial source of aluminum and its alloys in U.S. is the Aluminum Co. of America. See **Alcoa.** Other firms are beginning production, however.

Aluminum Alloy. General term covering aluminum-base alloys, containing up to 8% copper and other alloying elements, if desired.

Aluminum-Beryllium Alloy. See **Beryllium-Aluminum Alloys.**

Aluminum Brass. Brass, containing fractional percentages of aluminum, which helps to increase the oxidation resistance of the brass; when present in larger amounts it increases both hardness and strength, but decreases ductility.

Aluminum Bronze. Copper-base alloys, containing approximately 4 to 13% aluminum. Other metals, such as zinc, iron, tin, and manganese, may be present in small percentages.

Aluminum Bronze Powder. Actually, **Aluminum Powder** (see).

Aluminum Cladding. See **Cladding.**

Aluminum Foil. Aluminum in **Foil** (see) form.

Aluminum Gold. See **Aluminum Bronze.**

Aluminum-Magnesium Alloys. Aluminum-base alloys containing up to 10% magnesium; used commercially, even though the presence of over 3% magnesium causes brittleness. See **Alcoa.**

Aluminum-Manganese Alloys. Commercially used, aluminum-base alloys containing 1.5 manganese. See **Alcoa.**

Aluminum-Nickel Alloys. Aluminum-base alloys containing up to about 7% nickel, added to improve strength and resistance to corrosion.

Aluminum Powder. Aluminum, in the form of tiny flakes, made by stamping; used as a pigment in paints, inks, etc., usually after coating with a lubricant to gain luster and leafing characteristics.

Aluminum-Silicon Alloys. Aluminum-base alloys, containing up to 13% silicon, used primarily for light-metal castings.

Aluminum Silver. (a) Aluminum-base alloy, with about 5% silver. (b) Modification of **Nickel Silver** (see) containing several per cent of aluminum. A typical composition contains 20% nickel, 20% zinc and 3% aluminum.

Aluminum Solder. Hard Solder (see), containing small amounts of aluminum, varying up to 10%; used for soldering aluminum parts.

Aluminum Steel. Steel, containing alloyed metallic aluminum in fractional or low percentages; frequently used for **Nitriding** (see).

Aluminum-Zinc Alloys. Aluminum-base alloys containing up to 14% zinc, with optional low percentages of copper and silicon, used primarily for casting.

Alundum. Trade-marked fused alumina product, used for refractory and abrasive purposes (Norton Co.).

A.M.—. Series of magnesium-base metals and alloys, the latter most commonly with aluminum, manganese, and zinc (Aluminum Co. of America).

Amalgam. Alloy of mercury with any other metal, in liquid or solid state, depending on the nature and amount of the other component.

Amalgamating Barrel. Rotating cylindrical vessel, with pebbles or steel balls, used to effect contact between an ore and mercury. See **Amalgamation Process.**

Amalgamation. Formation of an alloy of mercury with another metal. See **Amalgam.**

Amalgamation Pan. Apparatus for effecting contact between an ore and mercury. See **Amalgamation Process.**

Amalgamation Process. Process of treating gold and silver ores, in which finely divided ore, suspended in water, is carried over a surface of liquid mercury, causing formation of amalgams. The resulting amalgam is later treated to separate the valuable metal, and the mercury returned to the process.

Amalgamator. See **Amalgamation Pan.**

Amalgam Gilding. Process of gilding in which the surface is coated with gold **Amalgam** (see), followed by driving off the mercury by heat.

Amalgam Retort. Iron retort for heating amalgams (see **Amalgamation Process**) to drive off the mercury and leave other (usually precious) metals behind.

Amalgam Silvering. Process of silvering in which the surface is coated with silver **Amalgam** (see), followed by driving off the mercury by heat.

Amalgam Treatment. See **Amalgamation Process.**

Amaloy. Lead-base alloy, containing 2.5 to 10% tin, used to form protective coatings on sheet steel and iron (American Machine & Foundry Co.)

Ambrac. Strong corrosion-resistant series of **Nickel-Silver** (see) alloys, containing approximately 75% copper, 20% nickel, and 5% zinc. The use of small percentages of tin in castings of Ambrac is optional. These alloys have a bright, nickel-like appearance and are available in sheet, wire and tube form, all easily machined (American Brass Co.).

Ambraloy. High-strength, corrosion-resistant **Bronze** (see) containing approximately 9.5% aluminum, 5% nickel, 2.5% iron and 1% manganese.

Amco Furnace. Modern type of **Soaking** (see) pit furnace (Amsler-Morten Co.).

Amercut (Steel). Series of free-cutting steels of simple steel composition; the free-cutting is accomplished by the presence of sulfur, up to 0.2% (United States Steel Corp.).

Amerhead (Steel). Series of simple steels, available in wire form for cold-heading (United States Steel Corp.).

American Alloy. Aluminum-base alloy, with 3% copper, 1% manganese, and 1% magnesium.

American Bloomery. See **Bloomery.**

American Forge. See **Champlain Forge.**

American Iron and Steel Institute Steels. See **A.I.S.I. Steels.**

American Malleable Iron. See **Black Heart (Malleable) Iron.**

American Petroleum Institute. See **A.P.I.**

American Screw Gage. See **Metal Gage.**

American Silver. White alloy of 50–60% copper content, with 20 to 24% zinc, 15 to 24% nickel, and optional low percentages of manganese, iron, lead, tin, and aluminum.

American Society for Testing Materials. Organization for issuing standard specifications on materials, including metals and alloys.

American Standard Pipe Thread. See **Briggs Pipe Thread.**

American Standard Silver. See **Standard Silver.**

American Steel & Wire Company Gage. See **Washburn & Moen Gage.**

American Water-Back Hearth. See **Water-Back Hearth.**

American Wire Gage. See **Metal Gage.**

AMF. High-nickel steel, with about 48% nickel, possessing great ductility and shock resistance at extremely low temperatures; consequently, useful in refrigeration and liquid air machinery (Midvale Co.).

Amola (Steel). Low-alloy steel of relatively high strength; the alloying elements are manganese and molybdenum (Chrysler Corp.).

Amorphous. Non-crystalline. In metallurgy, this term is usually limited to materials which appear non-crystalline under the microscope, though it is still a moot question if metals are ever truly non-crystalline.

Amorphous Cement. Thin film of metal, occurring at grain boundaries, in which the metal atoms are assumed to have a haphazard or non-crystalline arrangement.

Amorphous Phosphorus. Reddish-brown allotropic form of phosphorus, less active chemically than the white modification. It is the form commonly used in metallurgy for forming **Phosphorus** (see) alloys.

Ampcoloy. Series of copper-base alloys, used in cast form, with aluminum or tin up to 10%, and low percentages of iron, nickel, and lead (Ampco Metal, Inc.).

Ampco Metal. Aluminum bronze, containing from 82–88% copper, 9–14% aluminum, 2.5–5.5% iron, and small percentages of other elements if desired, corrosion-resistant to mild chemical agents such as dilute sulfuric or acetic acid (Ampco Metals, Inc.).

Amperage. Strength of electric current, measured in **Amperes** (see).

Ampere. Quantity of electric current flowing as a result of one volt electromotive force through one ohm resistance.

Ampere-Hour. Quantity of electricity resulting from passage of one ampere of current for one hour.

Amphoteric. Possessing both acidic and basic properties.

A.M.S. Symbol for "Aeronautical Material Specification" steels.

Amsco Alloy. Nickel-chromium-iron alloys of varying composition, used, particularly in cast form, for heat, fire, and chemical resistance (American Manganese Steel Co.). See **Fahralloy.**

Amsler-Laffon Hardness Tester. Device for measuring **Hardness** (see) of materials, in which a 90° cone is indented into the surface of the material, using a constant depth as standard.

Anaconda—. Series of copper-base alloys, of **Brass** (see) and **Bronze** (see) type, manufactured by American Brass Co.

Anatomical Alloy. See **Fusible Alloys.**

Angle of Bite. Maximum **Contact Angle** (see) at which a piece of material will enter a pair of rolls. For smooth rolls, the angle of bite is 30°; for roughened rolls, it is between 35 and 42°.

Anglo-Helvetium. See **Helvetium.**

Ångstrom Unit. See **A.**

Angular Fracture. See **Fracture.**

Angular Rolling. See **Diagonal Rolling.**

"A" Nickel. Commercially pure nickel, assaying about 99% metallic nickel (International Nickel Co.).

Anion. Negatively charged ion in any solution. Specifically, that portion of a chemical compound which, in solution, tends to flow toward the anode under the influence of direct current.

Anisotropy. Ability of certain substances to exhibit different properties along different lines of direction, e.g., variation of optical properties of a crystal with its axes.

Annealed-in-Process Wire. See **Bright Soft Wire.**

Annealed Wire. Softest grade of steel wire, wire made from basic open-hearth steel being softer than wire from Bessemer steel.

Annealing. Subjecting a metal to heating and cooling, in the solid state, generally in order to remove stresses and soften the metal. Full annealing, used principally on iron and steel, means heating the metal to about 100°F above the critical temperature range, followed by "soaking" at this point, and slow cooling below the critical temperature. Process or work annealing, also used mainly on iron-base alloys, means heating of the metal to the approximate lower limit of the critical temperature range, with subsequent cooling. Box annealing, commonly limited to iron-base alloys, is slow heating of the metal to a temperature below the critical, in a closed box or pot protected from oxidation, followed by slow cooling. Black annealing is box-annealing of sheets in order to impart a black color to the surface. Pot annealing is the same as box annealing. Blue annealing is the annealing of metal sheets after rolling. These sheets acquire a blue or black surface color because of exposure to the air. Pipe annealing is the same as box annealing. Bath annealing is immersion in a liquid bath (such as molten lead or fused salts) held at an assigned temperature. When a lead bath is used, the process is known as lead annealing.

Annealing Box. See **Annealing.**

Annealing Carbon. See **Cement Carbon.**

Annealing Color. Surface color on steel resulting from **Annealing** (see) in air.

Annealing Furnace. Batch or continuous furnace for **Annealing** (see) metals.

Annealing Oven. See **Annealing Furnace.**

Annealing Pot. See **Annealing.**

Annealing Twins. Twin Crystals (see) formed in many metals (most commonly, in copper-base metals) after cold-working and annealing.

Anode. Positive electrode of an electrolytic cell, at which the negative ions lose their charge. In electroplating processes, the anode is frequently made of the metal which is being plated, so that the continuous dissolving of the anode, during electroplating, replaces the metal, deposited at the cathode.

Anode Copper. Slab of copper, cast in crude form, prior to purification by electrolysis. See **Electrolytic Copper.**

Anode Effect. Tendency, in certain fused salt electrolyses, of the current to drop off sharply, owing to the sudden development of very high internal resistance which is apparently due to the formation of a continuous film of gas around the anode, preventing current flow.

Anode Mud. See **Anode Slime.**

Anode Slime. Insoluble residue, left after solution of the anode, in electroplating or electrorefining.

Anodic Cleaning. Removal of grease, oxide, etc., from metal, by making the metal anode in an electrolytic cleaning bath, usually alkaline. The evolution of oxygen helps remove foreign materials from the surface of the metal.

Anodic Coating. Coating which protects the base metal by being preferentially attacked chemically, as a result of the formation of an electrochemical couple. Zinc on steel is typical; as long as zinc is present, it will be attacked and not the steel base. Note difference from **Anodizing.**

Anodic Oxidation. See **Anodizing.**

Anodic Treatment. See **Anodizing.**

Anodizing. Forming a resistant oxide film on aluminum or similar metals, by electrolysis. The positive pole, or anode, is composed of aluminum and is contained in a bath of sulfuric, chromic, or oxalic acid, through which is passed a relatively high voltage. The formation of the oxide film on the aluminum prevents the development of high amperage during anodizing.

Anolyte. In electrolytic processes, the solution surrounding the **Anode** (see).

Antaciron. High-silicon cast iron, resistant to corrosion. Silicon content is about 14% (Antaciron, Inc.).

Anthracite Pig (Iron). Pig Iron (see), made with anthracite coal.

Anticorodel. Corrosion-resistant, aluminum-base alloys, containing low percentages of magnesium, silicon, manganese, and iron.

Anti-Friction Metals. Series of alloys, primarily those of lead, tin, antimony, and zinc, used for bearings, particularly in the form of linings on copper-base foundations.

Antimonial Lead. Lead with up to 25% antimony (about 7% is most common), acting as a hardener, hence the name "Hard Lead." Used for storage battery plates and connectors, chemical equipment, etc. With tin added, the alloys are used for type metal and bearings.

Antimonial Silver. Silver ore, containing variable quantities of antimony.

Antimonial Speiss. See **Speiss.**

Antimonide. Oxygen and sulfur-free compound containing antimony as the chemically negative constituent.

Antimony (chemical symbol, Sb). Element No. 51 of the periodic system, existing in several allotropic forms; atomic weight 121.76. It is a bluish-white, brittle metalloid; melting point 1167°F; boiling point 2993°F. Antimony has valences of three (antimonious) and five (antimonic). It is obtained commercially, principally from its sulfide ores, by means of smelting and reduction with carbon, or by reduction with iron. In metallic form, it is used almost exclusively as an alloying element in lead-base and tin-base alloys. It can be electrically deposited from both its acid and alkali compounds, however, this property has not been put to commercial use.

Antimony Hydrochloride Test. Chemical test for determining the quantity of zinc on galvanized iron or steel.

Antimony Regulus. See **Regulus of Antimony.**

Antimony Skimming. Removal of material, consisting principally of lead-arsenic-antimony-oxygen compounds, which collects on the surface of molten lead after completion of the **Tin Skimming** (see).

Antimony Star. See **Doubling Process.**

Antislip Metal. Metal, e.g., aluminum, bronze or iron, containing embedded abrasive particles, to prevent slippage when used on floors, stairs, etc.

Anvil. See **Drop Forging.**
Apex Zinc Alloy. Series of zinc-base alloys, with about 4% aluminum, up to 3% copper, and fractional percentages of magnesium (Apex Smelting Co.).
A.P.I. Symbol for American Petroleum Institute, which has issued standards for steel pipe, tubing, casing, and threads for these.
"A" Point. Temperature above which steel must be heated in order to achieve hardening effect.
"A_0" Point. See **Critical Point.**
"A_1" Point. See **Critical Point.**
"A_2" Point. See **Critical Point.**
"A_3" Point. See **Critical Point.**
"A_4" Point. See **Critical Point.**
"A_e" Point. See **Critical Point.**
"A_{cm}" Point. See **A_{cm} Line.**
Apolloy. Copper-bearing steel, containing approximately 0.08% carbon and 0.25% copper (Apollo Steel Co.).
Apparent Density. Term used in powder metallurgy, referring to the weight of a unit volume of powder, generally expressed as grams per cubic centimeter or cubic inch. Apparent density varies with the method of loading.
Apron. Set of guides for directing plates through a metal galvanizing or coating bath.
Aquadag. Graphite suspended, in colloid form, in water; used extensively as a wire-drawing lubricant (Acheson Colloids Corp.).
Aqua Regia. Mixture of 18% nitric and 82% hydrochloric acids, used to dissolve platinum, gold, and other noble metals immune to the action of individual acids.
Arborescent. See **Dendritic.**
Arc Furnace. Furnace in which an electric arc is the source of heat. Carbon or graphite normally acts as one electrode, and a similar unit, or the furnace charge, as the other electrode.
Architectural Bronze. Leaded yellow brass, containing approximately 40% zinc and 3% lead, possessing excellent machining characteristics combined with high strength and corrosion resistance.
Arcos. Welding Rods (see) made by Arcos Corp.
Arc Spectrum. Spectrum of non-ionized atoms of an element, obtained from an electric arc.
Arc Spraying. Method of **Metal Spraying** (see) in which the metal to be sprayed is vaporized by means of an electric arc.
Arc Welding. Welding (see) accomplished by means of an electric arc, formed between a metal (or carbon) electrode and the metal being welded.
Argall Furnace. Reverberatory **Roasting** (see) furnace with a reciprocating hearth, the ore moving continuously forward by the action of rabbles across the hearth.
Argental Mercury. Amalgam (see) of silver.
Argentan. Alloy of the **Nickel-Silver** type, containing 40–65% copper, 17–32% zinc, and 15–30% nickel. Iron is sometimes present in small amounts. Used mainly for ornamental purposes.
Argentiferous. Containing silver.
Argentiferous Lead. Lead containing small amounts of silver, usually as an impurity, resulting from the presence of silver in the original ore.
Argentine. (a) White metal coated with silver. (b) Spongy tin, obtained by precipitation of tin from solution by means of zinc.
Argentine Metal. Tin-base metal, containing about 15% antimony, used for ornamental purposes.
Argentite. See **Vitreous Silver.**
Argentum. Latin name for silver, used in medicine and pharmacy.
Armat Steel. Magnet steel, with about eutectoid carbon and 3% chromium content (Jessop Steel Co.).
Armature Grade Sheet. Electric steel sheet of about 0.5% silicon content, used for small motors and other purposes, requiring high magnetic permeability.
Armco Ingot Iron. See **Armco Iron.**
Armco Iron. Unusually pure ingot iron, containing approximately 0.1% carbon, 0.02% manganese, 0.005% phosphorus, and 0.025% sulfur (American Rolling Mill Co.).
Armco Radio (Steel). Series of **Silicon Steels** (see), used in the manufacture

of motor and electrical equipment (American Rolling Mill Co.).

Armco Steel. Series of steels manufactured by American Rolling Mill Co.

Armor Plate. Low-alloy steel, commonly surface-hardened, for use in military and naval installations (battleship plates, etc.). A typical armor plate contains 2% chromium and 3% nickel, and 0.20–0.40% carbon.

A_r Point. See **Critical Point.**

Arrest (Point). See **Critical Point.**

Arsenic (chemical symbol As). Element No. 33 of the periodic system, existing in several allotropic forms; atomic weight 74.91. "Alpha" arsenic is a brittle, crystalline solid, steel-gray in color. Under normal atmospheric pressure, it does not melt, but sublimes. Under a pressure of approximately 530 $lbs./in.^2$ the melting point is 1497°F. The ordinary metallic form has a specific gravity of 5.73. Chemically, arsenic is a metalloid and has valences of three (arsenious) and five (arsenic). Arsenic is usually obtained first as the trioxide, as a by-product of other metallurgical operations. A commercially pure grade is made by reduction of the trioxide with charcoal. In elementary form, its primary use is in forming alloys with lead and copper. It can be electrodeposited from both its acid and alkali compounds, but these have been only used as an addition to nickel plating baths for obtaining "black nickel" deposition.

Arsenical Admiralty. Admiralty Brass (see) to which approximately 0.05% arsenic has been added, to improve high-temperature characteristics.

Arsenical Copper. Copper containing approximately 0.30% of arsenic, usually deoxidized with phosphorus and, therefore, containing about 0.02% residual phosphorus. The addition of the arsenic improves the metal for use at elevated temperatures; consequently arsenical copper is frequently used for condenser tubes, etc.

Arsenical Nickel. Mineral, called niccolite, substantially an arsenide of nickel, approximating NiAs in chemical composition.

Arsenical Speiss. See **Speiss.**

Arsenic Kitchen. Chamber, normally of brick, for the condensation of arsenic trioxide. Usually, there is a series of such kitchens of progressively lower condensing temperatures.

Arsenide. Oxygen- and sulfur-free compound, containing arsenic as the chemically negative constituent. A number of metals are frequently found in nature as arsenide ores.

Arsenolamprite. Mineral **Arsenic** (see), occurring native, together with bismuth.

Arsenolite. See **White Arsenic.**

Arsentine Plate. See **Nickel Silver.**

Artificial Aging. See **Age Hardening.**

Artificial Gold. Stannic sulfide, SnS_2, a non-metal, used for imitation gilding.

As. Chemical symbol for **Arsenic** (see).

Asarcoloy. Cadmium-base bearing metal of high strength; contains about 1.3% nickel (American Smelting & Refining Co.).

Ascoloy. Series of chromium steels, with or without substantial amounts of nickel (Allegheny Steel Co.).

As Forged. Forgings (see), ordered and sold as such, with no treatment after completion of the forging operation.

Ashberry Metal. Tin-base alloy of the **Britannia Metal** (see) type, containing approximately 80% tin, 14–19% antimony, up to 3% copper and nickel, and up to 2% zinc.

Assay. Determination of metallic values in an ore by chemical analysis; usually "fire assay" is meant, with reduction of the metallic content of the ore to metal in a small analytical crucible.

Assay Value. Proportion of precious metal (gold, silver, etc.), usually indicated in ounces of metal per ton of ore, as revealed by assay.

Asterism. Radial streaks or bands which result in Laue X-ray patterns when a crystal, uniformly strained by slip, is subjected to X-ray examination.

A.S.T.M. See **American Society for Testing Materials.**

Aston Process. See **Byers Process.**

Aterite. Series of copper-base casting alloys, with 10–44% nickel, 5–20% iron, and, optionally, up to 23% zinc and 2% lead.

Atlas. Copper-base alloys of which a common constituent is aluminum, iron or lead (Ampco Metal, Inc.).

Atom. Smallest unit of a chemical element, taking part in a chemical change.

Atomic Arc Welding. See **Atomic Hydrogen Welding.**

Atomic Hydrogen Welding. Arc Welding (see), usually with tungsten electrodes, in a stream of hydrogen, which dissociates, in part, to form highly reducing atomic hydrogen.

Atomic Number. Number designating the position of an element in the **Periodic System** (see). Physically, it represents the number of excess positive charges present in the atom nucleus.

Atomic Weight. Weight of an atom of a given element in comparison to the weight of an atom of oxygen, set arbitrarily at 16.0000. Hydrogen has the lowest atomic weight, 1.0080, and uranium the highest known, 238.07. Theoretically, all atomic weights ought to approximate whole numbers and large deviations are due to the fact that, as found in nature, many elements represent a mixture of substances chemically identical, but varying slightly in their individual atomic weights. These varying-weight atoms of the same element are called isotopes.

Atomization. Dispersion of molten metal into small particles, by use of gas or liquid pressure.

Au. Chemical symbol for **Gold** (see).

Auer Metal. Alloy of iron and the cerium group of metals (such as Misch-metal). Typical compositions involve 35% of iron, with the remainder cerium and the other rare earth metals of the same family, the latter being obtained as an alloy, without attempts at separation into individual metals. Auer metal is pyrophoric, emitting showers of sparks when struck, and is, therefore, used for cigarette lighters and similar devices.

Auriferous. Containing gold.

Auromet. Series of **Aluminum Bronzes** (see) of varying composition (Aurora Metal Co.).

Aurum. Latin name for gold, used in medicine and pharmacy.

Austempering. Controlled quenching of steel, in which the formation of martensite is prevented by cooling in molten lead and then by air.

Austenite. Phase in certain steels, characterized as a solid solution, usually of carbon or iron carbide, in the gamma form of iron. Such steels are known as "austenitic." Austenite is stable only above 1333°F, but the presence of certain alloying elements, such as nickel and manganese, stabilizes the austenitic form, even at normal temperatures.

Austenitic Grain Size. Grain size developed in steels at solution or heat-treating temperatures, compared with grain sizes standardized by the **A.S.T.M.** (see). The austenitic grain size is found by etching quenched, cooled, or differentially quenched pieces.

Austenitic Nickel Steel. Steel in which the nickel content is sufficient to lower the critical ranges to points below room temperature, leaving the steel in the austenitic condition. Steels with about 18% nickel and 1% or more of carbon, and all steels with 25% or more nickel, are in this category.

Austenitic Steel. Steel which, because of the presence of alloying elements, such as manganese, nickel, chromium, etc., shows stability of **Austenite** (see) at normal temperatures.

Australian Gold. Standard Australian **Coinage Gold** (see), containing 8.33% silver and the remainder gold.

Auto Body Sheet. See **Automobile Body Sheet.**

Autogenous Soldering. See **Autogenous Welding.**

Autogenous Welding. Method of uniting two pieces of metal, without the use of solder or any added welding metal, by melting their edges together. Frequently used to eliminate the danger of electrolytic action of dissimilar metals, which may result from the use of solders or welding metals.

Automatic Arc Welding. See **Automatic Welding.**

Automatic Welding. Arc Welding (see), in which the arc is automatically moved along the welding line, with automatic control of the electrode feed and arc length.

Automobile Body Sheet. Cold-rolled and annealed steel sheet ordered and sold for use in the manufacture of automobile bodies, etc.

Available Base. In basic **Fluxes** (see), the percentage of base available for reaction with acidic constituents, after decomposition (as with carbonates), and after subtraction of acidic constituents already present in the flux.

Average Creep. Rate of **Creep** (see), determined by division of the total creep by the elapsed time.

Average Rate of Creep. See **Average Creep.**

Avesta (Steel). Series of high-chromium steels (chromium up to 22%), with carbon usually below 0.10% (A. Johnson & Co.).

Avialite. Aluminum bronze of high strength, hardness and chemical resistance, containing approximately 9% aluminum and small percentages of tin, nickel, and iron (American Brass Co.).

Awaruite. Mineral, alloy of nickel and iron, approximating $FeNi_2$ in chemical composition.

A.W.G. See **Metal Gage.**

Axis (of a Crystal). One of three or four imaginary lines passing through the center of a crystal, around which the mass of the crystal is symmetrically arranged.

B

B. Chemical symbol for **Boron** (see).

Ba. Chemical symbol for **Barium** (see).

Babbitt (Metal). Series of tin-base alloys, with antimony and copper. Up to 30% of lead may be added, but in high-tin alloys of this series, lead is normally limited by specification to 0.35%. Cadmium may be added up to 1%, to increase strength. Babbitt is extremely popular for lining and antifriction purposes.

Backing-Up Roll. See **Four-High Mill.**

Back-Pouring. Filling an ingot mold up to the **Sinkhead** (see), which is later filled with molten metal.

Back-Up Roll. See **Four-High Mill.**

Backward Creep. Tendency of the speed of metal, while rolled, to be slightly retarded on the entering side of the rolls, so that the movement of metal is slightly less than the roll peripheral speed.

Backward Slip. See **Backward Creep.**

Baffle Plate. Plate for deflecting flame or heated gas in a furnace, to achieve more even heating in the furnace.

Bag House. Large chamber for holding bags, used in the filtration of gases from a furnace, for the recovery of metal oxides and similar solids, suspended in the gases.

Bag Process. Recovering metal oxides, flue dusts, etc., from gas, by passing it through bags acting as filtering units.

Bag Room. See **Bag House.**

Bahnmetall. "Railway metal" (German). Lead-base bearing metal, introduced during the first World War as lining for locomotive and journal bearings. Approximate composition: Sodium 0.6%, lithium 0.04%, calcium 0.7%, remainder lead. It retains hardness at relatively high operating temperatures.

Baily Furnace. Early type of electric furnace, in which heat is generated by passage of current through crushed coke or graphite, lying between carbon electrodes.

Bainite. Decomposition product of **Austenite** (see), best developed at holding temperatures below those forming fine pearlite and above those giving martensite.

Baking. (a) Heating metal, for long periods, below the critical temperature, in order to remove adsorbed gases. (b) Heating sand molds or cores, prior to casting of metal, for several hours, at about 300°F.

Balanced Alloy Steel. Steel, containing alloying constituents adjusted to achieve desired characteristics, such as low creep, high temperature resistance, etc.

Balanced Mill. Roll mill, used instead of a roughing mill (see **Rolling Mill**), the top roll being always held against the chocks (see **Stand**), so as to move with the screws.

Balbach Process. Process for electrolytically separating gold from silver, using the alloy as anode, graphite plate cathodes and silver nitrate solution as bath.

Bale Ties. Steel wire of good pliability and high tensile strength, used for tying and bundling; most commonly, made from annealed, medium-carbon steel.

Baling Band. Light, narrow hoop (see **Hoop Iron**) used for tying and baling.

Ball. See **Puddling.**

Ballantine (Hardness) Test. Method of determining hardness of surfaces, in which a soft vertical metal cylinder, with a pointed hard top, is held against the surface and then struck by a weight, and the deformation of the cylinder measured as representing, inversely, the hardness of the surface.

Balling. Step in the **Puddling** (see) process, in which the metal is worked into a ball-like mass.

Balling Tool. Implement for collecting iron in a **Puddling** (see) furnace for removal to the squeezer.

Ball Squeezer. See **Puddling.**

Ball Steel. See **Bearing Steel.**

Ball and Wingham Process. Selective removal of sulfur from pig iron, in the fused state, by contacting with a molten mixture of sodium carbonate (soda ash), potassium carbonate, and potassium cyanide.

Banca Tin. Pig **Tin** (see) of about 99.95% purity, obtained from the Dutch East Indies.

Band. See **Hot-Rolled Strip Steel.**

Banded Structure. Appearance of a metal, under the microscope or viewed by the naked eye, on fractured or smoothed surfaces, with or without etching, showing parallel bands in the direction of rolling or working.

Band Steel. See **Hot-Rolled Strip Steel.**

Bank. Used in connection with a blast furnace, the term refers to closing down the furnace temporarily, feeding coke only until all ore has been consumed, and then withdrawing as much iron and slag as possible, followed by closing off all air flow and all openings.

Banka Tin. See **Banca Tin.**

Banking. See **Bank.**

Bar. Any flat piece of metal, narrower than about 12 inches and thicker than $\frac{3}{16}$ inch.

Barba's Law (of Similarity). In tensile and elongation testing of metals, similar results will be obtained only if test specimens are identical in proportions. Mathematically, the law may be given as: $P = L/\sqrt{a}$, where P is a proportionality factor, L the gage length under study, and a the cross-sectional area.

Barberite. Copper-base alloy (Barber Asphalt Co.), with about 5% tin, 5% nickel, 1.5% silicon, and fractional percentages of iron and manganese; used in cast form, it is resistant to marine corrosion.

Bare Chill. Metallic cooling section of a mold for casting **Chilled Rolls** (see) left bare to the roll, in contrast with "covered chill."

Bare Electrode. Electrode for use in **Arc Welding** (see), not coated with fluxing constituents. See **Coated Electrode.**

Barff Process. Forming a rust-resisting coating on iron-base metals, by oxidation with superheated steam.

Bar Furnace. First of a pair of **Sheet Furnaces** (see).

Barium (chemical symbol Ba). Element No. 56 of the periodic system, heaviest among the common members of the alkaline earth group; atomic weight 137.36. Silvery white metal, turning yellowish on exposure to air (due to nitride formation); the metal is ductile and malleable; it reacts with water, releasing hydrogen; melting point 1562°F; boiling point 2779°F; specific gravity 3.78. Chemically, barium is divalent; it is manufactured in the United States by interaction of barium oxide and aluminum, under high vacuum, and is used, as such or in the form of the magnesium alloy, as "getter" in the radio tube industry; the metal has also been used as a trace constituent of spark plug wire. Since it reacts readily with water, it cannot be electrodeposited. Barium is commercially available from Kemet Laboratories Co.

Barium Silicide. Theoretically, compound of barium and silicon, Ba_2Si. In commercial practice, however, the composition varies considerably through alloying in varying proportions. These alloys have some use as deoxidizing and desulfurizing agents for steel.

Barium Silicon (Alloy). See **Barium Silicide.**

Bark. Surface of metal, under the oxide-scale layer, resulting from heating in an oxidizing environment. In the case of steel, such bark always suffers from decarburization.

Barkhausen Effect. Discontinuous variation of induction in ferromagnetic substances in a field of gradually increasing strength.

Bar Mill. Merchant Mill (see) for rolling bars.

Barr Alloy. Series of **Bronzes** (see) of diversified composition, manufactured by McCallum-Hatch Bronze Co.

Barrel Amalgamation Process. See **Pan-Amalgamation Process.**

Barrel Copper. Small masses of native **Copper** (see).

Barrel Plating. Electroplating small articles (see **Plating**) by placing in a rotating perforated barrel, which is connected as cathode of the system.

Barrel Process. See **Pan-Amalgamation Process.**

Barrel Work. See **Barrel Copper.**

Barren Solution. Cyanide solution, after removal of silver and gold. See **Cyaniding** (b).

Bar Size. Structural shape (angle, channel, etc.) with greatest cross-section dimension less than three inches.

Bar Solder. Solder (see) cast into small bar form.

Bar Steel. (a) Straight lengths of steel, rolled from billets into rounds, hexagons, squares, flats, or shapes; used for reinforcing concrete. (b) **Blister Steel** (see) bars, hammered or rolled into flat form.

Bar Tin. See **Block Tin.**

Basaloy. See **Cerrobase.**

Basanite. See **Touchstone.**

Base. Chemical substance which forms hydroxyl ions, (OH^-), when dissolved, e.g., in water. Most metals are relatively immune to the action of bases in cold water solution; among the exceptions are aluminum, beryllium, gallium, silicon, and zinc. In metallurgical technology, materials are considered basic if they are susceptible of union with acidic substances, such as silica. Lime and magnesia are the most common basic materials of this type.

Base Box. Basis for ordering and selling of tin and terne plate. For tin plate, the base box, as a unit of measure, corresponds to 112 plates of 20 x 14 inch dimensions; it is used for tin plate up to 195 pounds per base box. For terne, the double base box is also frequently used, twice the above in size, corresponding to 112 plates of 20 x 28 inch dimensions.

Base Bullion. See **Work Lead.**

Base-Forming Element. Element, the hydroxy compounds of which are alkaline, the element itself being usually cationic in water solution (see **Cation**). The metallic elements are strictly base-forming; the metalloids are less so, and tend to become acid-forming.

Base Metals. More common metals, such as iron, lead, copper, etc., in contradistinction to the **Noble Metals** (see).

Basic Bessemer Pig Iron. Pig Iron (see) with relatively high (2–3%) phosphorus content, used for making basic-process Bessemer steel.

Basic Bessemer Process. See **Bessemer Steel.**

Basic Bessemer Slag. See **Basic Slag.**

Basic Bessemer Steel. See **Bessemer Steel.**

Basic Bottom. Bottom of furnace made of basic-reacting materials, such as dolomite, magnesia, etc.

Basic Cinder. See **Basic Slag.**

Basic Flux. Metallurgically basic material, such as limestone or dolomite, used as a **Flux** (see).

Basic Hearth. See **Basic Open-Hearth Process.**

Basic (Iron) Ore. Iron ore which yields pig iron, with phosphorus content not over 0.90%.

Basic Lining. Lining of furnace made of basic-reacting materials, such as dolomite, magnesia, etc.

Basic Lining Process. Modification of Bessemer process (see **Bessemer Steel**), using a basic lining in the converter and basic materials during the "blow" stage, resulting in removal of a substantial part of the phosphorus.

Basic Open-Hearth Iron. See **Ingot Iron.**

Basic Open-Hearth Process. See **Open-Hearth Process.**

Basic Open-Hearth Slag. See **Basic Slag.**

Basic Open-Hearth Steel. Steel made by the **Basic Open-Hearth Process** (see).

Basic Phosphate Slag. See **Basic Slag.**

Basic Pig (Iron). Pig Iron (see), used in the basic open-hearth or basic Bessemer process. For the open-hearth process, basic pig should have less than 1.5% silicon, less than 0.05% sulfur, manganese from 0.4 to 2.0%, and phosphorus from 0.11 to 0.90%. For the basic Bessemer (or Thomas) process, silicon should be in the 0.5 to 1.0% range, sulfur below 0.2%, manganese between 1.5 and 2.5%, and phosphorus between 1.9 and 2.5%.

Basic Process. See **Open-Hearth Process.**

Basic Refractory. See **Refractory.**

Basic Slag. Slag resulting from the operation of the basic Bessemer or basic open-hearth steel process. Because of its high phosphorus content, basic slag is of value as a phosphatic fertilizer.

Basic Steel. Steel made or melted in a base-reacting furnace lining. See **Basic Lining.**

Basis Box. British basis for ordering and selling tin plate, corresponding to 112 plates of 20 by 14 inch dimensions, weighing 112 pounds. Twenty basis boxes are equivalent to a gross ton (2240 pounds).

Basque. Mixture of clay and charcoal dust, used as a lining for furnaces and crucibles.

Bas Relief Polishing. Polishing a specimen, for metallographic examination, in any manner, to permit the softer constituents to wear down to a greater depth than the harder constituents.

Batch. Quantity of materials processed or produced in one operation.

Batch Furnace. Furnace operated on an intermittent charge basis as opposed to a **Continuous Furnace** (see).

Batch Pickling. See **Still (Tank) Pickling.**

Bath. (a) Mass of molten metal or other material in a furnace or holding vessel. (b) Solution for any chemical treatment, such as pickling, electrodeposition, etc.

Bath Annealing. See **Annealing.**

Bath Metal. Brass (see), with 17–45% zinc, frequently used for castings.

Batter. See **End Flow.**

Battery. (a) Commercially operable combination of two dissimilar metals (or metal and carbon), with an electrolyte between, which will yield an electric current when the two poles are connected. Primary batteries (or cells) cannot be recharged; **Storage Batteries** (see) can be discharged and recharged many times. The Daniell primary cell consists of copper and zinc electrodes, with zinc and copper sulfate solutions as electrolyte. The Lalande cell, extensively used in railroad signal work, uses zinc as anode, copper oxide as cathode, and sodium hydroxide solution as electrolyte. The Leclanche cell has zinc as cathode and carbon as anode, with ammonium chloride and manganese dioxide as electrolyte; practically all dry cells (dry batteries) are of this type. (b) Set of stamps in a stamp mill.

Battery Amalgamation. Amalgamation of the precious metals of an ore, in a stamp mill battery. See **Amalgamation Process.**

Bauxite. The only commercial ore of aluminum, corresponding essentially to the formula $Al_2O_3.xH_2O$.

Bayer Process. Standard process for purifying bauxite, prior to electrolysis, in the manufacture of aluminum. The bauxite is treated with sodium hydroxide (caustic soda) under heat and pressure to eliminate impurities.

B B Wire. See **Best Best Wire.**

Be. Chemical symbol for **Beryllium** (see).

Bead. (a) **Blister** (see) formed on wrought iron, when used in the manufacture of steel by the **Cementation Process** (see), caused by the carbon monoxide resulting from the reaction of the carbon with the slag oxides. (b) Globule of gold or silver obtained in cupellation. (c) Glassy mass of borax or similar chemical, when testing for the presence of certain metals, by heating together, usually in a blowpipe. (d) Line of metal, added during welding, usually raised above the plane of the metals being joined.

Beam. (a) Essentially parallel group of rays emanating from a common source. (b) Steel bar of standard shape, for structural purposes.

Bear. See **Salamander.**

Bearing Brass. See **Bearing Bronze.**

Bearing Bronze. Bronze usable for bearing purposes. Usually, bronzes containing dispersed lead; 5–10% tin and 5–25% lead copper-base alloys are typical. Zinc is an optional constituent.

Bearing Metals. General name, covering various alloys, primarily those of lead, copper, tin, antimony, and zinc; used for bearings.

Bearing Steel. Steel used in the manufacture of anti-friction bearing parts, such as balls, rollers, and races. Typical are those with carbon about 1.00%, chromium 0.80 to 1.50%, and vanadium optional up to 0.20%.

Bearite. Series of lead-base bearing metals, with about 17% antimony, and fractional percentages of copper and bismuth (A. W. Cadman Mfg. Co.).

Bearium. High-Lead Bronze (see) with 18 to 26% lead, used for bearings, etc. (Bearium Metals Corp.).

Beckett Metal. Copper-base bearing metal, with about 35% dispersed lead, and 5% tin (Beckett Bronze Co.).

Becquerel Effect. Electromotive force, resulting from unequal illumination of two identical electrodes, immersed in an electrolyte.

Bedplate. Iron plate, acting as the bottom of a furnace.

Bedstead Tubing. Light-weight steel pipe.

Beehive Coke. See **Coke.**

Beehive Oven. See **Coke.**

Belgian Furnace. Early type of **Zinc** (see) distillation furnace, in which small retorts, in four or five rows, are suspended in the combustion chamber.

Belgian Mill. See **Looping Mill.**

Belgian Process. See **Belgian Furnace.**

Belgian Zinc Furnace. See **Belgian Furnace.**

Bell and Hopper. Arrangement at the top of a blast furnace, for feeding raw materials. A hopper is mounted at the top, with the lower, smaller opening closed by a "bell," lowered whenever feeding raw materials into the furnace. It is also called "cup and cone."

Bell-Krupp Process. See **Bell Process** and **Krupp Process.**

Bell Metal. Copper-base hard alloy, with 15–40% tin, and optional additions of zinc, iron, and lead. Used primarily for bells and chimes.

Bell Process. Process for refining pig iron, in which the molten iron is stirred with molten iron oxide, to oxidize most of the silicon and phosphorus.

Bell's Dephosphorizing Process. See **Bell Process.**

Belly Pipe. See **Blowpipe** (a).

Bench. See **Wire-Drawing Frame.**

Bendalloy. See **Cerrobend.**

Bending Moment. Force, at any point, effective in bending an elastic rod.

Bending Rolls. Rolls for bending plate into round pipe, prior to welding the edges.

Bending Test. See **Bend Test.**

Bend Test. Test to determine deflection characteristics and soundness of metal, in which the material is bent around its axis or around an outside radius.

Benedict Metal. Copper-nickel alloy, with nickel in the 14–16% range, used in the manufacture of tubing.

Benedict Nickel. Special **Nickel-Silver** (see), now used only in cast form, with approximately 20% zinc, 12.5% nickel, 10.5% lead, and 2% tin.

Beneficiation. Treatment of an ore to make it more suitable for metallurgical use.

Beraloy. Series of copper-base alloys with up to 2.5% beryllium, added for age-hardening (Wilbur B. Driver Co.).

Berlin Iron. Relatively soft iron containing considerable proportions of phosphorus, used for smooth, small, ornamental castings.

Bertrand-Thiel Process. Two-step modification of the basic **Open-Hearth Process** (see) of steel manufacture, in which high-phosphorus pig iron is used as raw material. After removal of the high-phosphorus slag in the first treatment, the metal is subjected to a secondary purification.

Berylco Alloy. Series of copper-beryllium alloys, manufactured by Beryllium Corp. of Pennsylvania, in which age-hardening qualities are combined with good electrical conductivity. Available in strip or wire form; also as casting ingots.

Beryllium (chemical symbol Be). Element No. 4 of the periodic system; atomic weight 9.02; also known as glucinum (chemical symbol Gl). Grayish metal, of highly crystalline nature, it resists atmospheric corrosion much better than magnesium and aluminum; comparable with magnesium in lightness, having a specific gravity of only 1.85. Unlike this metal, however, it is quite brittle at all temperatures; small additions of zirconium or titanium increase malleability at higher temperatures. Beryllium is divalent; otherwise, its chemical reactions follow those of aluminum very closely; melting point 2345°F; boiling point uncertain but probably approximates 5400°F. Manufactured in this country by converting the oxide (made from beryl, a double aluminum silicate) into the anhydrous chloride, which is then electrolyzed in a fused bath of sodium chloride. Used primarily in the form of a 2–3% alloy with copper, on which it confers remarkable age-hardening and non-sparking characteristics. Light-weight alloys of beryllium have thus far proved disappointing. Beryllium cannot be electrodeposited. The metal and its alloys are available in the United States from Beryllium Corp. of Pennsylvania and Brush Beryllium Co.

Beryllium-Aluminum Alloys. Beryllium hardens aluminum, even in fractional percentages. Beryllium-aluminum alloys of higher beryllium content (up to 70% beryllium can be rolled into sheet form) have not been commercially successful to date, despite high tensile strength and hardness.

Beryllium Bronze. Copper-beryllium alloys, with 1 to 2.5% beryllium. See also **Beryllium Copper.**

Beryllium Copper. Copper containing approximately 2–3% beryllium and optionally fractional percentages of nickel or cobalt. Alloys of this series show remarkable age-hardening properties, and an ultimate hardness of about 400 Brinell. Because of such hardness and good electrical conductivity, beryllium copper is used in electrical switches, springs, etc.

Beryllium Manganese Bronze. Beryllium bronze of about 1% beryllium content, with about 3–10% manganese, added to increase strength and hardness.

Beryllium-Nickel. Nickel-base alloys, with up to 2.5% beryllium added to give age-hardening characteristics. Hardnesses up to 600 Brinell can be reached, with high tensile strength.

B.E.S.A. Abbreviation for British Engineering Standards Association. See **British Standards Institution.**

Bessemer Basic Pig (Iron). See **Basic Pig (Iron).**

Bessemer Converter. See **Bessemer Steel.**

Bessemer Iron. See **Bessemer Pig (Iron).**

Bessemer Iron Ore. See **Bessemer Ore.**

Bessemerizing. See **Bessemer Steel.**

Bessemer Ore. Iron ore, yielding pig iron with phosphorus content not exceeding 0.09%.

Bessemer Pig (Iron). Pig Iron (see), with phosphorus below 0.10%, used for making steel by the Bessemer or the acid open-hearth process.

Bessemer Process. See **Bessemer Steel.**

Bessemer Steel. Steel, manufactured by a process invented by Bessemer in England (and, simultaneously, by Kelly in the U.S.A.). Molten pig iron, held in a pear-shaped tilting converter of 5 to 25 tons capacity, is transformed into steel by the action of a blast of air passing up through the molten metal. This step is known as the blow, and the reaction between the oxygen of the air and the carbon and silicon of the metal can be followed by the changes in appearance of the flame at the mouth of the converter. At first, there is a shower of sparks, then dense brown fumes, followed by short reddish flames. After about five minutes, a short luminous flame appears around the mouth, gradually increasing in size and luminosity till it grows to thirty feet length, maintained for about eight minutes. This is known as the boil, during which the converter emits a roar caused by the agitation of the bath by the air blast and the carbon monoxide, resulting from the oxidation of the carbon. When the blow is about to end, the flame loses much of its luminosity. At this point, additions of anthracite coal, ferromanganese or spiegel, ferrosilicon, etc. may be made to recarburize the steel and deoxidize it simultaneously, bringing up the manganese and silicon contents to the desired values, after which the metal is poured, through the mouth, into a ladle. The adjusting additions can also be made in the ladle. The steel is then usually cast from the ladle through a bottom teeming hole into ingots which, when sufficiently cooled, are removed to soaking pits, preparatory to rolling. Acid Bessemer steel is made by using low-phosphorus pig iron as raw material, the phosphorus being under about 0.09% in the pig. Basic Bessemer steel utilizes pig iron with over 2% of phosphorus, removing this objectionable contaminant through the use of limestone during the conversion.

Best Bar. See **Double Rolled Iron.**

Best Best Wire. Steel wire of medium electrical resistance. See **Telephone & Telegraph Wire.**

Best Bright Finish. Highly glossy finish on cold-rolled strip or sheet metal, produced by highly polished finishing rolls.

Best Cokes. See **Coke Tin Plate.**

Beta Antimony. Allotropic modification of **Antimony** (see), amorphous **black powder,** obtained by oxidizing antimony hydride, at low temperatures.

Beta Arsenic. See **Black Arsenic.**

Beta Brass. Form of **Brass** (see) (body-centered cubic) of between 45.5 and 50.0% zinc content. Corresponds to the compound CuZn. Between 39.0 and 45.5% zinc, alpha and beta brasses are in equilibrium.

Beta Iron. Paramagnetic alpha iron, transformed from magnetic alpha iron when heated above 1414°F; since it has the same crystal structure and X-ray space lattice as magnetic alpha iron, its existence as a separate allotrope is no longer accepted by most metallurgists.

Beta Rays. Electrons emitted by radioactive materials.

Beta Tin. Metallic **Tin** (see) in its common, massive form.

Beta Transformation. Radioactive transformation characterized by the loss of an electron from the atomic nucleus.

Bethadur. Series of stainless steels, manufactured by Bethlehem Steel Co., with up to 21% chromium, 11% nickel, and the usual optional elements. The alloys are of high strength as well as corrosion and heat resistance.

Bethalon. Series of free-machining stainless steels, akin to the **Bethadur** (see) series, with fractional percentages of sulfur, added for greater machinability.

Bethanized Steel. Steel wire, electrolytically coated with zinc (Bethlehem Steel Co.).

Beth-cu-loy. Copper-bearing steel (Bethlehem Steel Co.), used in sheet form for sheet construction purposes.

Bethlehem Steel. Series of steels manufactured by Bethlehem Steel Co.

Betts Lead-Refining Process. See **Betts Process.**

Betts Process. Process of electrolytically purifying lead, in which lead fluosilicate is used in the electrolytic bath. Gold and silver are recovered from the anode residues.

BHN. Brinell hardness number. See **Brinell Hardness (Test).**

Bi. Chemical symbol for **Bismuth** (see).

Bidery Metal. Zinc-base alloy, with about 6% copper, 6% lead, and, optionally, 1% tin.

Bidri Metal. See **Bidery Metal.**

B.I.G. See **Metal Gage.**

Big-End-Down Mold. Ingot Mold (see) in which the bottom cross-section exceeds the top cross-section. In general, ingots of this type are easy to strip.

Big-End-Up Mold. Ingot Mold (see) in which the top cross-section exceeds the bottom cross-section. In conjunction with a hot top, the use of this type of a mold helps cut down losses by **Pipe** (see). In general, ingots of this type are difficult to strip.

Billet. Short thick bar, semi-finished in hot-rolled form from an ingot, for use in forging or rolling. Current usage limits the term to bars at least 1.5 inches in width and thickness, with cross-sections up to 36 square inches. See also **Bloom, Slab,** and **Sheet Bar.**

Billeting Roll. See **Billet Mill.**

Billet Mill. Rolling mill for the reduction of ingots to billets.

Billy. Small, idling roller, on the delivery side of a rolling mill, to help the catcher [see **Catcher** (a)] return the metal over the top roll to the **Roller** (see).

Bimetal. Two dissimilar metals adhering face-to-face, forming a composite. Normally, the term is limited to such a composition when used in temperature-measuring or regulating devices, operating by means of unequal rates of heat expansion, which causes the bimetal to curl and uncurl.

Bimetalism. Use of both gold and silver, at fixed relative values, as monetary standards.

Binary (Alloy). Alloy containing two major elements (exclusive of impurities).

Binary Diagram. Constitution Diagram (see) for a binary alloy system.

Binary System. See **Binary Diagram.**

Binder. In powder metallurgy, material used to help cement the metal powder together.

Birmingham Platina. Zinc-base alloy, with 20–47% copper.

Birmingham (Wire) Gage. See **Metal Gage.**

Bismuth (chemical symbol Bi). Element No. 83 of the periodic system; atomic weight 209.00. Unlike arsenic and antimony, it has only one modification: a relatively soft, crystalline, reddish-white metal of high luster and brittle nature; melting point 520°F; boiling point approximately 2840°F; specific gravity 9.78. It has valences of three (bismuthous) and five (bismuthic). Bismuth is usually obtained by reduction of its oxide or sulfide with carbon and iron. Metallurgically, its most common use is in conjunction with lead, antimony, and tin; it confers low-melting characteristics to alloys of this type and expansion on cooling. Bismuth can be electrodeposited from the fluosilicate or the acidified chloride; not used commercially.

Bismuth Bronze. Copper-base alloy, with high percentages of nickel (10% to 33%) tin (about 15%) and zinc (about 21%), and with about 1% bismuth.

Bismuth Silver. (a) Native silver ore, containing bismuth and usually approximating Ag_6Bi in chemical composition. (b) Non-metallic mineral, sulfide of lead, silver, and bismuth (schapbachite), approximating $PbAg_2Bi_2S_5$ in chemical composition.

Bismuth Solder. See **Fusible Alloys.**

Bite. Ability of rolls to engage metal for passage between the rolls.

Bivalent. Having a chemical **Valence** (see) of two.

Black Annealed Wire. See **Pot Annealed Wire.**

Black Annealing. See **Annealing.**

Black Antimony. See **Beta Antimony.**

Black Arsenic. Allotropic modification of **Arsenic** (see), obtained by condensing arsenic vapor under controlled conditions.

Black Body. Theoretically perfect surface, absorbing all radiation falling on it, without reflection or scattering.

Black Boil. See **Scaling Dip.**

Black Copper. (a) Impure metallic copper, containing several per cent of other elements, such as sulfur, iron, and lead; a by-product of blast-furnace reduction of copper ores. (b) See **Tenorite.**

Black Diamond. Non-gem-quality diamond, used for abrasive purposes.

Blackening. (a) Coating the inner surface of a mold with **Blacking** (see). (b) Treating steel in a chemical bath (alkaline and oxidizing) to form a black surface-film of iron oxide.

Black Heart (Malleable) Iron. Malleable iron with a completely decarbonized shell, white in color, and a dark core, containing finely precipitated temper carbon; this is the common form of malleable cast iron.

Black Heat. Temperature just below **Low Red Heat** (see).

Blacking. Graphite or similar coating applied to the surfaces of a **Mold** (see), prior to the casting of metal.

Blackjack. See **Mock Lead.**

Black Lead. Graphite. (The word "lead" here is misleading, since the material contains no lead and is completely non-metallic.)

Black Oxide Coating. Coating black iron oxide on iron or steel objects, usually by heating to about 1200°F for a long period, in a closed box, with steam under pressure, followed by cooling to about 300°F and dipping into linseed oil.

Black Pickling. Pickling (see) to remove hot mill scale.

Black Plate. Steel plate of 12 to 32 inches width, commonly of the acid Bessemer or basic open-hearth type, obtained by cold reduction, prior to any pickling or cleaning operation.

Black Red Heat. Temperature at which hot metal barely begins to be visible in daylight, about 1000°F.

Black Silver. Non-metal, sulfur-bearing ore of silver and antimony, approximating Ag_5SbS_2 in composition.

Black Slag. First slag forming in basic electric steel operations; black in color, because of the presence of free carbon.

Black Solder. Alloy of copper and zinc, with a small proportion of tin.

Black Tellurium. Non-metal, tellurium-containing ore, with gold, silver, and sulfur as other constituents.

Black Tin. Non-metal tin ore, after dressing, ready for smelting.

Black Wire. Steel wire of any shape, obtained directly by drawing or annealing, without pickling or cleaning.

Blackwork. Iron forgings, with surface as obtained directly from forging operations.

Blair Process. See **Chenot Process.**

Blake Furnace. Furnace with circular rotating hearth, terraced and rising toward the center.

Blanched Copper. Copper-base alloy, with arsenic as the alloying element.

Blanching. See **Tinning.**

Blank. (a) Piece of metal to be worked, by forging or stamping, into final form. (b) In powder metallurgy, a compacted mass of powdered metal, usually unfinished, which has been pressed, presintered, shaped, cut, or sintered. (c) Sheet metal with one or more edges curved to predetermined design.

Blast Cleaning. Cleaning surface of metal by air blast, using sand, grit, or metal shot as abrasive.

Blast Furnace. Standard apparatus for reducing iron ore to pig iron. The modern blast furnace is an essentially cylindrical structure, about 100 feet in height, and 12 feet in diameter. The bottom section, called the hearth or crucible, collects the molten iron. At a convenient point, in the crucible, is located the iron notch, or iron tapping hole, and slightly above this, the cinder notch for removal of the slag. Further up, evenly distributed around the upper circumference of the hearth, are the tuyeres, inlet passages for the hot air blast, furnished by blast engines and heated by hot blast stoves. Above the hearth, the blast furnace widens out gradually into the bosh, in which most of the interaction between ore, coke, and flux occurs. At the top of the bosh is the mantle, on which rests the remainder of the furnace, known as the shaft or stack, divided, imaginarily, into three, almost-equal parts, called the upper, middle and lower inwalls. From here, the reaction mixture feeds down into the body. In the top of the shaft are the appliances for continuous feeding in of the ore, coke, and flux, and for proper distribution. See **Bell and Hopper.** Similar apparatus are also used for many other metals, notably lead, copper, and nickel.

Blast Furnace Dust. Dust recovered from blast furnace gases, containing sufficient potash to be of interest as a fertilizer ingredient.

Blast Furnace Gas. Gas resulting from blast furnace operations; mixture of approximately 26% carbon monoxide, 4% hydrogen, 13% carbon dioxide, and 57% nitrogen. Generally used in the plant either directly, to drive engines, or as fuel.

Blast Furnace Process. See **Blast Furnace.**

Blast Furnace Slag. Slag formed in the operation of a **Blast Furnace** (see).

Blast Hearth. Hearth (see) operating with an air blast.

Blasting. See **Blast Cleaning.**

Blast Pipe. Pipe conducting air blast to a furnace.

Blast Roasting. Process of **Roasting** (see) ores, in which oxidation is achieved by internal combustion within a mass of ore by means of air currents forced through it.

Bleeding. (a) Phenomenon in some **Killed Steel** (see) ingots, from which a strong evolution of hydrogen or, sometimes, nitrogen, during the last stages of solidification, ejects molten metal through the top (see also **Rising**). (b) Shrinkage of castings, causing formation of small cavities.

Blende. See **Mock Lead.**

Blende Roasting. Roasting (see) of zinc ores, which can be either dead roasting or partial roasting.

Blick. Iridescence on silver or gold, appearing at the end of **Cupellation** (see).

Blind Roaster. See **Muffle Furnace.**

Bliss Mill. Rolling Mill (see), of the cluster mill type, consisting of six rolls, four of which act as back-up rolls.

Blister. (a) Defect in formed metal, on or near the surface, resulting from gas bubbles which failed to escape from the original molten metal. (b) See **Blister Copper.**

Blister Bar. See **Blister Steel.**

Blister Copper. Copper (about 96–99% in purity), after production by any smelting process; the blistered appearance is probably due to gas pockets.

Blister Refining. Refining **Blister Copper** (see), usually by a sequence of fire and electrolytic refining steps.

Blister Steel. Steel resulting from the **Cementation Process** (see) when wrought iron is used as the raw material. The surface of the steel is often covered with blisters due to carbon monoxide resulting from the reaction of carbon with the slag incorporated in the wrought iron.

Block. Apparatus for drawing wire; consists of a rotating (usually between 40 and 85 R.P.M.), tapered, hollow drum which draws the metal through the die and up toward the top of the drum.

Blocking. Maintaining carbon at a constant value in the manufacture or casting of steel, by adding pig iron, ferrosilicon, etc., to the fused metal.

Blocking Layer. Surface of contact between a conductor and semi-conductor, acting as a rectifier.

Block Spangle. Spangle on zinc-coated (galvanized) steel, in the form of regular blocks, usually obtained by conveying the sheets away from the bath on a wire-netting of the appropriate mesh.

Block Stripper. Hoist located above a drawing **Block** (see) to lift away coils of drawn metal.

Block Tin. Solid commercial tin, as distinguished from tinplate (tin-coated steel), commonly, though erroneously, called "tin."

Blood Red Heat. See **Dark Red Heat.**

Bloom. Semi-finished bar, hot-rolled from an ingot, for use in forging or rolling, with a rectangular cross-section of at least 36 square inches. See also **Billet, Slab,** and **Sheet Bar.**

Bloomery. Hearth type of furnace, formerly used for shaping wrought iron directly from the ore. The metal sides of the furnace were water-cooled and a hot blast was obtained by preheating. Alternate layers of charcoal and iron ore were added to the hearth, a pasty mass of iron, mixed with considerable slag, forming at the bottom. The sides of the furnace were called iron plates, the back hair plate, the front cinder plate. One side-plate, carrying the tuyere, was known as the tuyere plate, its upper part as the merrit plate; opposite this was the forespan plate, its top part known as the skew plate. At the bottom was the bottom plate.

Bloom Hook. See **Bloom Tongs.**

Blooming. Converting puddle balls into iron blooms. See **Puddling.**

Blooming Mill. Rolling Mill (see) for the reduction of ingots to blooms. There are two types of two-high blooming mills (one roll above another): the continuous one-directional type, and the reversing type. Three-high blooming mills are also in common use; in these, the top and center rolls flow the metal in one direction, whereas the center and lower rolls cause flow in the opposite direction.

Bloom Tongs. Device for holding and manipulating **Blooms** (see).

Blow. See **Bessemer Steel.**

Blowhole. Gas or air pocket in a metal casting, caused either by evolution of gas, released from the solidifying metal, or by mechanical entrapment in the mold.

Blowhole Segregation. Differences in composition occurring around blowholes, or in the filled zones formerly occupied by blowholes, in ingots. After working, blowhole segregates give rise to **Ghosts** (see).

Blowing Engine. See **Blast Furnace.**

Blowing In. Starting the operation of a **Blast Furnace** (see).

Blowing Out. Closing down the operation of a **Blast Furnace** (see).

Blowing-Up Furnace. Furnace for sintering lead and zinc ores and volatilizing the metals.

Blown Metal. Steel, obtained during the Bessemer process, after completion of the blow and before finishing, such as addition of ferrosilicon and recarburizers.

Blowoff Valve. Valve in **Blast Furnace** (see) system, for relieving pressure and allowing for escape of air when changing a hot-blast stove from air to gas, in the regenerative cycle.

Blowpipe. (a) In a **Blast Furnace** (see) a long horizontal pipe, conducting hot blast to the tuyeres; also known as belly pipe. (b) Torch, using an inflammable gas with air or oxygen. (c) Apparatus for creating a more intense flame, in which air is forced, through a separate tube, into the flame.

Blowtorch. See **Torch.**

Blue-Annealed Sheet. Hot-rolled heavy steel sheet, after blue **Annealing** (see) and flattening.

Blue Annealing. See **Annealing.**

Blue-Billy. Residue from the roasting of copper sulfide ores with salt.

Blue Brittleness. Lowered resistance to impact, exhibited by some steels when heated to about 400–600°F and also when so treated and then cooled to ordinary temperatures. It is probably due to precipitation of oxides and nitrides, since thoroughly **Killed Steel** (see) shows little effect of this sort.

Blue Brittle Range. Temperature range (about 400°F to 600°F) in which carbon and pearlitic alloy steels increase in tensile strength and yield point and decrease in ductility. See also **Blue Brittleness.**

Blue Chip (Tool) Steel. Series of alloy tool steels, with about 18% tungsten, 4% chromium, 1% vanadium, and, optionally, 3% cobalt (Firth Sterling Steel Co.).

Blued Plate. Black Plate (see) subjected to oxidizing by steam or air to produce a blue surface color.

Blued Sheet. Hot-rolled, annealed steel sheet, with final treatment of heating and exposure to steam or air, in order to obtain a blue film of oxide.

Blued Stock. See **Blued Plate.**

Blue Dust. By-product of zinc reduction, containing about 90% metallic zinc and 5–8% zinc oxide; a fine, bluish-gray powder.

Blue Gold. Alloy of gold with approximately 25–33% iron.

Blue Heat Range. See **Blue Brittle Range.**

Blueing. See **Bluing.**

Blue Lead. Non-metallic by-product obtained from the fumes in smelting lead ores; chemically, a mixture of lead-sulfur and lead-oxygen compounds.

Blue Metal. (a) See **Blue Dust.** (b) Copper matte of about 60% copper content.
Blue Powder. See **Blue Dust.**
Blue Room. First room in a **Bag House** (see).
Blue Shortness. See **Blue Brittleness.**
Bluing. (a) Producing an oxide film on steel, blue in color due to light interference, by low-temperature heating. (b) Heat treatment of steel springs, to relieve cold-working strains.
Bob. See **Drop Forging.**
Body. See **Roll Body.**
Body-Centered Cubic Space Lattice. See **Body-Centered (Cubic) Structure.**
Body-Centered (Cubic) Structure. Term describing internal crystal structures of materials, as determined by X-ray studies, in which the equivalent lattice points are at the corners of the unit cell and at the center of the cube.
Body Stock. Tin plate for the manufacture of the bodies of tin cans.
Boecker's Wire Mill. See **Double Belgian Mill.**
Bog Iron Ore. See **Limonite.**
Bog Manganese. Non-metallic, hydrated manganese ore, containing manganese and other oxides.
Bohn—. Series of copper-base alloys of highly diversified composition, manufactured by Bohn Aluminum & Brass Co.
Bohnalite. Aluminum-base alloy series (Bohn Aluminum & Brass Corp.), with optional copper up to 11%, silicon up to 13%, iron up to 1.5%, zinc up to 14%, cadmium up to 2%, and fractional percentages of manganese, chromium, and nickel. Extensively used for castings, particularly in the automotive and aeronautic fields.
Boil. Generation of steam by molten metal when running over a damp spot in a runner. See also **Bessemer Steel** and **Puddling.**
Boiler Iron. See **Boiler Plate.**
Boiler Plate. Soft steel plate or sheet of a quality (low-carbon) suited for boiler manufacture.
Boiler-Plug Alloys. See **Fusible Alloys.**
Boiler Steel. Steel, usually made in the open hearth, of low carbon content, and of a quality suited for boiler manufacture and cold-bending.
Boiling. Tendency of the ground coat of **Vitreous Enamel** (see), under faulty conditions of operation, to become covered with bubbles which penetrate through the cover coat. See also **Puddling.**
Boiling Point. Temperature at which the vapor pressure of a liquid equals the outside pressure, resulting in boiling.
Boilings. Slag, running out of the tap hole, during the "boil" steps of wrought iron manufacture. See **Puddling.**
Bonderite. Coating iron and steel with phosphate, for rust-proofing and to yield a better bonding surface for paint and lacquer.
Bonderizing. See **Bonderite.**
Bonding. See **Cladding.**
Bonding Treatment. (a) Chemical treatment of the surface of steel to improve the adhesion of paint, enamel, or similar coatings. (b) See **Cladding.**
Bonney-Floyd Steel. Series of low-alloy and high-alloy steels manufactured by Bonney-Floyd Co.
Boot. Joint, connecting a blast main with a **Tuyere** (see).
Booth Furnace. See **Booth-Hall Furnace.**
Booth-Hall Furnace. Indirect electric **Arc Furnace** (see), with single, double, or three phase connections and electrodes imbedded in the hearth, a hearth electrode for each incoming electrode.
Boot Leg. See **Tuyere Stock.**
Borcher's Metal. Series of nickel-chromium-iron alloys of broadly varying composition and high heat and corrosion resistance.
Borcher's Process. Electrolytic method for refining silver, using silver sheet cathodes and a dilute nitric acid electrolyte.
Borgenet Furnace. Belgian type zinc **Furnace** (see), with a single combustion chamber.
Borium. Hard-facing welding rod, composed principally of tungsten carbide (Stoody Co.).
Bornite. See **Peacock Copper.**

Borod. Hard-facing welding rod composed principally of tungsten carbide (Stoody Co.).

Borolon. Fused alumina product, used for refractory and abrasive purposes; manufactured by the Abrasive Co.

Boron (chemical symbol B). Element No. 5 of the periodic system; atomic weight 10.82; generally known as a brown, non-metallic powder obtained by interaction of boric oxide with magnesium; a more metallic allotropic modification can be obtained by electrolysis of fused complex boron fluorides, but this modification has not been studied; melting point approximately 4175°F; boiling point about 4625°F; specific gravity 2.54. Its normal chemical valence is three. Boron cannot be electrodeposited. Alloys with copper and iron are occasionally used as scavengers for these metals; alloys with other metals (manganese-boron and calcium-boron) are studied for the same general purpose. Recently, addition of minute amounts of boron (around 0.003%) has been gaining in popularity, to improve hardenability of steels. Boron alloys are available from Electro Metallurgical Co. and from Molybdenum Corp. of America.

Boron Alloys. Alloys of boron with the more common metals. Among those commercially available are **Ferro-Boron** (see) and **Manganese-Boron** (see).

Boron Carbide. Product of electric furnace interaction of carbon and boric oxide; the carbide corresponds to the formula B_4C, is extremely hard and consequently, is used for special abrasive purposes.

Borosil. Boron-containing iron master alloy, with silicon as an additional constituent, added to steel to increase hardenability (Ohio Ferro-Alloys Corp.).

Boroto Metal. Alloy of tin-lead-antimony base, in which colloidal graphite is incorporated. With Brinell hardness varying from 17 to 35, it is used for self-lubricating bearings.

Bort. Non-gem quality diamond, used for abrasive purposes, wire-drawing dies, tool tips or inserts, etc.

Bortam. Boron-containing master alloy for imparting deep hardenability to steel; consists of 16–18% titanium, 13–15% aluminum, 22–24% manganese, 20–25% silicon, and 1.5–2.0% boron, with carbon not exceeding 1% (Titanium Alloy Mfg. Co.).

Bosh. Portion of shaft furnace, above the hearth, in which it widens out to a maximum. See also **Blast Furnace.**

Bosh Breakout. Breakout of gas or coke through the brickwork of a **Blast Furnace** (see), caused by faulty operation.

Bosh Jacket. Water-cooled jacket for cooling the walls of a **Shaft Furnace** (see).

Bosh Plate. Iron casting of hollow flat design, water-cooled, to protect furnace walls.

Boss Process. Modification of the **Pan-Amalgamation Process** (see) in which the ore slurry flows continuously through a series of pans and settling tanks, making the process continuous.

Bott. (a) Short tapered iron rod, attached to a long steel rod handle, used to plug up the "cinder notch" in a **Blast Furnace** (see); it fits into the water-cooled "monkey." See **Cinder Cooler.** (b) Mass of clay, held at the end of a bar, to stop off the flow of molten metal or slag from any holding vessel.

Botting. Use of a **Bott** (see) to stop off the flow of molten metal or slag.

Bottled Gas. Propane or butane, under pressure in cylinders, for use as a source of heat when burned, as in a welding torch.

Bottle Top Mold. Ingot mold, with the top constricted; used in the manufacture of "capped steel," the metal in the constriction being covered with a cap fitting into the bottle neck, which stops "rimming" action by trapping escaping gases.

Bottom. See **Making Bottom.**

Bottom Blowing Converter. Bessemer converter (see **Bessemer Steel**), in which the air blast is forced upward through openings in the bottom of the converter.

Bottom Board. In casting technique, a board on which the bottom part of a pattern is laid.
Bottom Casting. Casting metal into a mold through the bottom of the mold by means of runners, through which the molten metal flows.
Bottom House. Building in which Bessemer converter bottoms (see **Making Bottom**) are repaired.
Bottom Plate. Plate supporting a **Mold** (see). See also **Bloomery.**
Bottom Pouring. (a) See **Bottom Casting.** (b) See **Teeming.**
Bottoms. Impure metallic copper, separating from the matte in smelting operations, when sulfur is inadequate to retain all the copper in the matte.
Bottone (Hardness) Test. Hardness test in which a small cylinder is pressed into the surface of the specimen to a specific distance, the load required being used as a measure of the hardness.
Bourne-Fuller Steel. Series of alloy tool steels, manufactured by Republic Steel Corp.
Bowco Steel. Series of complex alloy high-speed tool steels, manufactured by Craine-Schrage Steel Co.
Bower-Barff Process. Coating steel with a film of black oxide; the metal is heated to about 1200°F in a closed vessel, with steam under pressure, followed by cooling to 300°F and treatment with hot linseed oil.
Bowing. Lack of flatness in sheet or strip metal, the metal cross-section forming an arc.
Bowl Metal. Impure antimony, obtained from the **Doubling Process** (see).
Box. See **Flask.**
Box Annealing. See **Annealing.**
Box Groove. See **Box Pass.**
Box Hardening. Case-Hardening (see) conducted in an iron box.
Box Metal. Anti-friction or copper-base bearing alloy, used in the journal boxes of axles and shafting.
Box Pass. Rolling mill in which the combination of grooves in the rolls forms a square or rectangle, at the rolling cross-section.
Box Tuyere. Tuyere (see), consisting of an annular hollow iron casting, continuously cooled by water, passing through.
Boyd-Wagner Steel. Series of magnet and low-alloy tool steels, manufactured by Boyd-Wagner Co.
Braeburn Steel. Series of alloy, high-speed tool steels, of the high-tungsten type, manufactured by Braeburn Alloy Steel Co.
Bragg (X-ray) Method. Method of X-ray examination of crystals, using a single large crystal rotated through a small angle around an axis in a crystal face.
Brale. Diamond penetrator, essentially conical in shape, used with the Rockwell hardness tester, for hard metals. See **Rockwell Hardness (Test).**
Branner. Cleaning machine for removing oil from **Tin Plate** (see) and **Terne Plate** (see), after removal from the metal and palm oil baths.
Branner Mark. Flaw in **Terne** (see) and **Tin Plate** (see), caused by faulty operation of the **Branner** (see).
Brass. Alloy series of copper and zinc, with small amounts of other optional contituents, such as lead and tin. "Red Brass" contains about 77–86% of copper; it may be worked either hot or cold and is weldable by all standard procedures. "Yellow Brass" (also called "High Brass") contains approximately 63–76% of copper, has excellent cold-working and fair hot-working properties, and is weldable by the oxy-acetylene torch, resistance, and related methods.
Brass Foundry. See **Foundry.**
Brass Furnace. Furnace for melting brass, usually for casting purposes; usually of the reverberatory or crucible type.
Brass Ingot Metal. Brass, in ingot form, for casting purposes.
Brass Latten. See **Latten.**
Brassoid. Zinc sheet, coated with brass (American Nickeloid Co.).
Brass Powder. (a) Finely divided brass. (b) Mixture of finely divided copper and ochre.
Brass Special Zinc. Metallic zinc with total impurities not over 1.0%. Lead is

limited to 0.60%, iron to 0.03% and cadmium to 0.50%, with no aluminum allowed.

Brass Tin. Tin sheet, coated with brass (American Nickeloid Co.).

Braziers' Copper. Heavy copper sheet, used extensively in coppersmithing.

Brazing. Joining metals by means of alloys which melt at temperatures considerably below the melting points of the metals being joined. The more important brazing alloys include copper alloys of various formulae, and also the silver solders, the latter used where low melting and flow points are desired. Copper brazing, usually limited to steel, involves the use of pure copper as the brazing metal.

Brazing Metal. Copper-base alloy with about 15% zinc, frequently used for plumbing fixtures, hardware, etc.

Brazing Rod. Metal or alloy rod or wire, used in gas or electric **Brazing** (see).

Brazing Solder. See **Hard Solder.**

Breakdown. Steel **Sheet Bar** (see) after partial hot-rolling.

Breakdown Pass. See **Roughing Stand.**

Breakdown Rolls. See **Roughing Stand.**

Breakdown Stand. See **Roughing Stand.**

Breaker Pass. See **Scale-Breaking.**

Breaking Down Rolls. (a) See **Roughing Stand.** (b) See **Scale-Removing Rolls.**

Breaking Load. Load acting on a specimen under tensile test at the moment of fracture. The breaking load is less than the ultimate strength, because necking decreases the area under tension.

Breaking Strength. See **Breaking Load.**

Breaking Stress. See **Breaking Load.**

Breakout. Break in an open-hearth furnace, resulting in the metal and slag cutting through the furnace walls and breaking out of the furnace.

Breast. Side of hearth, in a shaft furnace, containing the **Metal Notch** (see).

Bridge. Cross wall in a furnace. In an **Air Furnace** (see), the wall near the fireplace is known as the "fire bridge" and the one near the stack is the "flue bridge."

Bridge Wire. High-quality, flexible galvanized steel wire, for use in cable suspension bridges.

Bridging. In powder metallurgy, the formation of arched cavities in the powder, resulting in voids, after the compact has been sintered.

Briggs Pipe Thread. Standard dimensions for pipe and tube threads; also known as American Standard Pipe Thread.

Briggs Standard (Gage). Pipe sizes and thicknesses, now generally accepted as standard.

Bright Annealed Wire. Steel wire, bright drawn and annealed by heating in sealed pots, in the presence of a non-oxidizing atmosphere.

Bright Annealing. See **Annealing.**

Bright Basic Wire. Bright steel wire, slightly softer than **Bright Bessemer Wire** (see), used for screws, bolts, rivets, etc.

Bright Bessemer Wire. Stiff bright steel wire of hard temper, normally drawn to size without annealing; used for nails, screws, cheap springs, etc.

Bright Cherry Heat. Temperature approximating 1470°F, as judged visually.

Bright Dip. Acid solution into which articles to be electroplated are dipped to obtain a clean, bright surface.

Brightener. Material added to an electroplating bath, to give the deposited metal a shiny surface.

Bright Finish. High-lustre and high-polish surface **Finish** (see) on metal, usually obtained by cold-finishing or by special polishing operations.

Bright-Finish Strip. Steel strip, with bright finish. See **Best Bright Finish** and **Regular Bright Finish.**

Bright Finish Tubing. See **Ellwood Bright Finish Tubing.**

Bright Hard Basic Wire. See **Bright Basic Wire.**

Bright Hard Wire. See **Bright Basic Wire.**

Bright Plating. Electroplating (see), with operating conditions and addition agents adjusted so as to yield a bright coating without need for elaborate polishing.

Bright Rope. Steel rope in which the metal wire has not been coated.
Bright Soft Basic Wire. Bright steel **Wire** (see), of soft temper, suitable for upsetting, forming, twisting, and other cold working operations; used for screws, chains, etc.
Bright Soft Bessemer Wire. Bright Bessemer steel wire of soft temper, suitable for difficult bending and flattening operations; used for tacks, chains, screens, etc.
Bright Soft Wire. Steel wire, drawn dry and annealed at some point during drawing, to yield a wire softer than **Bright Wire** (see).
Bright Wire. Steel wire, cold-drawn dry, without treatment subsequent to cold-drawing.
Brine. Solution of any salt in water. Commonly, sodium chloride (common table salt) is meant.
Brine Leaching. Recovering silver, lead, copper, or zinc values from ores, by contacting, after chloridizing roasting, with a strong solution of salt, frequently acidified with sulfuric acid.
Brinell Hardness (Test). Standard method of measuring the hardness of metals. The smooth surface of the metal is subjected to indentation by a hardened steel ball under pressure or load. The standard load and time are 500 kilograms for 60 seconds for soft metals, and 3000 kilograms for 30 seconds for steel and other hard metals. The diameter of the resultant indentation, in the metal surface, is measured by a special microscope and the Brinell hardness value read from a chart or calculated by formula.
Brinell Monotron Diamond. See **Monotron (Hardness) Test.**
Briquet. See **Compact.**
Briquette. See **Compact.**
Bristol Brass. Yellow **Brass** (see) (61–75% copper).
Britannia (Metal). Tin-base alloy, with antimony and other metals as hardeners; usually contains 80–95% of tin; used as the base for plated table ware and ornamental goods, and for bearings.
British Engineering Standards Association. See **British Standards Institution.**
British Imperial Gage. See **Metal Gage.**
British Plate. See **Albatra.**
British Standards Institution. Organization for issuing standard specifications, akin to the American Society for Testing Materials.
British Standard Specification. Specification on materials, as issued by **British Standards Institution** (see).
British Standard Wire Gage. See **Metal Gage.**
British Thermal Unit. See **B.T.U.**
British Wire Gage. See **British Standard Wire Gage.**
Brittleness. Characteristic of materials of fracturing when subjected to impact or heavy pressures. Coarse grained metals are prone to brittleness because of large cleavage planes. In general, the brittleness of alloys increases with hardness and grain size.
Broad Cold-Rolled Strip. See **Cold-Rolled Strip.**
Broad Hot-Rolled Strip. See **Hot-Rolled Strip.**
Broadside Mill. Rolling Mill (see), used on plate and strip, for rolling the slab crosswise; equipped with a turntable, to turn the metal slab for edgewise feeding into the mill. The broadside mill is usually followed by a "slab squeezer" to adjust the width and flatten the slab.
Broad Strip. See **Cold-Rolled Strip** and **Hot-Rolled Strip.**
Broken-Stick Fracture. See **Fracture.**
Bromocyanide Process. Method of recovering values from refractory or special gold ores, in which cyanogen bromide (CNBr), or a chemical mixture forming it, is used for treating the ore.
Bronzalum. Antislip Metal (see), with bronze as matrix (American Abrasive Metals Co.).
Bronze. Alloy series in which copper predominates (except the binary copper-zinc alloys, known as "brass"). Most commonly, and when not subjected to any special qualifications, "bronze" refers to the alloys of approximately 88% copper, with 8–10%

tin, and the remainder optionally zinc.
Bronze Gold. See **Gold Bronze.**
Bronze Powder. Powdered bronze, generally used as a pigment in paints and similar surfacing materials. Common bronze powder approximates 85% copper and 15% zinc, and therefore, corresponds to red brass.
Bronze Steel. See **Steel Bronze.**
Bronze Tuyere. Box Tuyere (see), fabricated of bronze.
Bronzing Liquid. Aluminum Bronze (see) powder suspended in an organic paint or lacquer vehicle.
Brown Iron Ore. See **Limonite.**
Brown Muffle Furnace. Mechanically rabbled **Roasting** (see) furnace, with longitudinal combustion flues under the hearth.
Brown-O'Hara Furnace. Long horizontal furnace with two hearths, for treating lead ores.
Brown Ore. See **Limonite.**
Brown & Sharpe Gage. See **Metal Gage.**
Brückner Cylinder. See **Brückner Furnace.**
Brückner Furnace. Furnace for roasting pulverized ore, consisting of a heated, horizontal, revolving cylinder.
Brunorizing. Normalizing process, used on steel rails, in which the rails, coming from the finishing rolls at 1750–1900°F, are cooled to about 1100°F, and then reheated in a continuous furnace to about 1525°F. Leaving the furnace, the railheads are hardened on the ends by dry quenching with cold compressed air issuing from special quenching nozzles, after which they return to hot beds where the heat, in the remainder of the rail, tempers the quenched portion.
Bru-Nor Process. See **Brunorizing.**
Brush. Blade pressed against the rolls in a tinning pot (see **Tin Plate**) to remove excess tin from the roll. At present, the blade is made of asbestos or similar material.
Brush Plating. Electroplating, in which the positive electrode (anode) is in the form of a brush wet with the electroplating solution, permitting plating of equipment, too large or clumsy to be immersed in a plating bath.
Brush Wire. High-carbon Bessemer steel wire (carbon from 0.30% to 1.20%) drawn dry to the second annealing step, then bright-annealed and drawn wet to finish size on continuous machines.
B. & S. G. See **Metal Gage.**
B.S.I. See **British Standards Institution.**
B.S.S. See **British Standard Specification.**
B.T.U. (British Thermal Unit). The heat necessary to raise the temperature of one pound of water one degree Fahrenheit.
Buckle. Defect in metal bar or sheet, represented by up-and-down waviness.
Buff. Polishing wheel usually consisting of large number of treated or untreated muslin discs sewn together.
Buffer. Substance, or mixture of substances, which in solution acts to maintain a constant hydrogen ion concentration despite addition of comparatively large amounts of acid or alkali.
Buffered Solution. Solution which tends to remain at a given acidity-alkalinity value, owing to the presence of a **Buffer** (see).
Buffing. Polishing of metals. See **Buff.**
Buflokast Iron. Low-alloy, gray cast iron, with carbon approximating 3.4%, and nickel and silicon up to 2% each (Buffalo Foundry & Machine Co.).
Building Sheet. Steel sheet of common grade, used for roofing, lath, etc.
Bulk Factor. Ratio of the volume of a given quantity of loose molding powder to the volume of the same material after molding.
Bulkhead. End of hearth, below the air and gas inlets, in an open-hearth furnace, consisting of an air-cooled, cast steel box covered with brick work. The back-walls of the vertical flues (uptake and downtake) are also known as bulkheads.
Bulk Modulus. Relationship of applied stress and volumetric strain produced when a stress is applied uniformly to all sides of a body.
Bull Block. Large size **Block** (see).

Bulldog. Roasted cinder from a **Puddling** (see) furnace, used for lining the furnace hearth. See also **Fettling.**

Bullion. Crude gold or silver, or silver-gold alloys prior to separation.

Bullion Bar. Crude gold or silver, in bar form.

Bung. Removable roof in **Air Furnace** (see), to permit charging large pieces of raw materials.

Burden. Ratio of amount of ore and flux, in a **Blast Furnace** (see), to the amount of coke. "Heavy burden" involves a high ratio of ore to coke; "light burden" involves a low ore/coke ratio.

Burdening. Regulating ratios of ore, flux, and coke in a **Blast Furnace** (see).

Burden Squeezer. Squeezer (see **Puddling**) in which one vertically mounted toothed cylinder revolves within a section of a large cylinder, set eccentrically to it.

Burloy (Steel). Series of corrosion-resistant, high-chromium and **Stainless Steels** (see), manufactured by Canada Electric Steel Castings, Ltd.

Burned. See **Burning.**

Burning. (a) Heating a metal beyond the temperature limits allowable for the desired heat treatment, or beyond the point where serious oxidation or other detrimental action begins. (b) Excessive pickling, resulting in deeply pitted areas on the metal. (c) See also **Calcining.**

Burnt Copper. Copper oxide, especially that formed by excessive heating of copper in air.

Burnt Deposit. Loose, dark metal powder, obtained in electroplating, instead of bright continuous films of metal; it results from the use of too high a rate of deposition. See **Plating.**

Burnt Edge. Defect in steel sheet or plate, due to excessive scaling of portions protruding out of the pile.

Burnt Iron. Metal, in steel manufacture by the Bessemer or open-hearth process, from which substantially all carbon has been oxidized, with iron oxide permitted to form in the metal. See also **Burning.**

Burnt Lime. **Limestone** (see), heated to drive off carbon dioxide, leaving substantially pure calcium oxide.

Burnt Metal. See **Burning.**

Burnt Steel. See **Burning.**

Burst. Rupture in metal due to improper rolling or forging.

Bursting. (a) Shattering of the outer layer of steel as a result of strains set up by passing through the recalescent point (see **Recalescence**). (b) Fracturing of metal due to improper forging.

Bus Bar. Copper, in strip form, used for conducting electric current, or any other metal so used.

Busheled Iron. Iron, made by heating pieces of scrap iron below the melting point and then forging or rolling together into massive form.

Busheling. Refining step in the puddling process of wrought iron manufacture; small pieces of **Muck Bar** (see) are heated to a white heat and collected into a ball, which is again squeezed and rolled or hammered into a bloom.

Bushing Bronze. Free-cutting phosphor bronze, approximating 88% copper, 4% zinc, 4% tin, and 4% lead; very easily machinable.

Bush Metal. Any alloy used for journals, bearings, etc.

Bustamente Furnace. Cylindrical shaft furnace, containing two compartments, for roasting mercury ores; the upper compartment contains the ore, the lower the fuel.

Buster Steel. Low-alloy, shock-resistant tool steel, with about 2.3% tungsten, 1.3% chromium, and 0.10% vanadium, in addition to 0.60% carbon and 0.30% manganese (Columbia Tool Steel Co.).

Bustle Pipe. Large pipe, surrounding a **Blast Furnace** (see), for the distribution of hot blast to the tuyeres.

Busy Metal. Metal, repeatedly subjected to cold rolling, vibration, or abrasion in service.

Butt Brass. Yellow brass (about 35% zinc), with about 1% lead, added to increase ease of machining.

Butterfly Method (of Rolling). Procedure for rolling steel into angles, in which the rolling passes are arranged

so that each pass rolls the metal further in the direction of the right angle shape, simultaneously bending and reducing it.
Button. Globule of metal, in assaying, remaining in a crucible or cupel after fusion has been completed. See **Cupellation** and **Button Metal.**
Button Brass. Brass (see) of about 10% zinc content, with about 0.5% tin as an additional constituent.
Button Metal. Zinc-base alloy, containing about 20% copper; used for buttons and ornaments.
Button Test. Test on galvanized steel wire, to determine adherence of the coating, in which the wire is wound around its own axis.
Butt-Welded Pipe. Pipe made from **Skelp** (see) by **Butt-Welding** (see).
Butt-Welding. Joining two edges, by placing one against the other, and **Welding** (see) the resulting seam.
Butt-Welding Furnace. Gas-fire furnace, for heating **Skelp** (see), prior to butt welding into pipe form.
Butt-Weld Process. See **Butt-Welded Pipe.**
B.W.G. See **Metal Gage.**
B & W Steel. Series of common and alloy steels, most commonly in tubular form, manufactured by Babcock & Wilcox Co.
Byers Process. Process for making wrought iron, in which the raw pig iron is melted in a cupola, desulfurized in a ladle, and then subjected to refinement in a Bessemer converter. After tapping into a ladle, the refined iron is poured at a regulated rate into another ladle containing molten slag, held several hundred degrees below the metal temperature, which causes freezing of the iron. The excess slag is poured away, and the metal sponge ball separately delivered to a squeezing press.
By-Pass. Pipe, in a blast furnace unit, leading from the cold blast main, around the stoves and into the hot blast main, for controlling the final temperature of the hot blast.
By-Product Coke. See **Coke.**
By-Product Oven. Oven, used for recovery of volatile substances, during the carbonization of coal to form **Coke** (see).

C

C. Chemical symbol for **Carbon** (see).
C. See **Centigrade.**
Ca. Chemical symbol for **Calcium** (see).
Cabbling. Breaking iron bars into pieces for **Faggoting** (see) and rolling.
Cadaloy Anodes. See **Cadalyte.**
Cadalyte. Process (Grasselli Chemical Dept. of duPont de Nemours) for cadmium plating. The anodes (Cadaloy) contain small percentages of mercury and zinc; the plating bath consists of cadmium oxide dissolved in sodium cyanide, together with brighteners and other addition agents.
Cadman Bearing Metal. See **Acorn (Babbitt)** and **Bearite.**
Cadmium (chemical symbol Cd). Element No. 48 of the periodic system; atomic weight 112.41. Soft, blue-white metal of good ductility, malleability and tarnish resistance; melting point 610°F; boiling point 1413°F; specific gravity 8.65. Chemically, cadmium is divalent. It is commercially obtained as a by-product of zinc distillation; being more volatile, it distills over, reacting with air to form the oxide, which is recovered and then reduced to metal by means of carbon. Used for electroplating, usually on steel, and as alloying element in bearing metals and low-melting alloys. Electrodeposition occurs, commercially, from a solution of a cadmium compound in sodium cyanide.
Cadmium Bronze. Copper-base alloys, with cadmium content up to 1%. Moderately-high strength, high electrical conductivity, good abrasion resistance.
Cadmium Copper. See **Cadmium Bronze.**
Cadmium-Nickel. Cadmium-base alloy, with about 1.80% nickel, used as a bearing alloy.
Cadmium Plate. Product of **Cadmium Plating** (see).
Cadmium Plating. Coating metal objects with cadmium or a high-cadmium

alloy. The object to be coated is made cathode (negative) in an electrolytic bath, containing a decomposable cadmium compound, usually a double cyanide or its equivalent. Cadmium metal or cadmium-base alloy anodes are most commonly used as the positive electrode.

Cadmium-Silver. Cadmium-base alloy, with about 1% silver and, commonly, about 0.50% copper, used for bearings.

Cadmium Solder. Lead-base solder, with 5–10% tin and 5–10% cadmium. Used as a substitute for solders of higher tin contents.

Caesium. See **Cesium.**

Cake. In powder metallurgy, the mass of weakly adherent metal powder obtained by reducing metal oxides.

Cake Copper. See **Tough Cake.**

Cake of Gold. Gold (see), in unmolten but compact form, resulting from distilling the mercury away from a gold amalgam.

Calcination. Heating of ores, usually to achieve partial decomposition.

Calcine. Product of **Calcination** (see).

Calciner. Furnace used in **Calcination** (see).

Calcining. See **Calcination.**

Calcium (chemical symbol Ca). Element No. 20 of the periodic system; atomic weight 40.08; principal member of the alkaline earth family of silvery-white metallic appearance; it is moderately ductile and malleable at normal temperatures; reacts readily with water, forming calcium hydroxide and developing hydrogen; moderately stable in air, due to crust formation on the surface; melting point 1548°F; boiling point approximately 2622°F; specific gravity 1.55. Chemically, calcium is mostly divalent. Since it reacts with water, it cannot be electrodeposited. Calcium is commercially isolated by electrolysis of the fused chloride; reaction of calcium-silicon with iron, under vacuum, is also reported. The metal is used, in relatively small quantities, for thermal reduction of metal oxides. In the United States, it is available from Chas. Hardy, Inc. Calcium-silicon alloys (Electro Metallurgical Co.) are used as deoxidizers and degasifiers for steel and cast iron.

Calcium-Aluminum-Silicon. Alloy used for deoxidizing and degasifying steel; contains 10–14% calcium, 8–12% aluminum, and 50–53% silicon.

Calcium Carbide. Chemically, CaC_2, formed when lime is reacted with coke in the electric furnace. With water, it reacts to evolve acetylene gas, which is extensively used in oxy-acetylene welding and cutting.

Calcium-Copper. Alloy of copper and calcium, used as master alloy or for desulfurizing copper-base and similar alloys.

Calcium-Lead. Lead hardened by a small percentage of metallic calcium, used occasionally as a bearing metal.

Calcium-Manganese-Silicon. Alloy for deoxidizing and degasifying steel; contains 16–20% calcium, 14–18% manganese, and 55–60% silicon, with iron up to 14%.

Calcium-Silicon. Alloy of calcium and silicon, usually with up to 6% iron; calcium content normally between 28–35%. Used as deoxidizer and degasifier for steel and cast iron.

Caldron Process. Early process for recovering silver from its ores, in which a slurry of the ore, contained in a copper vessel, was agitated with salt.

Calduro Steel. See **Warman Steel.**

Calido. Nickel-base, electric-resistance alloy, with about 25% iron, 8–16% chromium, and optionally, manganese up to 3% (Driver-Harris Co.).

California Intermittent Furnace. See **Leopoldi Furnace.**

Calite. Series of heat and oxidation resistant iron-nickel-chromium alloys, with 8–64% nickel and 5–20% chromium. Aluminum, manganese, and silicon, in low percentages, are optional (Calorizing Co.).

Calking Lead. Lead of not less than 99.73% purity, with the following maximum impurities: arsenic, antimony, and tin, together, 0.015%; copper 0.08%; zinc 0.02%; bismuth 0.25%; silver 0.02%.

Calmar Steel. See **Warman Steel.**

Calorie. A calorie (also called gram calorie and abbreviated cal.) is the heat necessary to raise the temperature of one gram of water one degree Centigrade. When written Calorie (abbreviated Cal.), the term means large calorie or kilogram calorie, i.e., the heat necessary to raise the temperature of one kilogram of water one degree Centigrade.

Calorific Intensity. Temperature which can be reached, under perfect operating conditions, by the combustion of a fuel.

Calorific Power. Heat in calories, evolved in the complete combustion of one gram of any material, more particularly, fuel.

Calorimeter. Apparatus for determining heat evolved or absorbed during the course of a chemical reaction (such as the burning of a fuel).

Calorite. Nickel-base, electric-resistance alloy, with about 12% chromium, 15–23% iron, and, optionally, 8% manganese (General Electric Co.).

Calorized Steel (or Alloys). See **Calorizing.**

Calorizing. Process of forming an oxidation-resistant coating of aluminum-iron on the surface of iron and steel; the iron object is heated at about 1740°F, while packed in aluminum powder, resulting in surface alloying.

Caloxo Steel. See **Warman Steel.**

Calsun Bronze. Copper-base alloy, with about 2% tin and 2.5% aluminum, used, primarily, in wire form (American Brass Co.).

Camber. Greatest deviation of a side edge of sheet or strip metal from a straight line, with measurement normally taken on the concave side.

Cambering. Formation of **Camber** (see).

Campaign. Period during which an equipment is actively operated or a continuous operation is carried on.

Campbell Process. Modification of the basic **Open-Hearth Process** (see) of steel manufacture, in which ore and pig iron are used as raw materials in a tilting furnace.

Can-End Stock. Tin plate for the manufacture of the tops and bottoms of tin can ends.

Cap. (a) See **Cope.** (b) See **Capped Steel.**

Cap Gilding (Brass). Red brass of about 10% zinc content, with optional addition of about 0.4% of lead for free machining qualities.

Capped Ingot. See **Bottle Top Mold.**

Capped Steel. See **Bottle Top Mold.**

Carat. The amount of gold in alloys is commonly expressed in parts per 24, called carats. Pure gold, therefore is 24 carats; 14-carat, a common composition, represents 14/24 of gold.

Carbide. Compound of carbon with one or more metals.

Carbide Divorcement. See **Ferrite Divorcement.**

Carbide of Iron. See **Cementite.**

Carbofrax. Silicon carbide product used for highly refractory installations (Carborundum Co.).

Carboloy. Tungsten carbide cemented carbide, with cobalt as the binding metal, used for cutting tools, dies, etc. Carbides of other metals, such as tantalum and titanium, may also be present (Carboloy Co.).

Carbomang. Low-alloy tool steel, cast into shape, with approximately 1% of carbon, 1% of manganese, 0.5% of chromium and 0.5% of tungsten (Detroit Alloy Steel Co.).

Carbometer. Device for determining the quantity of carbon in steel, by measuring the magnetic characteristics of a standard-size sample, as cast at the furnace.

Carbon (chemical symbol C). Element No. 6 of the periodic system; atomic weight 12.01; has three allotropic modifications, all non-metallic: amorphous carbon, graphite, diamond. Its chemical relationships are so complex as to give rise to a separate branch of chemistry ("organic"). Its chemical valence is normally four; occasionally it is also divalent, as in carbon monoxide; melting point uncertain, but above 6300°F; boiling point approximately 7600°F; specific gravity: graphite 2.25; amphorous carbon about 1.9. It cannot be electrodeposited. All three forms of carbon are found in nature; graphite can be made synthetically from amorphous carbon by electric furnace operations. Carbon is present in practically all ferrous alloys, and has tremendous effect on the properties of the resultant metal. Carbon is also an essential component of the cemented carbides. Its metallurgical use, in the form of coke, for reduction of oxides, is very extensive.

Carbonaceous. Containing carbon. Term, commonly used in describing non-metallic, natural materials.

Carbonado. See **Black Diamond.**

Carbon Arc Welding. Arc **Welding** (see), in which an electric arc is formed between a carbon electrode and the metal being welded.

Carbon Carburizing Steel. Carbon steel, susceptible to **Carburizing** (see). Usually, steel with 0.10–0.25% carbon is used for this purpose, with manganese less than 1.60%, to prevent embrittlement. The steel may be either of the fine-grain or coarse-grain type, the former giving a slightly thinner case, with higher carbon concentration.

Carbon of Cementation. See **Cement Carbon.**

Carbon-Chromium Steel. Steel with chromium as the only alloying element present in substantial quantity, most commonly with about 1% carbon. See **Chrome Steel.**

Carbon Flame. Luminous, white flame, resulting when a carbon-containing fuel burns in the absence of oxygen sufficient for complete combustion. See also **Carbonizing Flame.**

Carbon Gradient. Variation in carbon content from the outer surface of carburized or decarburized iron to the central, unaltered core.

Carbonization. Heating organic or carbonaceous substances in order to remove volatile constituents or products, leaving non-volatile residues high in carbon.

Carbonizing. See **Carburizing.**

Carbonizing Flame. In gas torch welding with a carbon-containing fuel, a flame with fuel gas in excess of that needed to produce a "neutral" flame, i.e., perfect combustion.

Carbon-Manganese Steel. See **Medium Manganese Steel.**

Carbon-Molybdenum Steel. Steel with molybdenum as the only alloying element present in substantial quantity. See **Molybdenum Steel.**

Carbon Steel. Common or ordinary steel, as contrasted with special or alloy steels, which contain other alloying metals, in addition to the usual constituents of steel in their common percentages. In carbon steel, manganese does not exceed 1.65%, silicon 0.60%, and copper 0.60%.

Carbon-Steel Strip. Hot-rolled strip (see **Hot-Rolled Strip Steel**) fabricated from **Simple Steel** (see).

Carbon-Steel Wire. Steel wire, made by cold-reduction from hot-rolled rods, after removal of scale.

Carbon Tool Steel. High-carbon and very-high-carbon steel, with carbon ranging from about 0.70 to 1.60% and with no substantial amount of alloying elements; used in tool manufacture.

Carbon-Vanadium Steel. Low-alloy steel, frequently used for locomotive and automotive forgings, with vanadium as the sole alloying element. The vanadium minimum is 0.15%; carbon usually ranges from 0.30% to 0.55%.

Carbon-Vanadium Tool Steel. Carbon tool steel, with 0.08 to 0.40% vanadium added for deoxidation, to prevent grain growth and increase shock resistance.

Carbonyl. Compound of carbon monoxide with a metal, e.g., $Ni(CO)_4$, $Fe(CO)_4$, $Co(CO)_3$, etc.

Carbonyl Iron. Extremely pure iron (total impurities approximately 0.05%) obtained by thermal decomposition of iron carbonyl.

Carbonyl Particles. See **Carbonyl Powder.**

Carbonyl Powder. In powder metallurgy, metal powder manufactured by decomposition of a metal carbonyl compound.

Carborundum. Silicon carbide product, used for refractory and abrasive purposes (Carborundum Co.).

Carbotan. Boron-containing master alloy for imparting deep-hardenability to steel, consisting of 16–17% titanium, 2.5–3.0% silicon, 1–1.25% boron, and

6.5–7.5% carbon (Titanium Alloy Mfg. Co.).

Carburization. See **Carburizing.**

Carburizing. Process of increasing the carbon content of the surface of iron-base alloys by exposure, at a temperature below the melting point of the metal, to a source of carbon. Appropriate heat treatment will harden the surface, while the interior core remains in its original soft state. The degree of carburizing and the depth of the hardened zone, frequently called the case, will vary with temperature, nature of carburizer, and length of exposure. When only certain parts of the exposed surface are to be carburized, the other areas are stopped-off, or protected, frequently by copper plating or by a special cement, which prevents access of the carburizing material to the iron. Pack carburizing calls for heating the metal with a solid carburizing material in a closed container called carburizing pot or box. Gas carburizing relies upon a carbon-containing gas, to act as a source of carbon. Liquid carburizing uses a fused carburizing compound, to furnish the carbon for the hardening action. Often the compound acts also as a source of nitrogen.

Carburizing Box. See **Carburizing.**

Carburizing Compound. Mixture usable for **Case Hardening** (see) or carburizing.

Carburizing Flame. See **Carbonizing Flame.**

Carburizing Gas. See **Carburizing.**

Carburizing Pot. See **Carburizing.**

Carburizing Steel. Steel suitable for **Carburizing** (see). Either carbon steel (see **Carbon Carburizing Steel**) or low-alloy steel may be used for this purpose.

Carburizing Test. See **McQuaid-Ehn Test.**

Carinthian Furnace. (a) Small reverberatory furnace, having an inclined hearth, for smelting lead ores. (b) Furnace for distilling zinc, equipped with a series of small vertical retorts.

Carinthian Process. Process of reducing lead ores, in which a small charge is roasted slowly at a low temperature, the lead being collected outside the furnace by means of an inclined hearth. See **Carinthian Furnace** (a).

Carmichel-Bradford Process. Modification of the **Huntington-Heberlein Process** (see), using gypsum instead of lime.

Carnegie-Illinois Steel. Series of low-alloy steels, manufactured by United States Steel Corp.

Carpaloy Steel. Series of high-chromium steels, manufactured by Carpenter Steel Co.

Carpenter Impact (Torsion) Tester. Instrument for determining impact strength (see **Impact Test**). The tested specimen, clamped on one end, supports a cross-arm on the other end, which is struck by a rotating head during the test. When the head is given sufficient speed to twist the specimen to the point of fracture, the energy absorbed is calculated from the loss in momentum of the rotating head.

Carpenter Stainless Steel. Series of stainless steels, manufactured by Carpenter Steel Co., with chromium up to about 20%, nickel up to about 10%, and the usual optional additional alloying elements. The alloys are of high strength, and highly resistant to corrosion and heat.

Carpenter Steel. Series of low-alloy and high-alloy steels of varying composition, manufactured by Carpenter Steel Co.

Carrot. Mass of metal, carrot-like in shape; calcium is obtained in this form when manufactured by electrolysis.

Cartridge Brass. Yellow brass of about 30% zinc; because of its deep-drawing qualities, it is used for cartridges, shells, and similar purposes.

Case. Surface of iron or steel, after case-hardening. See also **Core.** In casting technology, case is synonymous with **Cope** (see).

Case Carburizing. See **Case-Hardening.**

Case Carburizing Quality Steel. See **Case-Hardening Steel.**

Case Depth. Thickness of case formed on steel by any **Case-Hardening** (see).

Cased Tin. Non-metallic, finely-divided tin ore.

Case-Hardening. Heat treatment of iron or steel in which the surface only is hardened by altering its composition, usually by chemical addition of carbon or nitrogen or both, followed by the appropriate heat treatment. Usually, low-carbon or alloy steels are used for case-hardening; about 0.30% carbon represents the maximum. Silicon is normally kept below 0.30%, manganese below 0.50%. Among the simpler materials effective in case-hardening are charcoal, leather, bones, etc. It is believed, however, that the actual interaction is between the iron and a carbon-bearing gas, such as carbon monoxide, cyanogen, etc.

Case-Hardening Steel. Steel susceptible to **Case-Hardening** (see) or **Carburizing** (see).

Casing. Steel tubing, used in oil- or gas-well operations.

Cassiopeium. See **Lutecium.**

Cassiterite. Principal ore of tin, chemically substantially SnO_2.

Cast Gate. Channel through which molten metal is poured into a mold, in **Casting** (see) operations.

Cast House. Building for **Casting** (see) pigs or ingots.

Casting. (a) Pouring molten metal from any container, such as a furnace, cupola, ladle, etc., into a **Mold** (see). (b) Metal object resulting from casting.

Casting Box. See **Flask.**

Casting Copper. Relatively impure copper, adequate for casting purposes, but not for working into wire or sheet.

Casting Ladle. Ladle (see) for pouring molten metal directly into a mold.

Casting Machine. Arrangement for rapid casting of **Pig** (see) metal, in which the molds are carried by an endless metal belt in front of the furnace ladle. See also **Casting Wheel.**

Casting Pit. Space for holding molds, during **Casting** (see).

Casting Strains. Deformations, resulting from the cooling of cast metal.

Casting Wheel. Arrangement for rapid casting of **Pig** (see) metal, in which the molds are placed, either radially or tangentially, at the circumference of a wheel rotated so as to bring each mold under the furnace ladle.

Cast Iron. Iron, normally not malleable at any temperature, with carbon ranging from about 2.0 to 4.5%; usually contains considerable silicon. Low melting temperature and excellent fluidity are the great advantages of cast iron; its greatest handicap is non-malleability (except after special treatment). In white cast iron, the carbon is largely combined as cementite (the iron carbide of the formula Fe_3C), making the metal hard and brittle. In gray cast iron, the combined carbon is limited to eutectoid values, with the remainder of the carbon present in the form of flakes of graphite disseminated through the metal. Most castings are of gray iron, since these provide ready machinability. Malleable cast iron represents white cast iron in which the combined carbon has been changed to free carbon, which then appears in the form of nodules dispersed through the metal. Chemically, it is essentially ferrite and carbon.

Castner Cell. See **Castner Process.**

Castner Process. Method of producing sodium metal from fused sodium hydroxide, by means of an iron electrolytic cell, electrically connected with heavy iron anodes, dipping into the bath from above, which surround a vertical central cathode rising from below. An iron gauze diaphragm, located midway between anode and cathode acts to collect the liquid sodium and prevent its flowing back to the anode. Simultaneously with the deposition of the sodium, hydrogen is liberated at the cathode and oxygen at the anode.

Castoloy (Steel). High-carbon, high-chromium die steel, of about 1.5% carbon, 13% chromium, and fractional percentages of manganese and molybdenum (Detroit Alloy Steel Co.).

Cast Refractory. See **Electrocast Refractory.**

Cast Scrap. Cast iron **Scrap** (see).

Cast Steel. (a) Steel which has been cast, from the molten state, into a mold. (b) See **Ingot Iron.**

Catalan Forge. **Forge** (see) for reducing iron ore to wrought iron by means of charcoal.

Catalan Hearth. See **Catalan Forge.**

Catalysis. See **Catalyst.**

Catalyst (or Catalytic Agent). Substance whose presence changes the rate of a chemical reaction without being itself changed.

Catalytic Poison. Material which prevents normal action of a **Catalyst** (see).

Cataract Metal. Series of copper-nickel alloys (Niagara Falls Smelting & Refining Co.).

Catarinite. Alloy of iron and nickel, occurring native and corresponding approximately to Fe_2Ni, in chemical composition.

Catcher. (a) Worker in rolling mill operations, whose duty is to receive the metal bar or pack after it has passed through the rolls. (b) Device, in tin and terne plate operations, for lifting the coated plate away from the exit rolls and placing it on the conveyor.

Catcher Mark. Flaw in **Terne Plate** (see) and **Tin Plate** (see), due to faulty operation of plate-catching devices.

Catcher Rolls. Pair of rolls, in the manufacture of terne and tin plate by the palm oil process, for removing the coated metal from the oil bath.

Catenary Furnace. Furnace for the continuous heat-treating of steel sheet and strip, in coil form; the metal is drawn through by its own tension, without rollers or conveying devices of any sort.

Cat Gold. Mica of golden color; a non-metallic mineral.

Cathion. See **Cation.**

Cathode. Negative electrode in an electrolytic cell. In electroplating, the metal being plated on is usually made the cathode of the electrolytic system.

Cathode Copper. Copper which has been electrodeposited on starting sheets, connected as negative electrodes (cathodes), in the electrolytic cell method of refining copper.

Cathode Deposit. Metal deposited on the negative electrode (cathode), in electroplating. See **Plating.**

Cathodic Coating. Electroplated or mechanical coating on base metals, in which protection is based on mechanical surface-coverage only. Any break in the surface results in the corrosion of the base metal.

Catholyte. In electrolysis, the solution located near the **Cathode** (see).

Cation. Positively-charged ion in any solution; that portion of a chemical compound which, when dissolved (usually in water), tends to flow toward the cathode (negative pole) under the influence of a direct electric current.

Cat's Eyes. Surface appearance of steel, during the molten stage of the crucible process (see **Crucible Steel**).

Cat Silver. Mica of silvery lustre, a non-metallic mineral.

Causal Metal. Corrosion-resistant cast iron, containing nickel, copper, and chromium as alloying constituents (Lunkenheimer Co.).

Caustic (Alkali). Water-soluble, strong **Base** (see).

Caustic Embrittlement. Cracking in mild-steel boiler-plates, particularly where rivetted, due to localized concentration of the OH^- ion (fundamental to all alkalies), breaking down the cohesion between the ferrite grains.

Caustic Silver. Non-metallic chemical compound; silver nitrate, $AgNO_3$.

Cavity Corrosion. Corrosion localizing on the sides of a narrow crack.

Cavity Corrosion Test. Test to determine susceptibility of a metal to **Cavity Corrosion** (see) by subjecting a composite of two pieces of the metal, separated only by several thousandths of an inch, to a corroding solution.

Cazo Process. See **Caldron Process.**

Cb. Chemical symbol for **Columbium** (see).

Cd. Chemical symbol for **Cadmium** (see).

Ce. Chemical symbol for **Cerium** (see).

Cecolloy. Shock-resistant, low-alloy iron, with carbon approximating 3.0%, nickel 0.6–1.5%, molybdenum about 0.5%, and chromium, optionally, up to 0.35% (Chambersburg Engineering Co.).

Cell. Combination made up of electrodes of two dissimilar elements (usually metals) and an electrically conducting solution between them, or the equivalent of such a combination. Such a cell will set up an electromotive force (voltage), and, if the electrodes are connected externally, an electric current will flow.

Cellular Structure. See **Network Structure.**

Celtium. Non-existent element, formerly believed to occupy position No. 72 of the periodic system, and to be the last of the rare earth-metal series. The discovery of hafnium as element No. 72, congener of zirconium and not a rare earth-metal, proved the nonexistence of celtium.

Cement. (a) Finely divided metal obtained by chemical precipitation or **Cementation** (see). (b) Material in which iron and similar base metals are packed for surface cementation. See **Metallic Cementation.**

Cementation. (a) Process of obtaining a metal from a solution of a compound of that metal by displacing it by a more electropositive element. (b) See **Carburizing.** (c) See **Metallic Cementation.** (d) See **Cementation Process.**

Cementation Process. Obsolete, seldom-used process for making steel, in which wrought iron, or other low-carbon iron, is heated to red heat (below melting) in contact with charcoal or similar carbonaceous matter, the carbon absorption being proportional to the time and temperature of contact.

Cement Bar. See **Blister Steel.**

Cement Carbon. Carbon united with iron in **Cementite** (see).

Cement Copper. Copper in impure, finely-divided form, resulting from the action of copper-bearing solutions on a metal more electropositive than copper, such as iron, which will necessarily contaminate the cement copper.

Cemented Carbide (Tools). See **Sintered Carbide.**

Cement Gold. Gold obtained by **Cementation** (see) from solution.

Cementing Furnace. Furnace used in **Metallic Cementation** (see).

Cementing Oven. See **Cementing Furnace.**

Cementite. Carbide of iron, Fe_3C, containing 6.69% of carbon. Hard and brittle, it is the hard constituent of cast iron, and the normal form in which carbon is present in steel. It is magnetizable, but not as readily as ferrite.

Cementitic Steel. See **Ferritic Steel.**

Cement Silver. Silver obtained by **Cementation** (see) from solution.

Cement Steel. Steel made by the **Cementation Process** (see).

Center Combustion Stove. Blast Furnace (see) stove, in which the combustion chamber is located in the center.

Centered. See **Body-Centered.**

Centigrade. Temperature scale in which the freezing point of water is taken as zero, its boiling point as 100. It is standard throughout the world in scientific work, and also in common daily use everywhere, except the English-speaking countries.

Centrifugal Casting. Casting of metal into a rotating mold; the process yields unusually sound castings and is frequently used in the manufacture of large-diameter pipe.

Cerium (chemical symbol Ce). Element No. 58 of the periodic system; atomic weight 140.13. Most common of the rare earth metals, it can be separated with relative ease from the others. Steel-gray in color, soft, ductile, and malleable, and stable against normal atmospheric conditions and water; melting point 1499°F; boiling point approximately 2550°F. Its more common valence is three (cerous), but may also be tetravalent (ceric). Like all rare earth metals, it cannot be electrodeposited. Cerium is usually isolated in alloyed combination with other rare earth metals, in the form of "mischmetall," by electrolysis of the fused mixed chlorides; the pure metal can be made in analogous manner, using pure anhydrous cerous chloride. Metallurgically, cerium is mostly used as **Mischmetall** (see), in combination with iron, for pyrophoric purposes.

Cermak-Spirek Furnace. Rectangular, reverberatory furnace, divided into two sections by a wall; used for **Roasting** (see) zinc and mercury ores.
Cerrobase. Lead-bismuth alloy (Cerro de Pasco Copper Corp.) with about 58% bismuth. It has a relatively low melting point (255°F).
Cerrobend. White metal, main constituent bismuth, with approximately 27% lead, 13% tin, and 10% cadmium. Because of its low melting point (160°F), it is used as a filler for bending tubing, without causing buckling or flattening (Cerro de Pasco Copper Corp.).
Cerromatrix. White metal, with bismuth as main constituent, manufactured by Cerro de Pasco Copper Co.; contains approximately 15% tin, 29% lead, and 9% antimony. Because of slight expansion on solidification, it is used as a matrix for holding mechanical parts, etc.
Cerrosafe. Low-melting alloy (Cerro de Pasco Copper Corp.), with about 40% bismuth and 40% lead, 11% tin, and 9% cadmium. Because shrinkage is offset after solidication, by growth, it is used for small castings where precision of final measurements is desired.
Cesium (chemical symbol Cs). Element No. 55 of the periodic system; atomic weight 132.91. It is the most highly electropositive metal isolated, reacting violently with water and explosively with carbon tetrachloride. It resembles potassium in most characteristics, being extremely soft and malleable and corroding rapidly in air (a freely-cut surface is silvery white); melting point 83°F; boiling point 1238°F; specific gravity 1.87. Chemically, it is mostly monovalent. It may be isolated by reaction of the hydroxide with aluminum in vacuo. It finds some use in electronic, television, and similar tubework. Since it reacts with water, it cannot be electrodeposited.
Chace Bimetal. Series of **Bimetal** (see) composites, manufactured by W. M. Chace Co.
Chafery. Forge (see), used for the reheating of metal.
Chain Grate. Continuous grate, in a furnace, for feeding raw materials, fuel, etc.
Chalcocite. See **Vitreous Copper.**
Chalybite. See **Siderite.**
Chamber Dust. See **Flue Dust.**
Chamet Bronze. Copper-base alloy, with about 39% zinc and 1% tin (Chase Brass & Copper Co.).
Champion (Tool) Steel. Series of low-alloy and high-alloy tool steels, manufactured by Crucible Steel Co.
Champlain Forge. Continuous **Forge** (see) for the direct production of wrought iron from finely divided ore.
Chance-Claus Process. Process for recovering sulfur from sulfide waste, by treatment with carbon dioxide and oxidation of the resulting hydrogen sulfide with air in the presence of a catalyst.
Change of State. Change in the physical form of a material, from solid to liquid, from liquid to gas, or from solid to gas (or the reverse changes).
Chaotic Orientation. See **Random Orientation.**
Chaplet. Metal support on which one end of a casting **Core** (see) may be placed; the chaplet is absorbed in the casting when the metal is poured.
Chapmanizing. Modified **Nitriding** (see) process, in which dissociated ammonia gas is forced, at about 1400°F, into a fused cyanide bath in which the articles under treatment are heated.
Charcoal. Amorphous **Carbon** (see), made by heating wood to red heat in the absence of air.
Charcoal Finery. See **Finery.**
Charcoal Iron. Pig Iron (see) made in a charcoal furnace; because of the relative purity of charcoal, iron so made is of higher quality, higher density and closer structure.
Charcoal Pig (Iron). See **Charcoal Iron.**
Charcoal Plate. See **Charcoal Tin Plate.**
Charcoal Tin Plate. Tin Plate (see) with a relatively heavy coating of tin (higher than the "Coke Tin Plate" grades).

Charge. Materials fed into a furnace or similar apparatus at one time or during one cycle.
Charging. Feeding raw material into an apparatus, e.g., into a furnace, for treatment or conversion.
Charging Box. Steel box, used for holding raw materials fed into an open-hearth furnace.
Charging Machine. Machine, used in conjunction with open-hearth furnaces, manipulating **Charging Boxes** (see) in feeding material to the furnace.
Charging Scale. Scale for weighing ingredients of a **Charge** (see).
Charpy (Impact) Test. Method of testing resistance to impact. A notched bar, lying on a split anvil, 1.575 inches between supports, is broken by a pendulum-hammer blow carrying 217 foot-pounds of energy; the rise of the pendulum past the anvil, after breaking the bar, is used as a measure of the impact strength of the metal.
Charpy Keyhole Specimen. See **Keyhole Specimen.**
Chase—. Series of copper-base alloys, of highly diversified composition, manufactured by Chase Brass & Copper Co.
Chasing. (a) Finishing surface of castings by polishing and removing minor surface flaws. (b) Engraving for ornamental purposes, particularly on silverware.
Chateaugay (Pig) Iron. Low-phosphorus pig iron, produced from the magnetite ores of New York State; sulfur and phosphorus have maximum limits of 0.035%.
Check. Crack in the surface of an ingot, caused by improper mold surfaces. After rolling the ingot, checks will frequently appear as seams or slivers.
Check Analysis. Analysis of metal after formation into finished or semi-finished form.
Check Analysis Quality Strip Steel. See **Forging Quality Hot-Rolled Strip Steel.**
Checkerboard Spangle. Spangle on zinc-coated (galvanized) steel of checkerboard appearance, usually produced by carrying the sheets away from the bath on a conveying wire-netting of the appropriate mesh.
Checkers. See **Regenerator.**
Checkerwork. See **Regenerator.**
Checking. Temporarily reducing the temperature or the volume of the hot air blast in a **Blast Furnace** (see). See also **Check.**
Cheek. Intermediate portion of a casting **Pattern** (see), lying between the top and bottom.
Chemical Compound. See **Compound.**
Chemical Deposition. Precipitation of a metal from a salt solution by a metal higher in the electromotive series or by addition of a reducing agent. Metal so precipitated normally appears in powdered form.
Chemical Dip Brazing. Dip Brazing (see), in which an inert chemical bath is used for heating the metal. Usually, the filler metal is added before immersion into the bath.
Chemical Element. See **Element.**
Chemical Lead. Lead with low fractional percentages of copper and silver, as originally obtained from the ore; used for the manufacture of storage battery plates, chemical piping, etc.
Chemically Bonded Brick. Refractory brick, usually of ignited magnesite or of chromite, formed into shape under high pressure and bonded by the presence of a refractory bonding material.
Chemical Metallurgy. Science of the composition and methods of isolation of metals from their ores.
Chemical Symbol. For each chemical element, there is an abbreviated symbol, consisting of a capital letter, sometimes followed by a small letter. Such an abbreviation is always used in chemical formulas and equations.
Chemical Valence. See **Valence.**
Chempure Tin. The purest commercially available tin of 99.9+% tin content.
Chenot Process. Process of making iron sponge by heating a mixture of ore and coal-dust in vertical retorts.
Chessy Copper. Non-metallic mineral, corresponding approximately to basic copper carbonate, $CuCO_3.Cu(OH)_2$, in composition.

Chile Bar. Bar of impure **Copper** (see), of about 98% copper content, weighing about 200 pounds.

Chill. (a) Hard outer surface of white iron formed on cast iron when cooled rapidly, the cooling rate being adjusted so as to leave the interior in the form of gray cast iron. (b) Metallic portion of a mold, to speed cooling of a casting. (c) Freezing of metal in a hearth, due to sudden addition of comparatively cold raw material (as in the faulty operation of a blast furnace).

Chill Cast Pig. Pig iron which has been cast into metallic molds and, therefore, cooled rapidly.

Chill Crystals. Small crystals, resulting from quick freezing of cast metal at the surface of a cold mold.

Chilled Casting. Cast iron, hardened by chilling, i.e., either cast against a chill [see **Chill** (b)] or rapidly cooled in air or water.

Chilled Cast Iron Roll. See **Chilled Roll.**

Chilled Hearth. See **Chill** (c).

Chilled Iron. Iron, cast into metallic molds, with consequent rapid cooling; the surface of such metal is of **White Cast Iron** (see), the interior of **Gray Cast Iron** (see).

Chilled Roll. Roll made of cast iron, in which the interior is of gray cast iron, while the outer layer, the "chill," is of hard white iron, as the result of rapid cooling. The chill may be classed as "mild," "medium," or "hard." Mild chill rolls are used for hot-rolling, medium chill rolls, for strand and finishing passes, while hard chill rolls are used for planishing.

Chill Hardening. See **Chill** (b).

Chilling. Casting metal in contact with a metal chill [see **Chill** (b)].

Chill Roll. See **Chilled Roll.**

Chimneying. Faulty operation of a blast furnace, causing the coarser portion of the raw materials to segregate near the center of the furnace, with the result that the hot gases pass more readily through this region, creating a disparity in the rate of reaction across the cross-section.

Chimney Valve. Valve, in the base of stove of **Blast Furnace** (see), preventing escape of air through the chimney when the hot blast stove is on blast.

Chimo (Steel). Tough, low-alloy, tool steel, with approximately 2% molybdenum, 2% silicon, 1% manganese, and 0.4% vanadium (Firth-Sterling Steel Co.).

China Metal. Not a metal, but porcelain.

Chinese Bronze. See **Nickel Bronze.**

Chipping. (a) Removing surface defects from ingots, etc., by chiselling (usually with the aid of mechanical equipment). (b) Tendency of faulty **Vitreous Enamel** (see) to peel away from the base metal, in large flakes.

Chisel Steel. Tool Steel (see) used for the manufacture of chisels and similar tools. Carbon is commonly in the 0.50–0.75% range. Small amounts of alloying elements may be added: tungsten up to 2.50% (particularly for hot-working tools), chromium up to 2.00%, manganese up to 0.80%, silicon up to 2.25%, and vanadium up to 0.30%.

Chloanthite. Naturally-occurring, mineral alloy, chemically essentially nickel diarsenide, $NiAs_2$.

Chloridizing (Roasting). Roasting (see) ores, particularly sulfides, with salt (sodium chloride), to transform the metallic values into relatively soluble chlorides.

Chock. See **Stand.**

Chromaloid. Zinc sheet, coated with chromium (American Nickeloid Co.).

Chromax. Series of alloy cast irons, manufactured by Driver-Harris Co., with 20–35% nickel and 15–20% chromium. Because of high temperature and abrasion resistance, the alloys are used for furnace parts, chains subjected to heat, etc.

Chrome. Abbreviated name for **Chromium** (see), used to designate alloys, e.g., ferrochrome.

Chrome Brick. Refractory, in brick form, made by heating **Chromite** (see), with or without bonding materials.

Chrome Checks. See **Flakes.**

Chromel. Series of nickel-chromium alloys, manufactured by Hoskins Mfg. Co.; usually nickel is dominant, with iron optional; commonly used for heating elements and heat-resistant applications.

Chrome-Molybdenum Steel. Steel, usually made by the basic open-hearth or by the electric furnace process; contains both chromium and molybdenum. Typical of low-alloy-content steels are those of 0.3–1.1% chromium and 0.15–0.30% molybdenum. Molybdenum high-speed steels are also generally chrome-molybdenum in type, often with tungsten and vanadium as additional elements.

Chrome-Nickel Cast Steel. See **Nickel-Chromium Cast Steel.**

Chrome-Nickel Pig (Iron). Pig Iron (see), made from ores containing chromium and nickel.

Chrome-Nickel Steel. Steel, usually made by the basic open-hearth process, in which chromium and nickel predominate as alloying elements. Typical of low-alloy-content compositions are those in which nickel ranges from 0.45% to 1.75%; the stainless steels of the 18% chromium 8% nickel type are the best known of the chrome-nickel type.

Chrome Ore. See **Chromite.**

Chrome-Pickle Process. Process of forming a corrosion-resistant oxide film on the surface of magnesium-base metals by immersion in a bath of an alkali-bichromate, with either acid or alkali added.

Chrome Plate. See **Chromium Plate.**

Chrome-Silicon Steel. Low-alloy steels, usually approximating 1% chromium, 1% silicon, and 0.40% carbon; frequently used for the manufacture of springs. Steels with much higher chromium content (up to 19%) are also known, commonly with carbon down to about 0.10%; used for heat and corrosion resistance work, as in automobile engine valves.

Chrome Steel. Steel, usually made by the basic open-hearth, electric-furnace, or crucible process, containing over 0.5% chromium; used for tool manufacture and similar purposes requiring increased strength and toughness. The lower ranges of chromium, from 0.5% to 2%, define the pearlitic chromium steels; 2% to 17% chromium steels frequently are martensitic; ferritic chromium steels contain 17–35% chromium. The important group of austenitic chromium steels contains chromium in the 10–30% range.

Chrome-Vanadium Steel. Steel, made by the basic open-hearth or electric-furnace process; contains up to 1.1% chromium and about 0.15% vanadium; frequently used in the manufacture of high-quality springs.

Chromite. Ore corresponding approximately to the chemical formula $Fe(CrO_2)_2$; used as a neutral refractory and a source of chromium metal and its alloys.

Chromium—. See also **Chrome.**

Chromium (chemical symbol Cr). Element No. 24 of the periodic system; atomic weight 52.01. It is of bright silvery color, relatively hard (hardness depending, however, on method of preparation), and strongly resistant to atmospheric and other oxidation; melting point 3281°F; boiling point about 4820°F, specific gravity 7.1. Chemically, chromium commonly exhibits valences of two (chromous), three (chromic), and six (chromates and bichromates). The purest grade of metal may be made by the action of ultra-pure hydrogen on chromic oxide. It is much more common, however, in the form of **Ferrochrome** (see), and in iron-base alloys of great value (such as stainless steel). Chromium plating has also become a large outlet for the metal, deposition being made from a solution of chromium trioxide (chromic acid). Chromium-nickel alloys are extensively used as electric resistance-heating elements, and chromium-cobalt alloys are also met in commercial practice (see **Stellite**). American suppliers of the metal include Electro Metallurgical Co. and Vanadium Corp. of America.

Chromium-Carbon Tool Steel. Carbon Tool Steel (see), with chromium as an alloying element. Three chromium ranges are standard: (1) 0.10–0.25%, (2) 0.25–0.50%, and (3) 0.50–0.90%.

Chromium Cast Steel. Cast steel of high strength [see **Cast Steel** (a)] with about 0.50% carbon content, and chromium content ranging from 0.50 to 1.5%.

Chromium Copper. Copper-base alloy, containing approximately 8–11% chromium; used primarily for the introduction of chromium into non-ferrous metals of low melting point. See also **Copper-Chromium Alloys.**

Chromium - Manganese - Molybdenum Steel. Modified low-alloy **Chrome-Molybdenum Steel** (see), with 1.00–1.30% manganese.

Chromium - Manganese - Vanadium Steel. Modified **Chrome - Vanadium Steel** (see), with 1.00–1.60% manganese, 0.30–0.65% chromium, 0.06–0.12% vanadium, and 0.15–0.50% carbon.

Chromium-Molybdenum Cast Steel. Cast Steel (see) of high strength and ductility, with carbon up to 1%, chromium 0.7–1.1%, and molybdenum 0.2–0.4%.

Chromium - Molybdenum - Vanadium Steel. Low-alloy steel, containing about 0.40% carbon, 1.50% chromium, 0.40% molybdenum, and 0.20% vanadium; used for uniformity of strength and hardness in large sections.

Chromium-Nickel Steel. See **Nickel-Chromium Steel.**

Chromium Plate. Product of **Chromium Plating** (see).

Chromium Plating. Process of coating metal objects with a thin film of chromium, in which the object is made cathode (negative electrode) in an electrolytic bath, containing a decomposable chromium compound. Commercially, the latter is nearly always chromium trioxide (so-called chromic acid) together with small amounts of additional agents. For decorative purposes, a highly polished surface is deposited; hard chromium is a dull deposit, usually much thicker, used for its high wear-resisting qualities. Current density and bath composition determine which type of deposit will be obtained.

Chromium Tool Steel. See **Low-Chromium Carbon Tool Steel.**

Chromium-Tungsten Cast Steel. Cast steel of high strength and corrosion resistance, particularly at elevated temperatures [see **Cast Steel** (a)]. Carbon is normally about 0.20%, chromium 4.5–6.5%, and tungsten 0.75–1.0%.

Chromium-Tungsten Steel. Low-alloy steel, with chromium and tungsten as the alloying elements. Most commonly, chromium is about 5% and tungsten 1%; characterized by resistance to heat and chemical corrosion.

Chromizing. Surface-alloying of metals, most commonly steel, with chromium. The most practical method is passing vaporized chromium chloride over the steel, at about 1800°F.

Chrom-X. Exothermic Ferrochrome (see) of high carbon content, made by Chromium Mining & Smelting Corp.

Churn Pickling. See **Agitator Pickling.**

Cimet. Corrosion-resistant iron-chromium-nickel alloy series, with chromium approximately 27% and nickel 10% (Driver-Harris Co.).

Cinder. See **Slag.**

Cinder Bank. See **Cinder Dump.**

Cinder Block. Block containing the cinder notch in a **Blast Furnace** (see).

Cinder Cooler. Hollow, water-cooled, cone-shaped copper casting, fitted into the cinder notch of a blast furnace. The opening is further reduced by insertion of another water-cooled unit called the intermediate cooler or monkey cooler. Inside the latter is a still smaller unit of the same construction, called a monkey, which reduces the opening diameter to about two inches.

Cinder Dump. Area for dumping **Slag** (see).

Cinder Fall. Dam over which slag flows from a furnace cinder-notch. See **Slag Hole.**

Cinder Notch. See **Slag Hole.**

Cinder Outbreak. Escape of slag, from a furnace, due to erosion or chemical corrosion of the fire brick.

Cinder Pig (Iron). Pig iron made by reworking slag. Because of high phosphorus, such iron is of inferior quality.

Cinder Pit. Pit containing water, into which molten slag is cast, causing it to become granulated.

Cinder Plate. See **Bloomery.**

Cinder Runner. Trough for carrying away slag from the cinder-notch of a furnace.
Cinder Tap. See **Cinder Notch.**
Cinder Wool. See **Mineral Wool.**
Circle "C" Steel. High-speed alloy tool steel (Firth-Sterling Steel Co.), with about 18% tungsten, 9% cobalt, 4% chromium, 2% vanadium, and 1% molybdenum.
Circle "L" Steel. Series of low-alloy and high-alloy steels, including stainless steels, maunfactured by Lebanon Steel Foundry.
Circular Mil. Area of a circle of one "mil" diameter (a mil is 0.001 inch).
Circular Mill Gage. See **Metal Gage.**
Cladding. Process of covering one metal by another metal, the latter usually more valuable or chemically more resistant, in substantial thickness. For example, steel may be clad with stainless steel or nickel, copper with silver, or duralumin with pure aluminum. Usually, the faces of fairly thick slabs of two metals are carefully contacted and then subjected to co-rolling, so that a clad composite results.
Clad Metal. See **Cladding.**
Clad Steel. See **Cladding.**
Clamp. Pile of ore for **Roasting** (see).
Clarite (Tool) Steel. High-carbon alloy tool steel (Columbia Tool Steel Co.), with about 18% tungsten, 4% chromium, 1% vanadium, and 0.70% carbon.
Clark Cell. Standard cell for measuring electrical potential, using mercury and zinc amalgam electrodes in zinc sulfate solution.
Class. Referring to steel, class indicates the general group of service or size of the steel.
Classification. Separation of materials according to particle size, while suspended in a liquid medium.
Clay. Natural plastic mixture of silica and alumina, containing other constituents as impurities. Clay hardens when heated and loses all plastic properties; it is the basis of all common brick, tile, etc. In metallurgy, it is frequently used where its relative refractoriness is of value.
Clay Gun. Apparatus for ramming wet clay into tap-holes, to prevent metal running out of the hole, before tapping is desired.
Cleaner. See **Branner.**
Cleaning. Removing foreign substances from the surface of metal, prior to chemical or mechanical treatment of the surface.
Cleanness. Relative absence, in metal, of non-metallic inclusions.
Clean Pass. Rolling Pass (see), free from surface cracks and defects.
Clear Chill. Portion of **Chill** (see) nearest the surface, which is white and not mottled. It extends to ⅛–2 inches, depending upon size of section, cooling rate, etc.
Clearing. In the **Puddling** (see) process, the step of stirring the molten metal after covering with a thin layer of slag, in order to oxidize out the impurities. Clearing is followed by bringing the metal to a boil.
Cleavage. Property of certain materials of splitting easily along one or more crystallographic planes.
Cleavage Plane. Any crystal plane along which a material tends to fracture. See **Cleavage.**
Cletaloy. Series of copper-tungsten and silver-tungsten composites, used in the fabrication of resistance-welding electrodes (Cleveland Tungsten, Inc.).
Climax (Metal). Iron-base electric-resistance and heat-resistant alloy, with nickel about 24% and manganese about 2% (Driver-Harris Co.).
Clink. Tiny crack, formed within cast steel ingots by different rates of contraction of surface and core.
Clinker. Iron scale, formed on iron when heated in air.
Clipper. Apparatus for simultaneously shearing and bending **Skelp** (see) into a form ready for grasping by tongs, for passage through the welding bell.
Clock Brass. Free-cutting **Brass** (see) of about 37% zinc and 1.5% lead; used frequently for the manufacture of clock parts.
Close Annealing. See **Box Annealing.**
Closed Box Pass. Rolling mill in which the combination of grooves in the rolls forms a square or rectangle,

the two rolls interlocking, in part, so as to form a completely closed cross-section.

Closed Front. See **Lurmann Front.**

Closed Pass. Rolling mill in which the rolls interlock, in part, so as to form a completely closed cross-section.

Closed Topped Housing. Housing for rolling mill rolls, in which the base, legs, and top are made as a single unit.

Close Grain. Freedom from porosity and other unsoundness, particularly in steel.

Close Mold. Mold consisting of two **Flasks** (see) and filled by pouring through ingates.

Close-Packed. Arrangement of atoms or ions, in which each atom or ion faces twelve neighbors.

Clotting. Sintering of ore during **Roasting** (see).

Cloudburst (Hardness) Test. Test designed to determine soft spots in case-hardened surfaces. Small steel balls are permitted to cascade on the surface, with the result that soft spots are affected, showing up as matte areas against the normally bright background.

Cluster Mill. Six-roll rolling mill, in which each working roll (see **Four-High Mill**) is supported by two back-up rolls.

CMSZ Alloy. Alloy manufactured by Electro Metallurgical Co., containing 30–35% chromium, 7–10% manganese, 30–35% silicon, 4–6% zirconium, with the balance iron. Used as an addition (in the ladle) to cast iron, to increase hardness and wear resistance.

C. M. (Tap Steel). Non-deforming, oil-hardening tool steel, of approximately 1.2% carbon, 0.5% chromium and 0.5% manganese (Vanadium Alloys Steel Co.).

Co. Chemical symbol for **Cobalt** (see).

Coal Blacking. Blacking (see), made of powdered coal.

Coal Brass. Non-metallic, iron pyrites as found in coal seams.

Coalesced Pearlite. See **Granular Pearlite.**

Coalescence. Union of one or more crystals or aggregates into a simple, larger unit.

Coarse Grain. Appearance of metal with crystals of relatively large size.

Coarse-Grained Steel. Steel which, on the basis of the A.S.T.M. austenitic grain size standards, shows a grain size from #1 to #5.

Coarse Metal. Crude copper matte of 20–40% copper content, obtained from the smelting of sulfide-copper ores.

Coarsening Temperature (Range). Temperature range in which steel shows substantial grain growth.

Coarse Round Wire. See **Coarse Wire.**

Coarse Wire. Round steel wire, coarser than #20 gage (Washburn & Moen steel wire gage).

Coated Electrode. Electrode for use in **Arc Welding** (see), in which the metal wire is coated with fluxing constituents.

Coated-Finish Steel. Steel strip or sheet, coated with a non-ferrous material, such as copper, terne, lacquer, etc.

Coating. Material, metallic or non-metallic, which forms a continuous film over another surface.

Cobalt (chemical symbol Co). Element No. 27 of the periodic system; atomic weight 58.94. A gray magnetic metal, of medium hardness; it resists corrosion like nickel, which it resembles closely; melting point 2696°F; boiling point about 5250°F; specific gravity 8.9. Chemically, its normal valence is two (cobaltous); trivalent in some compounds (cobaltic). The metal can be isolated, in pure form, by hydrogen-reduction of the oxide. It is used as the matrix metal in most cemented carbides and is occasionally electroplated instead of nickel, the sulfate being used as electrolyte. Larger quantities are used in the form of ferrous alloys, particularly high-grade tool and magnet steels. A cobalt-chromium alloy, Stellite, is in common use, also for high-speed tools.

Cobaltcrom. Series of air-hardening steels, of about 14–18% chromium and 3.5% cobalt content, with fractional percentages of molybdenum, vanadium, nickel, manganese, and silicon (Darwin & Milner, Inc.).

Cobalt Glance. See **White Cobalt** (b).

Cobaltite. See **White Cobalt** (b).

Cobalt Plating. See **Nickel Plating.**

Cobalt Steel. Steel, usually made by the crucible or electric-furnace method, containing substantial percentages of cobalt in addition to other alloying elements. Cobalt high-speed steels contain 5–12% cobalt, commonly, 14–20% tungsten, about 4% chromium, 1–2% vanadium, and fractional percentages of molybdenum. Carbon is normally about 0.80%. Such high-speed cobalt steels are of particular value for cutting hard, abrasive materials.

Cobble. (a) Faulty puddle ball (see **Puddling**), which breaks up when subjected to squeezer action. (b) See **Cobbles.**

Cobbles. (a) Coke tin plates so poor as to require complete scrapping. (b) See **Cobble.**

Cobbling. Twisting or turning down of metal while being rolled in a rolling mill. Usually, such metal has to be scrapped.

Cock Metal. Soft metal conglomerate, consisting essentially of two parts of copper and one part of lead; used for the manufacture of taps and cocks.

Co-Co Steel. High-speed, alloy tool steel (Colonial Steel Co.), with approximately 18% tungsten, 4% chromium, 4% cobalt, 1% vanadium, and fractional percentages of molybdenum.

Coefficient of Expansion. Change of a substance in unit length (or volume) per degree of temperature change.

Coercive Force. Tenacity with which a magnet retains its magnetism.

Cogging. Converting ingots, by rolling or forging, into blooms.

Cogging Hammer. Forging hammer used in **Cogging** (see).

Cogging Mill. See **Blooming Mill.**

Coil-Break. Slight reduction in thickness, across a metal strip in narrow streaks, occasionally occurring when the metal has been coiled hot and uncoiled cold.

Coinage Alloys. See **Coinage Metals.**

Coinage Bronze. Copper-base alloy, commonly used for copper coins. Tin is approximately 3–4%, and zinc 1–2%.

Coinage Copper. See **Coinage Bronze** and **Coinage Metals.**

Coinage Gold. See **Coinage Metals.**

Coinage Metals. Metals used for minting coins vary from country to country, but gold, silver, copper, and nickel predominate. Usually, the metal used is not pure, but is hardened by alloying. American coinage gold and silver, for example, contain 10% copper. Coinage copper (more properly called coinage bronze) contains approximately 4% tin and 1% zinc. American coinage nickel is in reality, a copper-base alloy with only 25% nickel.

Coinage Nickel. See **Coinage Metals.**

Coinage Silver. See **Coinage Metals.**

Coining. See **Sizing.**

Coin Silver. See **Silver.**

Coke. Porous masses, consisting chiefly of carbon, resulting from the heating bituminous coals in limited air supply, or in the absence of air. In the beehive-oven procedure for making coke, just enough air is admitted to burn away the organic matter volatilized out of the coal; consequently, no by-products are obtained from the beehive oven. In the by-product or retort coking process, air is completely excluded, and the approximately 25% of volatile matter in the coal is recovered as ammonia, tar, aromatics (benzol, toluol), etc.

Coke Breeze. Coke of less than three-fourth inch size.

Coke Iron. Iron reduced from the ore by means of coke.

Coke Oven. Oven used to produce coke by the carbonization of coal.

Coke Oven Gas. See **Retort Gas.**

Coke Pig (Iron). The most common type of **Pig Iron** (see), made with coke as the reducing agent.

Coke Plate. See **Coke Tin Plate.**

Cokes. See **Coke Tin Plate.**

Coke Tin Plate. Standard tin plate, with the lightest commercial tin coat, used for food containers, oil canning, etc. A higher grade is the best cokes, with special cokes representing the best of the coke tin variety. For higher qualities and heavier containers, see **Charcoal Tin Plate.**

Cold Bed. Platform for the storage of cold bars, blooms, etc., during rolling.

Cold Bend. Bend Test (see), conducted on cold metal.

Cold-Blast. (a) Air, under about 15 pounds per square inch pressure, for passage into the hot-blast stoves in the **Blast Furnace** (see) process. (b) Air, not subjected to pre-heat, for use in a furnace or similar heating apparatus.

Cold-Blast Pig. Low-silicon pig iron, commonly made by means of cold-blast procedure.

Cold-Blast Valve. Valve, in **Blast Furnace** (see), located in the air line, prior to the entry of the cold main into the hot-blast stove.

Cold-Drawing. Drawing metal, without application of outside heat. See **Drawing** (b).

Cold-Drawn. See **Cold-Drawing.**

Cold-Drawn Steel. Steel given a final treatment by drawing through dies while cold, resulting in a high luster. See also **Cold-Rolled Steel.**

Cold-Finishing. Cold-working on metal, as a finishing operation, after the greater portion of working has been accomplished hot.

Cold Galvanizing. See **Electrogalvanizing.**

Cold-Hardening. See **Cold-Working.**

Cold-Heading. Forming the heads of rivets, bolts, screws, etc. by cold plastic flow, commonly by pressure in dies.

Cold-Heading Wire. Steel wire ordered or sold on the basis of suitability for cold-heading and cold-forging.

Cold Melt Process. Any process of melting metal, in which the charge consists entirely of cold, solid material.

Cold Metal Process. See **Cold Melt Process.**

Cold-Pressing. In powder metallurgy, the formation of compacts at room temperature.

Cold-Pressing Quality (Steel Plate). Steel plate ordered or sold on the basis of specifications governing suitability for cold-bending or forming, both longitudinally and transversely, at ordinary temperatures.

Cold-Reduced Plate. Steel plate, for use in tin plate manufacture, made from hot-rolled strip, by rolling on cold reduction mills.

Cold-Reduced Products. See **Cold-Rolled Products.**

Cold-Reduced Sheet. See **Cold-Reduced Plate.**

Cold-Reduced Strip. Steel strip, made from hot-rolled strip, by rolling on cold-reduction mills.

Cold-Reduction. Reduction of metal size, particularly thickness, while the metal is maintained at room temperature or below 250°F.

Cold-Reduction Mill. Rolling Mill (see) for the cold-rolling of wide and broad strip. See **Cold-Rolled Strip.**

Cold-Rolled Bar. See **Cold-Rolled Steel.**

Cold-Rolled Commercial Quality Steel Sheet. Cold-rolled steel sheet ordered or sold without special specifications. Good surface and moderate drawing ability are characteristic of this grade of steel sheet.

Cold-Rolled Deep Drawing Quality Steel Sheet. Cold-rolled steel sheet, ordered or sold on the basis of specifications governing deep-drawing quality.

Cold-Rolled Luster Surface Quality Steel Sheet. Cold-rolled steel sheet, produced on ground and polished cold-rolls, yielding a highly lustrous surface on the sheet.

Cold-Rolled Mill Run Steel Sheet. Cold-rolled steel sheet, ordered or sold on the basis of no separation of prime sheets (see **Cold-Rolled Prime Steel Sheet**) from seconds.

Cold-Rolled Plate. See **Cold-Rolled Steel.**

Cold-Rolled Prime Steel Sheet. Steel sheet, cold-rolled, ordered or sold on the basis of specifications requiring specific surface condition.

Cold-Rolled Products. Steel sheet, strip, and similar flat stock, rolled at temperatures between 50 and 250°F.

Cold-Rolled Seconds (Steel Sheet). Cold-rolled steel sheet, with surface defects of such nature that the metal may be used after a reasonable amount of finishing by the user.

Cold-Rolled Sheet. See **Cold-Rolled Steel.**

Cold-Rolled Steel. Steel with carbon below about 0.30%, made in the open-hearth furnace, and cold-rolled into strip, sheet, bar, or shafting form. It is formed hot, then annealed and pickled, and finally passed, cold, through a final finishing roll, which results in a smooth, highly polished surface. Available in tempers varying from "hard" to "dead soft," the latter being used for heavy or deep drawing.

Cold-Rolled (Steel) Sheets with Seconds Arising. Cold-rolled steel sheet, ordered or sold on the basis of a limited percentage of seconds. See **Cold-Rolled Seconds.**

Cold-Rolled Strip. See **Cold-Rolled Steel.** "Narrow cold-rolled strip" width up to 24 in.; "wide cold-rolled strip" width 24–46 in.; "broad cold-rolled strip" finished width above 46 in.

Cold-Rolling. Rolling metal at a temperature sufficiently low to create strain hardening (work hardening). See **Cold-Reduction.**

Cold-Rolling Mill. Rolling Mill (see) for the cold-rolling of narrow strip. See **Cold-Rolled Strip.**

Cold Short. Certain metals are brittle at ordinary temperatures; this characteristic is known as cold shortness. Phosphorus is a frequent cause of cold shortness in iron and steel.

Cold Shot. Round particles occasionally found in imperfectly-cast, chilled cast iron.

Cold-Shut. Defect, resulting from faulty technique in the casting of metals, when complete unity of the metal flow is somehow broken, so that a discontinuity appears in the cast metal.

Cold-Working. Plastic deformation, such as rolling, hammering, etc., at a temperature sufficiently low to create strain-hardening (work-hardening). Commonly, the term refers to such deformation at normal temperatures. See **Cold-Reduction** and **Cold Rolling.**

Colhed (Steel). Low-alloy tool steel, with about 0.5% vanadium, and 1% carbon (Vanadium Alloys Steel Co.).

Collar. Outside rims on rolls in rolling mills, limiting the lateral flow of metal being rolled.

Collar Marks. Defects on the surface of rolled metal, due to the metal going beyond the rolling groove and getting under the collars of the rolls.

Collaurin. Colloidal gold.

Collecting Electrode. Electrode, in the Cottrell precipitation process, on which the dust collects.

Collecting Plate. See **Collecting Electrode.**

Colloidal Metal. Colloidal dispersion of a metal.

Colloid Solution. See **Colloid (State).**

Colloid (State). Intermediate, in particle size, between molecular solution and coarse suspension, the particle size being 1 to 100μμ (1μμ is 10^{-6}mm). Colloidal solutions differ greatly from coarse suspensions, owing to the enormous surface exposed by the colloidal particles.

Colmonoy. Steel welding rod of low-alloy and high-alloy compositions (Wall-Colmonoy Corp.).

Colonial Steel. Series of low-alloy and high-alloy steels, including stainless steels, manufactured by Vanadium Alloys Steel Co.

Color. Particle of gold, obtained in panning.

Color Bluing. See **Bluing.**

Color Coating. See **Color Plating.**

Colorimetric Analysis. Analysis based on the law that the intensity of color of certain solutions is proportional to the amount of substance in the solution.

Colorimetry. See **Colorimetric Analysis.**

Color Marking. Indicating grade of metal, most commonly steel, by color identification, marked on the end of the metal bar, strip, etc.

Color Plating. Process of forming a colored film on a metal, most commonly by means of electrolysis in special, chemical baths.

Colors. Minute particles of metals, which occur native, as found in minerals.

Colorstrip. Strip steel, colored by chemical or japanning methods (Acme Steel Co.).
Columbia Steel. Series of low-alloy and carbon tool steels, manufactured by Columbia Tool Steel Co.
Columbium (chemical symbol Cb). Element No. 41 of the periodic system; atomic weight 92.91; also known as niobium (chemical symbol Nb). A steel-gray, malleable and ductile metal; it is extremely resistant to corrosion and most types of chemical attack; melting point 3542°F; boiling point unknown, but above 6000°F; specific gravity 8.4. Chemical valence is normally five (columbic), in some compounds four (columbous). The metal can be isolated in pure form by electrolysis of the fused double potassium fluoride. It has found use in commerce primarily as a stabilizer in stainless steels, to prevent intercrystalline precipitation when the metal is heated to 930–1300°F. The electrodeposition of columbium has not been definitely achieved. The pure metal may be obtained in the United States from Fansteel Metallurgical Corp.
Columnar Fracture. See **Fracture.**
Columnar Structure. Coarse structure of slender parallel columns, in an ingot, caused by slow solidification; likely to cause cracks.
Combination Mill. Rolling Mill (see), in which the greater portion of the reduction is done on continuous rolls, with the final shaping in a guide or loop mill.
Combination Plate. Plate (usually steel), as rolled, intended for shearing into two or more standard-size plates.
Combination Pot. See **Combination Process.**
Combination Process. Process of **Terne** (see) or **Tin Coating** (see) in which one coating machine, using the flux process, is followed by another, using palm oil.
Combination Quenching. Quenching in a bath consisting of water covered with oil to approximately equal depth. The metal piece, passing through the oil, is partly cooled and maintains an oil film when it reaches the water, which retards the quenching action.
Combined Carbon. Carbon in steel or cast iron, which is chemically combined with the iron (as Fe_3C) or with alloying elements, as simple or double carbides. All of the carbon in white cast iron is in the form of combined carbon.

Combining Heat. See **Heat of Formation.**
Combustion. Chemical union with oxygen. The term is commonly limited to the burning of fuels.
Combustion Chamber. Space in a furnace for the burning of the fuel.
Combustion Method. Method of quantitatively determining carbon, sulfur, etc., in metals, particularly iron-base metals, by burning a weighed sample and weighing the resulting gases.
Combustion Tube. Tube, used in the **Combustion Method** (see) of chemical analysis.
Comet (Metal). Iron-base, high-nickel (about 30%) resistance alloy, with up to 5% chromium (Driver-Harris Co.).
Coming to Nature. Stage in the **Puddling** (see) process, just prior to balling, in which the metal appears as small globules.
Commercial Brass. Yellow **Brass** (see), with 60–65% copper and up to 3.75% lead, added for greater machinability.
Commercial Bronze. Copper-base alloys (copper 87–97%, the alloying element being zinc), with excellent hot and cold working properties. In reality, the alloys are red brasses, and not bronzes at all.
Commercial Dust. Impure **Gold Dust** (see).
Commercial Finish Long Terne Sheet. See **Long Terne Sheet.**
Commercial Forging. Forging made without conforming to standard specifications as regards composition or grade of steel.
Commercial Hard. Term descriptive of the hard temper of commercially available copper and brass materials.
Commercial Quality Hot-Rolled Strip Steel. Hot-Rolled Strip Steel (see) not subjected to pickling and ordered or sold without special specifications.

Comminution. See **Pulverization.**
Common Brass. See **High Brass.**
Common Iron. See **Simple Steel.**
Common Lead. Lead of lower grade (more impurities) than **Corroding Lead** (see) or **Chemical Lead** (see).
Common Mill Edge. Convex edge, the most common form of edge finish on hot-rolled steel bands. See **Hot-Rolled Strip Steel.**
Common Solder. Soft solder, of about 60% lead and 40% tin content. Antimony is optional to about 1.5%.
Common Steel. See **Simple Steel.**
Common Terne. Terne (see) plate with 6–8 pounds of terne coating per double **Base Box** (see).
Common Wire. See **Coarse Wire.**
Compact. In powder metallurgy, the product resulting from compressing one or more powders, to form a semi-adherent mass.
Complete Roasting. See **Dead Roasting.**
Complex Alloy Steel. See **Complex Steel.**
Complex Steel. Alloy steel of more than two alloying elements, i.e., more complex than a **Quartenary Steel** (see).
Compo. Graphited tin bronze, made by sintering; tin content is about 10% and graphite about 1.5% (Bound Brook Oil-Less Bearing Co.).
Components. In the **Phase Rule** (see), the least number of chemical substances which can create the system under study.
Composite Metal. See **Cladding.**
Composite Roll. Roll for rolling metal; consists of a very hard iron alloy as the outer surface and a very strong, tough central section, combining hardness and strength.
Composition Brass. Copper-base alloy series, with 4–6% each of lead, tin, and zinc, used extensively for valves, hardware, etc. Fractional percentages of nickel and iron are optional.
Composition Face. Face of actual contact in **Twin Crystals** (see).
Composition Leaf. See **Dutch Metal Leaf.**
Composition Plane. See **Composition Face.**
Compound. Substance, containing two or more elements, always in definite proportions.
Compound Blowpipe. See **Oxy-Hydrogen Blowpipe.**
Compressibility. In powder metallurgy, the ability of the mass of powder to yield to pressure, in greater or lesser degree.
Compression Ratio. In powder metallurgy, the ratio of volumes of the loose powder, before compression, and the resulting compact, after application of any specific pressure.
Compression Stress. Stress that tends to compress the specimen under consideration in one or more directions.
Compression Test. Testing materials by subjecting to compression. Generally, the same type of equipment is used as for tensile testing. See **Tensile Strength.**
Compressive Strength. Maximum stress which a material can withstand, without deformation, when subjected to compression; commonly expressed as pounds, or tons, per square inch, or, in the metric system, as kilograms per square millimeter.
Concentrate. Product of **Concentration** (see), ready for chemical and heat treatment.
Concentration. Separation of valuable portions of an ore from the gangue, i.e., undesirable portions, by physical methods.
Concentration Cell. Electric cell with two similar electrodes, immersed in solutions of different concentrations of the same compound.
Concentric Converter. Bessemer converter (see **Bessemer Steel**), with its mouth at the center of the top, concentric and parallel to the bottom.
Concrete Bar. Steel bar used to reinforce concrete.
Concrete Reenforcement. Woven or electrically welded, steel-wire fabric used to reenforce concrete, for added tensile and bending strengths.
Condensed System. Phase Rule (see) system in which the vapor phase is substantially absent, thereby eliminating pressure as a factor, as expressed by the modified equation: $P+F=C+1$.

Most metallurgical processes are of this type.

Condenser. Chamber or zone, in which volatile material is condensed to a liquid or solid, by cooling action, most commonly of water or air.

Condensing Kitchen. See **Arsenic Kitchen.**

Conductance. See **Electrical Conductivity.**

Conducting Bottom Furnace. Electric **Arc Furnace** (see) with a bottom of electrically conducting materials; few electric furnaces now in use are of this type.

Conduction. See **Electrical Conductivity** and **Thermal Conductivity.**

Conductivity. See **Electrical Conductivity** and **Thermal Conductivity.**

Conductivity Bronze. Copper-base alloy, strengthened with approximately 1% cadmium and 0.6% tin. Despite the alloying, it maintains 50% of the electrical conductivity of pure copper.

Conductivity Copper. Electrolytically purified copper or arsenic-free Lake Superior copper of standardized conductivity.

Conductor. Substance capable of conducting electricity.

Conduit Pipe. Pipe, frequently of wrought iron, for protection of electric wire.

Congener. Member of the same family of the chemical **Periodic System** (see).

Consolute Temperature. See **Critical Solution Temperature.**

Constantan. Copper-nickel alloy, containing 10–55% nickel; characterized by relatively constant electrical resistivity irrespective of temperature; used for electrical resistance, and also for thermocouples, usually in conjunction with copper.

Constant Creep Period. See **Secondary Creep.**

Constant Permeability Nickel. Alloy of equal parts of nickel and iron, used as **Permeability Alloy** (see).

Constituent. Component of an alloy.

Constitutional Change. Change, in a solid alloy, affecting the nature or proportion of the constituents of the alloy.

Constitution Diagram. Diagram, indicating phase stability relationships between metal (or other) constituents and temperature.

Construction Line. See **Pitch Line.**

Contact Angle. Angle formed between the line of centers of the rolls and a radius to the top of the metal piece being rolled. The arc, equivalent to this angle, is known as the "contact arc."

Contact Arc. See **Contact Angle.**

Contact Area. Area of steel under the "contact arc" in a roll. See **Contact Angle.**

Contact Metal. Corrosion and arc resistant metal used for the fabrication of electric contacts. Silver, platinum, tungsten, and their alloys are among the most common materials used for this purpose.

Continental Process. Reducing copper ores to metal in a shaft furnace, after preliminary roasting.

Continuous Annealing Furnace. See **Continuous Furnace.** (a).

Continuous Billet Furnace. See **Continuous Furnace** (a).

Continuous Billet Mill. Mill, in which the metal passes from the ingot stage to a finished billet through a series of horizontal roll stands in tandem. See **Rolling Mill.**

Continuous Blooming Mill. See **Blooming Mill.**

Continuous Drawing. Wire-drawing through a succession of dies without interruption, eliminating intermediate handling of the wire.

Continuous Electrode. Carbon electrode, used in electric furnaces, formed continuously above the furnace, from raw materials, and fed downward as needed.

Continuous Furnace. (a) Furnace, in which the material being heated moves steadily through the furnace, often on a conveyor or in cars. (b) **Recuperative Furnace** (see).

Continuous Hot Strip Mill. See **Continuous Strip Mill.**

Continuous Lead Annealing. Process of annealing steel wire continuously by drawing the wire through a bath of molten lead.

Continuous Mill. Series of rolling mill stands, in which billet metal entering the first pass moves in a straight line and is continuously reduced in cross-section by the subsequent passes, being drawn by each set of rolls from the previous set.

Continuous Phase. In metals or alloys containing more than one phase, the phase which forms the matrix or background in which the other phase or phases are distributed as isolated units.

Continuous Pickling. Passing sheet or strip metal continuously through a series of **Pickling** (see) and washing tanks.

Continuous Rod Mill. See **Continuous Mill.**

Continuous Rolling. Process for converting slabs of metal, particularly steel, into sheet by continuous, progressive reduction of the thickness through a series of rolls, properly synchronized. See **Continuous Mill.**

Continuous Sheet Mill. Series of rolling mill stands for continuous rolling of steel sheet. See **Continuous Mill.**

Continuous Sintering. Sintering in a **Continuous Furnace** (see).

Continuous Slab (Heating) Furnace. Furnace for heating (or reheating) metal slabs, in which the metal is carried on a conveyor or the slabs are progressively pushed, one after another, through the furnace.

Continuous Strip Mill. Series of rolling mill stands for continuous rolling of steel strip. See **Continuous Mill.**

Continuous Wire Drawing. See **Continuous Drawing.**

Contraction of Area. See **Reduction of Area.**

Contraction Cavity. See **Pipe Cavity.**

Contraction Rule. See **Molders' Rule.**

Controlled Cooling. See **Insulated Cooling.**

Controlled Loop. In continuous wire-drawing, slippage of wire is avoided by permitting a number of idle, or controlled, loops, the number and length fluctuating with wear of the die.

Controlled Spangle. Spangle on zinc-coated (galvanized) steel of any pre-assigned geometrical appearance, usually accomplished by carrying the sheets away from the bath on a conveying wire netting of the appropriate mesh. See **Galvanizing.**

Convection. See **Thermal Convection.**

Conversion Iron. See **Conversion Pig Iron.**

Conversion Pig Iron. Pig Iron (see), intended as raw material for the manufacture of steel or wrought iron.

Converted Steel. See **Cement Steel.**

Converter. See **Bessemer Steel.** The converter may also be used for other metals, notably copper and nickel.

Converting. See **Bessemer Steel** and **Converter.**

Conveyor Furnace. Continuous Furnace (see) through which the material is carried on a mechanical conveyor.

Conveyor Sheet. Metal sheet, on which steel sheets are carried through a conveyor-type, continuous furnace, most commonly for normalizing purposes.

Cooler Arch. Opening in tuyere breast of a furnace, holding the **Tuyere** (see) cooler.

Cooling Bed. Equipment for holding metal sheet and strip for cooling after completion of rolling mill operations.

Cooling Crack. Crack in metal caused by too rapid cooling or by faulty cooling technique.

Cooling Curve. Curve resulting from plotting time and temperature for a slow-cooling metal sample. Since most changes in internal constitution involve heat development or absorption, there will be a break in the curve at every change of this kind. Cooling curves are extensively used in determining alloy constitution diagrams.

Cooling Floor. In the rolling of metal, the space between a roll train and the cutting shears.

Cooling Lag. Tendency of physical changes to lag behind temperature changes during cooling operations.

Cooling Medium. See **Quenching Medium.**

Cooling Stress. Stress, in a material, resulting from various parts of the material cooling at an unequal rate.

Cooling Zone. Zone in a **Continuous Furnace** (see) in which the metal cools, totally or in part, after heating.

Cool Metal. Non-metal; stratum in the earth in which a coal seam occurs.
Cooper Alloy. Series of low-alloy and high-alloy, ferrous casting compositions (Cooper Alloy Foundry Co.).
Cooper Stainless Steel. Series of high-chromium and **Stainless Steels** (see), in cast form, manufactured by Cooper Alloy Foundry Co.
Coordination Lattice. Crystal lattice, in which each ion bears the same relation to its adjacent ions, in all three coordinate directions.
Cope. Upper half of a casting mold.
Copel. Electric-resistance wire, for use up to 800°F only, with about 55% copper and 45% nickel.
Copper—. See also **Cupro—.**
Copper (chemical symbol Cu). Element No. 29 of the periodic system; atomic weight 63.57. A characteristically reddish metal of bright luster, highly malleable and ductile, and having high electrical and heat conductivity; melting point 1981°F; boiling point 4237°F; specific gravity 8.94. Chemically, it is both monovalent (cuprous) and divalent (cupric). Copper is obtained from its ores by roasting and smelting, followed by electrolytic refining. Extensively used in the pure form wherever high electrical conductivity is needed, and as the base for the many commercially valuable brasses and bronzes. Copper-plating is commercially wide-spread; both sulfate and double alkali cyanide baths are widely used.
Copper-Aluminum. Alloy, containing approximately equal parts of copper and aluminum, used as **Master Alloy** (see).
Copper Amalgam. Alloy of the copper-mercury series.
Copper Bath. Solution of an electrolyzable copper compound, such as the sulfate or alkali copper cyanide, for use in the electrodeposition or electrorefining of copper.
Copper-Bearing Steel. See **Copper Steel.**
Copper Bottoms. Product, essentially metallic in nature, resulting from smelting rich copper ores in a reverberatory furnace.
Copper Cast Steel. Cast Steel (see), with copper ranging from 0.90 to 1.50%.
Copper-Chromium Alloys. Copper-base alloys, with 0.5–1% chromium, combining high strength with about 80% of the electrical and heat conductivity of standard-conductivity copper.
Copper-Cladding. See **Cladding.**
Copper-Clad Steel. See **Cladding.**
Copper-Coated Steel. See **Coated Finish Steel.**
Coppered Wire. Steel wire, with a thin coating of copper, obtained by wet-drawing the wire while immersed in a solution of copper sulfate.
Copper-Head. Red-brown spot, appearing on the **Ground-Coat** (see) of vitreous enamels.
Copper-Lead Alloys. Series of copper compositions, with up to 3% of lead; used for bearings and to increase machinability. See also **Leaded Copper.**
Copper Loss. Power loss in electrical apparatus, due to resistance of the conductors.
Copper Matte. Matte (see), intermediate product in the smelting of copper ores; normally contains about 35% sulfur.
Copper Nickel. (a) See **Arsenical Nickel.** (b) See **Copper-Nickel Alloy.**
Copper-Nickel Alloy. Series of **Master Alloys** (see), containing up to 60% nickel. See also **Cupro-Nickel.**
Copperoid. Zinc sheet, electrocoated with copper (American Nickeloid Co.).
Copper Phosphide. The compound, Cu_3P_2, resulting from reaction of copper and phosphorus in chemically proper proportions. See also **Phosphor Copper.**
Copper Plate. (a) Product of **Copper Plating** (see). (b) See **Copper Plates.**
Copper Plates. Plates of copper, amalgamated on the surface with mercury, over which crushed gold ore is passed, in the appropriate type of **Amalgamation Process** (see), in order to recover the gold as an amalgam.
Copper Plating. Process of coating metal objects with copper, in which the object is made cathode (negative electrode) in an electrolytic bath, containing a decomposable copper salt,

acid copper baths being mostly made with copper sulfate, alkali baths with cyanide. Pure copper metal is normally used as anode (positive electrode). Nickel and chromium plating on steel is usually preceded by a light copper-plating (called a flash), with the nickel or chromium then deposited on the copper.

Copper Powder. Essentially-pure copper in powder form. Such material, much used in powder metallurgy, may be manufactured either by mechanical means or by electrodeposition under special conditions.

Copper Pyrites. Ore, double sulfide of copper and iron, corresponding essentially to the formula $CuFeS_2$. It is used as a source of copper, being smelted with considerable silica or silicaceous material, which unites with the iron oxide formed by oxidation of the iron content, while the sulfur content is also oxidized, to gaseous sulfur dioxide.

Copper-Rich Brass. See **Rich Gold Brass.**

Copper Scale. Coating formed on copper (a mixture of cuprous oxide and cupric oxide, the latter predominating), as a result of heating.

Copper Silicide. See **Silicon-Copper.**

Copper-Silicon. See **Silicon-Copper.**

Coppersmiths' Copper. Hot-rolled copper sheet, in soft temper and relatively heavy thickness.

Copper Smoke. Gas, resulting from the roasting of copper sulfide ores; the preponderant material is sulfur dioxide.

Copper Sponge. Copper in porous form, made by **Powder Metallurgy** (see) techniques. Frequently, the copper, after being formed into finished form, is filled with lead and the resultant material used for bearings.

Copper Steel. Low-carbon steel, containing approximately 0.25% copper, which slightly increases resistance to atmospheric corrosion.

Copper Sulfate Test. See **Preece Test.**

Copper-Titanium. Master alloy of titanium with copper.

Copperweld. Copper-coated steel wire, manufactured by pouring copper around a round steel billet, followed by hot-rolling the resultant composite billet and drawing into wire (Copperweld Steel Co.).

Cop-R-Loy. Copper-bearing ingot iron (Wheeling Steel Corp.) of low carbon content, with fractional percentages of copper, to increase resistance to atmospheric corrosion.

Cordirie Process. Removing oxidizable impurities from molten lead by passing steam through the metal.

Core. (a) The sub-surface portion of iron or steel after the surface has been case-hardened (the core is much softer than the case). (b) Form, made of sand or similar material, inside of a mold; prevents the metal from filling that space during casting.

Core Box. Box in which cores [see **Core** (b)] are made and held during operations.

Cored Bar. In powder metallurgy, a compacted bar, electrically heated to a temperature sufficient to melt its center.

Cored Electrode. Metal electrode with an internal core of flux, used in electric **Arc Welding** (see).

Cored Mold. Mold, for casting metal, containing one or more cores. See **Core** (b).

Cored Structure. See **Zonal Structure.**

Coreless (Induction) Furnace. Type of induction furnace, operating at high frequencies, which eliminates the iron core necessary for lower frequencies.

Core Loss. Electric energy, lost as heat within a core of magnetic material during magnetization; equal to the sum of eddy current loss and hysteresis loss.

Core Plate. Plate on which cores are formed, in metal casting operations, by ramming sand-binder mixtures into a mold or flask. See **Core** (b).

Core Plating. Coating of silicon steel sheets with varnish, for use in electrical machinery.

Core Rod. In powder metallurgy, the die part, used to form any hole in a compact.

Core Sand. Mixture, predominantly sand and clay, used in making cores. See **Core** (b).

Core-Type Induction Furnace. See **Induction Furnace.**

Corinthian Process. See **Carinthian Process.**

Corner Segregation. Tendency for segregation in the corners of an ingot having sharp corners or angles.

Cornice Copper. Hard-rolled sheet copper used in construction work, usually pre-formed to the final desired shape.

Corroding Lead. Lead of high purity (99.93–99.99% analysis), used in the manufacture of white lead.

Corronizing. Process for rustproofing by electroplating, in which a layer of nickel is first plated on the base metal, followed by a very thin second layer of zinc or tin and heating to 500–750° F, to accelerate surface alloying.

Corrosion. Generally, gradual chemical attack on a metal at normal temperatures by atmosphere, moisture, and similar common agents; usually an oxide is formed.

Corrosion Cracking. See **Season Cracking.**

Corrosion Efficiency. Total loss of metal by anodic oxidation in an electrolytic process, divided by the loss of metal to be expected theoretically.

Corrosion Fatigue. See **Fatigue.**

Corrosion-Resistant Alloy. Alloy showing better-than-normal resistance to corrosion (usually, atmospheric corrosion). More specifically, the term is often limited to ferrous alloys high in chromium, particularly the stainless steels of the 18% chromium-8% nickel variety.

Corrosion-Resistant Steel. See **Corrosion-Resistant Alloy.**

Corrosiron. High-silicon cast iron, containing approximately 14% silicon, made by Pacific Foundry Co. and by Bethlehem Foundry & Machine Co. Because of its acid corrosion resistance, it is commonly used for chemical equipment.

Corrosive. Chemical material causing **Corrosion** (see).

Cor-Ten Steel. Series of low-alloy steels (United States Steel Corp.), with chromium in the 0.50–1.50% range and optionally fractional percentages of copper and silicon, in addition to normal steel constituents. Carbon does not exceed 0.10%. They show better-than-average corrosion resistance and are used for pipes, structural work, etc.

Corvic Bronze. Phosphor Bronze (see) with about 1.5% tin and 0.30% residual phosphorus (Chase Brass & Copper Co.).

Coslettized Steel. See **Coslettizing.**

Coslettizing. Process of forming rust-resistant coating on iron or steel by immersion in a hot solution of acid iron phosphate.

Cotton Tie Band. Light, narrow hoop (see **Hoop Iron**) used for tying and baling cotton.

Cotton Ties. See **Cotton Tie Band.**

Cottrell (Electric Precipitation) Process. Removing suspended solids from smokes by passage through a strong electric field. The suspended solids, being electrically charged, discharge and deposit on one or the other of the electrodes, depending on the charge of the suspended particles.

Coulomb. Quantity of electricity equivalent to one ampere per second.

Country Heat Steel. Steel, with about 0.60% carbon, made by carburization of wrought iron.

Couple. Two dissimilar metals in contact; will generate voltage under proper electrolytic influences or on heating.

Coupling Box. Loose-fitting connection, in **Rolling Mill** (see), between the roll and the spindle, which transmits power from the pinion to the roll.

Coupon. Extra piece of metal, e.g., on a casting, from which a specimen for physical testing is cut or machined.

Cover Coat. In processes, such as enameling, where more than one coat of material is used, any coat which is put on after the ground-coat.

Covered Chill. Metallic cooling section of a mold for casting **Chilled Rolls** (see) covered with sand to reduce chilling effect. This is in contrast with bare chill in which the chill is left exposed.

Covered Electrode. See **Coated Electrode.**

Cover Sheet. Metal sheet laid on a steel sheet during normalizing operations, in a conveyor-type continuous furnace.

Cowles Alloy. Series of copper-aluminum alloys, with 1–11% aluminum. See **Cowles Process.**

Cowles Process. Process for the direct manufacture of aluminum alloys, such as copper-aluminum, from aluminum ores, by electric-furnace reaction with carbon in presence of the alloying metal.

Cowper-Siemens Stove. Brick hot-blast stove (see **Blast Furnace**), operating by regeneration.

C.P. Chemically pure; designates substances of analytical grade.

Cr. Chemical symbol for **Chromium** (see).

Crack of Gas. Small explosion, or puff, of gas in a **Blast Furnace** (see) during operation.

Cramp Alloy. Series of copper-base alloys of varying composition, manufactured by Baldwin Locomotive Works.

Cramp's Elfur Iron. Series of low-alloy and high-alloy cast irons (Baldwin Locomotive Works).

Crane Ladle. Ladle (see), hanging from a crane, for pouring fused metal into molds.

Crankcase Alloys. Aluminum alloys used for automotive purposes, primarily for crankcase castings, approximating 90% aluminum, with copper as the primary alloying agent. Iron and, optionally, zinc are also present in small percentages.

Crasco Steel. Series of alloy tool steels, manufactured by Craine-Schrage Steel Co.

Crazed Corner. Surface defect on steel sheet, resulting from faulty cold-working technique.

Creep. (a) Tendency of metals to deform permanently when stressed for long periods, even when such stresses are below the normal yield point, particularly at higher temperatures. As short-time tensile values are not safe for design purposes at higher temperatures, creep test values must be used instead, determined by subjecting specimens to constant load and temperatures for long periods (up to as long as several years' duration). At first, creep is quite rapid (called initial creep), then a constant secondary creep takes place. (b) In rolling metal, the difference in the speeds of a point on the roll surface and any point in the metal.

Creep Curve. Curve, in creep test, indicating relationship between elongation of a metal sample (at a stated high temperature) and total elapsed testing time.

Creep Limit. Maximum tensile stress, at any assigned temperature, measured before significant **Creep** (see) is in evidence, or one which will maintain creep below any assigned rate.

Creep Recovery. Tendency of polycrystalline materials to recover part of their creep deformation, after removal of the creep-causing stresses.

Creep Strength. See **Creep.**

Creep Stress. See **Creep.**

Creep Test. See **Creep.**

Crescent Steel. High-carbon tool steel (Crucible Steel Co.) of approximately 7% tungsten, 3% manganese and 2.0% carbon content.

Crilley Metal. Bronze containing mercury, used for bearing purposes.

Cristite. High-alloy iron resistant to wear, chemical action, abrasion, and high temperatures. Tungsten content is about 17%, with chromium about 11%, and carbon about 3%. Low percentages of molybdenum and titanium are optional (Commercial Alloys Co.).

Critical Cooling Rate. Rate of cooling barely sufficient to prevent undesired transformations. In steel, usually the rate necessary for the formation of martensite.

Critical Point. Temperature at which internal changes take place within a metal, as evidenced by absorption of heat on heating and liberation of heat during cooling. The resulting break in the heating or cooling curve is known as an arrest. In iron and steel, any arrest occurring in the heating curve is indicated by the letters Ac, whereas those occurring during cooling (usually somewhat lower) receive the designation Ar. These letters are usually followed by a number; O representing an unimportant magnetic change occurring in cementite at 415°F; 1 representing the most important of these changes, the eutectoid transformation at 1333°F; 2 (Curie point) representing a magnetic change otherwise not observable as a phase change at 1414°F; 3 representing formation of ferrite from austenite in the 1333–1670°F range; 4 representing the transformation of gamma iron to delta iron at 2552°F. Changes in chemical composition shift the critical points; generally, increased impurities lower their values. When nothing specific is indicated, the term critical point generally refers to the A-1 recalescence point.

Critical Point Test. Test for determining critical temperatures (see **Critical Point**) in a metal specimen. Most common among the methods are drawing time-temperature curves while heating or cooling, temperature-expansion curves (called the dilatometric method), temperature-magnetic curves (called the magnetic method), and measuring the electrical resistance of samples after specified heat treatments.

Critical Range. Temperature range at which internal changes take place within a metal. See also **Critical Point.**

Critical Shearing Stress. Stress at which rate of slip can just be measured with standard equipment.

Critical Solution Temperature. Temperature below which two mutually soluble liquids become partially soluble in each other.

Critical Strain. The percentage strain below which no recrystallization will take place on annealing and above which large grain growth occurs on annealing.

Critical Temperature. See **Critical Point.**

Crocar. Iron-base alloy, manufactured by Vanadium Alloys Steel Co., approximating 12% chromium, 2% carbon, 0.3% silicon, 0.75% vanadium and 0.75% cobalt; used where heat and abrasion resistance is required.

Croloy. Series of low-alloy and high-alloy steels, available in rolled and tube form; chromium ranges from 4 to 26%, with nickel up to 20% and molybdenum up to 1.5% (Babcock & Wilcox Tube Co.).

Cromal. Aluminum-base alloy, with about 3% chromium and optional low percentages of nickel and manganese, adding to strength and corrosion resistance.

Cromansil (Steel). Series of low-alloy, high-strength steels, with carbon usually below 0.14%, 1.0–1.4% manganese, 0.3–0.7% chromium and 0.6–0.9% silicon.

Cromic. Series of nickel-chromium resistance alloys; with nickel ranging from 25% to 80% and iron optional up to 50% (George W. Prentiss & Co.).

Cromoco. High-carbon die steel containing about 12% chromium, 1% cobalt, 1% molybdenum, and 1.6% carbon (Firth-Sterling Steel Co.).

Cromodizing. Process for rust-proofing of steel and forming a paint-bonding surface, in which a solution of chromic acid and sodium chloride is sprayed on the metal at about 180°F.

Crop. Discarded end of an ingot, cut away from the body because of pipe and similar defects.

Crop End. See **Crop.**

Cropping. Cutting off **Crop** (see) from an ingot.

Cropping Shears. Hydraulic shears for cutting off ends of blooms, to remove unsatisfactory metal prior to passing into rolling mills.

Cro-Sil. Series of corrosion-resistant, high-chromium alloys (10–14% chromium), manufactured by Crucible Steel Co.

Cross-Break. See **Coil-Break.**

Cross Country Mill. Group of roll stands arranged in trains or in tandem (see **Rolling Mill**); the metal moves semi-continuously from one mill to another; from each set of rolls it goes to stationary roll tables before entering the next.

Cross Rolls. Set of two rolls, with axes at an angle of about 30°, for straightening of pipe which is revolved by the rolls while slowly moved forward.

Cro-Tung. Alloy steel series (Industrial Steels, Inc.), with about 5% chromium and 1% tungsten; carbon maximum, 0.30%. Used for corrosion-resistant products.

Crouse's Alloy (or Metal). See **Fusible Alloys.**

Crown. Increased thickness in the center of metal sheet or strip, as compared with thickness at the edge.

Crown Gold. Gold (see) of 91.7% purity, standard in Great Britain for coinage.

Crucible. (a) Pot, shaped like a small barrel, for melting metals. The most common type, the graphite crucible, is made of a mixture of graphite and clay, in approximately equal proportions, with sand frequently added. (b) Lower part of a shaft furnace, in which fusion takes place. See also **Blast Furnace** and **Cupola.**

Crucible Furnace. Furnace for melting of metals in crucibles.

Crucible Practice. Technique and care given steel produced by the crucible process. See **Crucible Steel.**

Crucible Process. See **Crucible Steel.**

Crucible Steel. Steel made and melted in graphite **Crucibles** (see). Because of its small production capacity and consequent costliness, the crucible process is limited to the better steels, and, in particular, to tool steels; although it permits great precision in steel composition and quality, it is being gradually replaced by the electric-furnace method, where the same precision can be obtained with much larger production capacity. Note that the Crucible Steel Co. does not limit its production to steel made by the crucible process, i.e., not all steel labeled "Crucible" is necessarily manufactured by the crucible process.

Crude Antimony. Antimony sulfide ore; a non-metal.

Crushed Steel. Small fragments of steel, used as an abrasive.

Crushing Hardness. See **Compressive Strength.**

Crushing Strength. See **Compressive Strength.**

Cry. See **Tin Cry.**

Cryocarbide. See **Pearlite.**

Cryolite. Double fluoride of sodium and aluminum, chemically Na_3AlF_6, occurring naturally in Greenland or made synthetically; used as fused bath from which aluminum is electrolyzed when aluminum oxide is added.

Cryptocrystalline. Composed of crystals too small to be resolved by the microscope or other resolving apparatus.

Crystal. (a) Essentially homogeneous particle in which the atoms, ions, or molecules are arranged in a three-dimensional periodic pattern. Plane surfaces and definite angles between the crystals are normally the consequence of such structure. Most metals are either macrocrystalline or microcrystalline in nature; it is still a moot question whether, sub-microscopically, this is not also true of metals called amorphous, or non-crystalline. See also **Idiomorphic Crystals.** (b) See **Spangle.**

Crystal Aggregates. Grouping of crystals and grains within a metal.

Crystal Analysis. Science of determining crystal structure of materials, usually by X-ray methods.

Crystal Boundary. Line or plane, separating crystals within a metal.

Crystal Face. Recurring, characteristic surface of a crystal, or a plane parallel to it.

Crystal Face Index. See **Indices of Crystal Face.**

Crystalline Form. Geometrical shape of a crystal.

Crystalline Fracture. See **Fracture.**

Crystallite. Small, imperfect crystal.

Crystallization. Crystal formation; in metal technology, this usually occurs during solidification from molten metal. The term is sometimes applied to the growth of larger crystals from smaller ones. A common error is to charge fatigue failure of metals as being due to crystallization of the latter type.

Crystallization System. See **Crystal System.**

Crystallized Tin Plate. Tin Plate (see) with surface subjected to the action of a dilute solution of hydrochloric and nitric acids, yielding a crystalline surface appearance.

Crystallogram. Photograph of X-ray diffraction pattern of a crystal.

Crystallographic Axes. Principal axes of symmetry common to all crystals belonging to a given system.

Crystallographic Systems. See **Crystal System.**

Crystallography. Science of the properties and structure of crystals. See **Crystal** (a).

Crystalloid. Substance capable of easy crystallization, as opposed to colloid. See **Colloid (State).**

Crystal Nucleus. Small crystal acting as focal point for further crystal growth.

Crystalon. Silicon carbide product, manufactured by Norton Co., used for refractory and abrasive purposes.

Crystal Pattern. Pattern formed by action of X-rays on a crystal.

Crystal Structure. Size and shape of the unit cell of crystalline materials and the arrangement of atoms, ions, or molecules in the unit cell.

Crystal System. Crystals are divided into six basic systems, according to their symmetry: cubic, hexagonal, orthorhombic, tetragonal, monoclinic, and triclinic. See Individual Systems.

Cs. Chemical symbol for **Cesium** (see).

Cu. Chemical symbol for **Copper** (see).

Cubbling. Breaking flat iron, preliminary to piling, faggoting, or rolling.

Cubical Cleavage. Cleavage (see) of equal facility in three directions perpendicular to one another.

Cubic (Crystal System). Crystal arrangement in which the three axes are equal and perpendicular to one another.

Cufenium. Copper-base alloy series, with about 21% nickel and 6–20% iron; used for the same purposes as ordinary **Nickel Silver** (see).

Cu-Lead-Ite. Series of high-lead bearing bronzes, with 20–50% lead, and tin optional up to 10% (Scientific Alloys, Inc.).

Culvert Stock. Steel sheet ordered or sold on the basis of specifications governing use in the manufacture of culverts.

Cumloy. See **West (Steel).**

Cupaloy. Copper-base alloy, manufactured by Westinghouse Electric & Mfg. Co., containing fractional percentages of chromium and silver.

Cup and Cone. See **Bell and Hopper.**

Cup-and-Cone Fracture. See **Fracture.**

Cupel. See **Cupellation.**

Cupellation. (a) Process of melting metals, such as silver-rich lead, in a reverberatory furnace with exposure to a blast of air, the base metals being oxidized, leaving behind the non-oxidizing metals (such as silver). (b) Treatment of silver- or gold-bearing lead in a small, shallow, porous cup, called cupel, to oxidize the lead, leaving the precious metals, for assaying.

Cup Fracture. See **Cup-and-Cone Fracture.**

Cupola. Tall cylindrical shaft furnace for melting metal. The bottom part is the crucible, with metal and slag tapping holes; above the slag tapping line there is a line of tuyeres, inlets for air. Above this are alternate layers of metal and fuel (such as coke), reaching up to a charging door at the top.

Cupola Furnace. See **Cupola.**

Cupped Fracture. See **Cup-Shaped Fracture.**

Cupped Wire. Wire with internal cavities, resulting from faulty wire-drawing operations.

Cupping. Tendency of defective wire to break prematurely, with a cup fracture; generally caused by segregation, with the hard center more brittle than the ductile exterior.

Cupping Process. Making seamless pipe tubing, cylinders, etc., from a flat plate or disc of metal, by forming into a cup-like shape by a series of forging operations.

Cupping Test. Test of the deep-drawing capacity of sheet metal. A standardized plunger is forced into the sheet and the depth of the cup is measured just at the moment of fracture. See **Olsen Ductility Test** and **Ericksen Ductility Test.**

Cupriferous. Containing copper.

Cupro—. See also **Copper—.**

Cupromagnesium. Alloy of copper and magnesium, the latter usually about 10%.

Cupromanganese. See **Manganese Copper.**

Cupron. Alloy of about 55% copper and 45% nickel (Wilbur B. Driver Co.). See **Constantan.**

Cupro-Nickel. Alloys of copper and nickel, with up to about 20% nickel, used for hardware in the lower nickel ranges, condenser tubes, and similar products in the higher nickel ranges.

Cuprous. (a) Pertaining to copper. (b) Chemically, compounds of monovalent copper.

Cuprous Manganese. Variety of **Bog Manganese** (see), with appreciable percentages of copper oxide, frequently containing also cobalt oxide.

Cup-Shaped Fracture. Fracture of metal, particularly in a tensile test, in which the metal shows, at the point of fracture, a cup-like face.

Curie Point. Temperature at which a ferromagnetic substance becomes paramagnetic.

Curie's Law. Magnetic susceptibility is inversely proportional to the absolute temperature.

Current Density. Electric current flowing per unit cross-section.

Current Efficiency. In electroplating or electrorefining, the ratio of metal deposited to theoretical values corresponding to Faraday's laws.

Cut. In powder metallurgy, the proportion of metal powder which lies, in particle size, between any two limits, usually as judged from screen mesh values.

Cutanit. Cemented carbide cutting tool, comprising tungsten carbide and other carbides, such as those of titanium and molybdenum.

Cutting Alloys. Alloys (excluding steels and other iron-base alloys), used for the manufacture of cutting tools, such as **Stellite** (see).

Cutting Flame. Flame in a **Cutting Torch** (see), in which oxygen is present in excess of that needed for complete combustion. Also any other flame containing excess oxygen.

Cutting Hardness. Resistance of a material to cutting.

Cutting-Out. (a) Enlargement of the clay nozzle during tapping of molten metal from a ladle. (b) Enlargement of the diameter of a die through wear.

Cutting Test. Determination of serviceability, or "tool endurance" of a cutting tool.

Cutting Torch. Oxy-acetylene or oxy-hydrogen torch, with excess oxygen, used for cutting through metal in sheet or plate form.

Cyanicide. Substance reacting with aqueous solutions of cyanides, in the extraction of gold or silver, resulting in loss of cyanide activity. See **Cyaniding** (b).

Cyanidation. See **Cyaniding.**

Cyanide Bath. (a) See **Cyaniding** (a). (b) Bath containing fused alkali or alkaline earth cyanide, for the cleaning of metal.

Cyanide Dip. Treatment of steel, in which the metal is first heated below the critical temperature range, then transferred to an alkali or alkaline earth cyanide bath above the critical range for up to a minute, and finally quenched.

Cyanide Hardening. See **Cyaniding** (a).

Cyanide Process. See **Cyaniding** (b).

Cyanide Reheat. Treatment of steel, in which the finished metal piece is plunged, cold, into an alkali or alkaline earth cyanide bath for twenty to forty minutes.

Cyanide Wash. Treatment of steel, in which the metal is first heated below the critical temperature range, then transferred to an alkali or alkaline earth cyanide bath above the critical range, for a period of one to five minutes, and finally quenched.

Cyaniding. (a) Form of case-hardening in which a molten cyanide is used in the hardening bath, to increase surface carbon and nitrogen. (b) Process of extracting gold and silver from their ores by treatment with a dilute solution of alkali cyanide.

Cyclops (Steel). Series of high-alloy steels, manufactured by Cyclops Steel Co. Usually chromium is present in ranges from about 4 to 8%, with relatively large amounts of other alloying elements, such as nickel, tungsten, vanadium, cobalt, and molybdenum. Because of their rust- and heat-resistance, the alloys are used as tools, high-temperature tubing, etc.

Cylinder Finish. Highly smooth and polished finish, commonly used on the inside of cold-drawn steel tubing, intended for pump and engine cylinders, etc.

Cylinder Iron. Hard, close-grained cast iron, used for engine cylinders and pistons.

Cylinder Metal. Cast iron, with manganese up to 2%, used for the manufacture of engine cylinders because of low friction characteristics.

Cyprus Bronze. High-Lead Bronze (see), with about 30% lead, used for bearings, etc.

CYW Steel. Low-alloy, hot-die steel (Firth-Sterling Steel Co.), with about 3.5% chromium and 1% carbon.

D

Dalton's Alloy (or Metal). See **Fusible Alloys.**

Dam. Brick or refractory wall, built on a supporting iron plate, called the "dam plate," acting as the front of the fore-hearth in a **Blast Furnace** (see). The metal is drawn through a hole in the dam and the slag through a notch in the top.

Damask. Etching formed on polished steel, as a result of corrosion.

Damping. See **Damping Capacity.**

Damping Capacity. Ability of a material to absorb vibrations.

Damping Test. See **Damping Capacity** and **Foeppl-Pertz Damping Tester.**

Dam Plate. Water-cooled metal plate, at the tapping hole in a **Blast Furnace** (see) to protect the hearth jacket when metal is being tapped.

Dam Stone. Wall of fire brick or refractory, enclosing the hearth front in a **Blast Furnace** (see).

Dancer Rolls. Rolls, in continuous pickling-tank operations, which control the rate of travel of the coiled metal through the bath.

Dandelion Metal. Lead-base alloy, with about 18% antimony and 10% tin, used for bearings in heavy duty machinery.

Daniell Cell. See **Battery.**

Darcet's Metal (or Alloy). See **Fusible Alloys.**

Dark Cherry Heat. Temperature approximating 1175°F, as judged visually.

Dark Orange Heat. Temperature approximating 1650°F, as judged visually.

Dark Red Heat. Temperature range from about 1020 to 1160°F, as judged visually.

Dark Ruby Silver. Non-metallic mineral, pyrargyrite, a silver-antimony sulfide approximating Ag_3SbS_3 in composition.

Dark Spot. See **Flux Spot.**

Darling Cell. See **Darling Process.**

Darling Process. Obsolete process for manufacturing sodium by electrolysis of fused sodium nitrate.

Darwin (Steel). Series of alloy tool steels, manufactured by Darwin & Milner, Inc.

Davis Furnace. Reverberatory furnace, with a long hearth and heated by several side fireplaces, used for **Roasting** (see) sulfide ores.

Davis Metal. Copper-base alloy series (Chapman Valve Mfg. Co.), with 25–29% nickel, and low percentages of iron, carbon, silicon, manganese, and lead. Because of corrosion resistance, the alloys are used in valves and fittings.

Dead-Dipping. Forming a dull surface on metal, usually copper-base alloys, by dipping in acid, followed by washing and drying.

Deadened Mercury. See **Floured Mercury.**

Deadhead. See **Riser.**

Dead Hole. Shallow hole in a casting.

Dead Load. Static Stress (see) in a structure, caused by the weight of the structure itself.

Dead Metal. See **Dead Steel.**

Dead Riches. See **Base Bullion.**

Dead Roasting. Roasting (see) carried to the lowest practical point of sulfur content.

Dead Soft Annealing. Heating metal to above the critical range, to develop the greatest possible commercial softness or ductility.

Dead Soft Steel. Steel, normally made in the basic open-hearth furnace, with carbon less than 0.10% and manganese in the 0.20–0.50% range, completely annealed.

Dead Soft Temper. Condition of maximum softness commercially attainable in wire, strip, or sheet metal.

Dead Steel. Steel, during manufacture, which no longer reacts with the carbon, resulting in a cessation of carbon monoxide evolution.

Dead Surface. See **Dead-Dipping.**

Debye-Scherrer (X-ray) Method. X-ray examination of crystals, using refracted monochromatic X-rays and powdered material, with crystals lying at random.

Decalescence. Absorption of heat by iron and iron alloys during heating through the so-called decalescence point (metallographically known as the Ac_1 point), equivalent to the recalescence point, on cooling.

Decantation. Pouring away supernatent liquid from a sediment or precipitate.

Decarbonize. See **Decarburization.**

Decarburization. Removal of carbon from the outer surface of iron or steel, usually by heating in an oxidizing or reducing atmosphere. Water vapor, oxygen and carbon dioxide are strong decarburizers. Reheating with adhering scale is also strongly decarburizing in action.

Decarburizing. See **Decarburization.**

Decay. See **Weld Decay.**

Decomposition Potential. See **Decomposition Voltage.**

Decomposition Voltage. Minimum voltage necessary to maintain an appreciable flow of current through an electrolytically decomposable solution.

Decrepitation. Cracking of materials, with a crackling noise, when subjected to heat, as in roasting operations.

Deep-Drawing. Drawing of sheet metal into forms involving great amounts of deformation. Metal sheets for deep-drawing, consequently, require high ductility.

Deep-Drawing Brass. Brass (see) of 66–70% copper content, used for deep-drawing, forming, spinning, and other cold-working operations.

Deep-Drawing Dull Finish Auto Body Sheet. Cold-rolled and annealed steel sheet, used in the manufacture of door and back panels of automobiles. See **Automobile (Body) Sheet.**

Deep-Drawing Quality Hot-Rolled Strip Steel. Hot-Rolled Strip Steel (see), ordered or sold on the basis of suitability for deep-drawing.

Deep-Drawing Steel. Steel ordered or sold on the basis of suitability for drastic drawing operations. See **Drawing** (b).

Deep Etching. See **Etching.**

Deep Etch Test. Test, conducted on a smoothened section of steel, by immersion in a 50% hydrochloric acid solution at 180°F. Because of preferential attack on different components of the steel, an apparent porosity results, and flaws may be detected visually.

Deep-Hardening Steel. Steel capable of attaining a thick **Case** (see), after standard hardening operations.

Deep Seam. In butt-welded pipe, a defect in the form of depressions on the inside and outside of the pipe.

Deep-Seated Blowhole. Blowhole (see) located at an appreciable distance from the surface of the metal.

Defender Metal. Lead-base alloy, with 10–12% antimony and 5.5–7.5% tin, used as a substitute for tin-base bearings (Magnolia Metal Co.).

Defiheat (Rustless) Iron. High-chromium steel, manufactured by Rustless Iron & Steel Corp., resistant to high temperatures and corrosion, with 25–30% chromium, about 0.30% carbon, and the usual fractional percentages of manganese and silicon.

Defirust (Rustless) Iron. Series of chromium steels, manufactured by Rustless Iron & Steel Corp., highly resistant to corrosion. The chromium content may vary from 12 to 18%, and carbon is kept below 0.12%, with manganese and silicon in the usual, fractional percentages.

Defistain (Rustless) Iron. Stainless steel, manufactured by Rustless Iron & Steel Corp., with 17–19% chromium, 7–9% nickel, 0.50% manganese, 0.50% silicon and carbon below 0.20%; used for the same purposes as other stainless steels. See **Stainless Steel.**

Deformation Curve. Curve showing percentage of deformation, with time, under specified conditions.

Deformation Resistance. In rolling metal, the ratio between roll pressure and contact area.

Deformed Bar. Metal bar, particularly steel bar, with projections or depressions regularly spaced; used for special purposes, such as concrete reinforcement.

Deforming Groove. Groove, in rolling mill, for rolling **Deformed Bars** (see).

Deforming Test. Determination of deformation, or non-deformation, of a metal, when subjected to heat treatment.

Degreasing. Removing grease, oil, etc., from metal surfaces, prior to finishing operations (such as plating). Usually, degreasing is accomplished by an organic solvent. In vapor-phase degreasing, chlorinated solvents, such as trichlorethylene, are extensively used.

Degrees of Freedom. Number of independently variable quantities such as temperature, pressure and concentration, which may be altered at will, without causing alteration in the number of phases in a system, or the number of such variables which must be fixed arbitrarily to completely define the system. See **Phase Rule.**

Dehydrated Blast. See **Gayley Process.**

Delay Table. Table for holding partially rolled metal, before further rolling.

Delhi (Steel). Series of heat and corrosion resistant chromium steels, manufactured by Ludlum Steel Co., containing 12–18% chromium and approximately 1% silicon. Carbon is normally kept below 0.12%.

Delivery Guide. Device for supporting metal in the proper position while delivered out of a rolling mill.

Delta Brass. See **Delta Metal.**

Delta Bronze. See **Delta Metal.**

Delta Iron. Allotropic modification of iron, stable above 2552°F. It is of body-centered, cubic crystal structure.

Delta Metal. Brass containing about 36% zinc and low percentages of lead, tin, phosphorus, and iron.

Demagnetization. Removal of magnetic polarity in magnetizable metal.

Dendriform. See **Dendrite.**

Dendrite. Crystalline structure, within a metal, resembling a tree in appearance, with thick branches coming off a main trunk. Because of this appearance, a dendritic structure is also called pine tree and fir tree.

Dendritic. See **Dendrite.**

Dendritic Pattern. See **Dendrite.**

Dendritic Structure. See **Columnar Structure.**

Denitriding. Removing the hard nitride surface from steel, caused by **Nitriding** (see). This can be accomplished by heating in a bath of fused salts at about 1500°F, or by direct heating and long annealing to 1800°F.

Density. Weight per unit volume; its numerical value is equal to that of **Specific Gravity** (see).

Density Ratio. In powder metallurgy, the ratio of the **Apparent Density** (see) to the true density of the metal.

Dental Alloys. Alloys used for bridges, fillings, braces, and similar work. Gold and silver base alloys are generally used, frequently with addition of the nobler metals, such as platinum and palladium.

Dental Amalgams. Alloys of mercury with one or more of the metals: tin, cadmium, silver, and gold; used in dentistry.

Dental Gold. Gold-base alloys, with 5–12% silver and 4–12% copper; used in dental practice. Occasionally, members of the platinum family of metals are also present.

Deoxidation. See **Deoxidizing.**

Deoxidized Copper. Essentially, pure copper treated with a chemical agent, usually phosphorus, to reduce any oxide. Typical deoxidized copper contains about 0.02% residual phosphorus.

Deoxidized Sheet. Steel sheet, usually box-annealed (see **Annealing**) and treated with a deoxidizing gas while cooling, followed by cold-rolling and/or roller-levelling.

Deoxidizer. Material which, added to fused metal, will react with free or combined oxygen, and eliminate it.

Deoxidizing. Removal of oxygen. Frequently the term is used to indicate the removal of oxygen from molten metal by addition of active elements, such as **Scavengers** (see). In steel sheet technology, the term refers to heating sheets in a reducing atmosphere, to lessen the amount of **Scale** (see).

Dephosphorization. Removal of phosphorus; the term is particularly used in the basic Bessemer process. See **Bessemer Steel.**

Depolarizing. Chemical method of eliminating or curtailing **Polarization (see).**

Deposit. Product of **Electrolytic Deposition** (see).

Deposition. See **Electrolytic Deposition.**

Deposition Potential. Minimum difference in potential between an electrode and its surrounding electrolyte necessary to deposit a particular ion.

Descaler. Apparatus for treating rolled steel, after **Scale Breaking** (see), in which high-pressure jets of water or steam impinge on the surface at an angle, to wash or blow away the scale.

Descaling. Procedure, usually mechanical, for removing adherent scale from metal (see **Scaling**).

Deseamer. Machine for removing surface defects from ingot or semi-finished metal.

Deseaming. See **Scarfing.**

Desiliconization. Removal of silicon. In the case of metals, the term refers to removal of elementary silicon; in the case of minerals, to removal of silica, SiO_2.

Desilverization. Removing silver from crude lead. See **Desilverized Lead.**

Desilverized Lead. Lead which has been subjected to desilverizing, as by the **Parkes Process** (see) or **Pattinson Process** (see).

Desilverizing Kettle. Round kettle used in the desilverization of base bullion.

Destruction Limit. Limiting shearing stress at which a crystal begins to lose its lattice structure, as shown by its Laue X-ray pattern change.

Desulfurization. Removal of sulfur, usually by chemical treatment.

Detail Etching. See **Etching.**

Detail Fracture. See **Progressive Fracture.**

Detinning. Removing the tin coating from scrap **Tin Plate** (see). The most common process involves the use of chlorine, which causes the formation of liquid tin tetrachloride, while the iron remains unattacked. See **Goldschmidt Process** (a).

Detroit Furnace. See **Detroit Rocking Furnace.**

Detroit Rocking Furnace. Indirect-Arc Furnace (see) in which graphite electrodes enter horizontally from opposite ends; the furnace, supported on rollers, is rocked continuously during operation.

De Vathaire Process. Process of selectively removing sulfur from pig iron, in the fused state, by contacting with a mixture of barium cyanide, lime, and carbon.

Deward (Steel). Non-Deforming Steel (see) with about 1.6% manganese, 0.30% silicon and 1.00% carbon (Ludlum Steel Co.).

Dezincification. Corrosive action on brass which removes all or part of the zinc from the surface, leaving copper or a high-copper alloy. Dezincification is usually accompanied by increase in red color.

Diagonal (Method of) Rolling. Rolling in which the shaping begins with the first pass and all rolling action is in compression, without lateral spreading.

Diagram. See **Constitution Diagram.**

Diamagnetism. Characteristic of certain metals, of being repelled from a magnet. In sequence of decreasing repulsion, the metals are: bismuth, antimony, zinc, tin, cadmium, mercury, lead, silver, copper, arsenic, uranium, iridium, tungsten. See also **Ferromagnetism.**

Diamite. Alloy white **Cast Iron** (see), containing about 3.2% total carbon, 3–5% nickel, 1–2% chromium, up to

1.5% silicon, and 0.90% manganese (Weatherly Foundry & Mfg. Co.).

Diamond (Cubic) Structure. Term used to designate crystal structures analogous to that of diamond, with two face-centered cubic atom-centers displaced from the other by one-quarter of the diagonal of the unit cube, in the direction of the cubic diagonal, as determined by X-ray test.

Diamond Pass. Rolling mill in which the combination of grooves in the rolls forms a diamond-shaped area at the cross-section.

Diamond Pyramid Hardness (Test). See **Vickers Hardness.**

Diamond Tin. Non-metallic ore of tin, cassiterite, SnO_2, in the form of large, bright crystals.

Didymium. Mixture of two rare earth metals, **Neodymium** (see) and **Praseodymium** (see), which are so similar that they were thought, at first, to be a single element.

Die. (a) In powder metallurgy, mold in which metal powder is subjected to pressure for compacting. (b) In wire drawing, a metal unit or diamond with a hole through which the metal is drawn in order to achieve reduction of cross-section.

Die-Casting. Casting alloys, usually relatively low-melting point zinc- or aluminum-base alloys, into molds. If the molten alloy is forced into the mold, the process is called pressure die-casting; otherwise, gravity die-casting.

Die-Casting Alloy. See **Die-Casting Metal.**

Die-Casting Metal. Aluminum-base, magnesium-base, or zinc-base alloys, used for **Die-Casting** (see). Occasionally, lead-base, and even copper-base alloys, are used for this purpose.

Diehl Process. Modification of the cyanide process [see **Cyaniding** (b)], using cyanogen bromide as reagent.

Die Holder. See **Wire-Drawing Frame.**

Die Steel. Steel suitable for the fabrication of steel dies; usually, a plain 0.50–0.90% carbon steel or one of the low-alloy varieties, although hot-working die steels are generally of high-alloy type.

Die Temper Steel. **Tool Steel** (see), with a carbon content of about 0.75%.

Differential Hardening. Quenching part of an object, so that various portions of the object cool at different rates.

Differential Heating. Heating an object so as to bring various parts to different temperatures, thus subjecting them to different types of heat treatment.

Differential Quenching. See **Differential Hardening.**

Diffraction Pattern. Pattern resulting from the action of X-rays on a photographic plate, after being scattered by a crystalline material.

Diffusion. Spreading of gaseous or other mobile molecules, due to molecular movement.

Diffusion Alloying. Alloy formation resulting from pressing and heating mixed metal powders below the melting point. Alloying results from the diffusion of the metals into one another.

Diffusion Chamber. See **Eutectrol Process.**

Dilatation. Changes in the dimensions of materials when heated or cooled.

Dilatometer. Instrument for measuring relatively small changes in the length of a specimen under test.

Dilatometric Method. See **Critical Point Test.**

Dilver. See **Platinite.**

Dimondite. Cemented tungsten carbide, manufactured by Firth-Sterling Steel Co., with 5–13% cobalt as the cementing material.

Dip Brazing. **Brazing** (see), in which the object is heated by immersion in a bath of molten metal or inert chemical. When metal is used for the bath, the filler metal may also come from this source.

Direct Arc Furnace. **Arc Furnace** (see), with one electrode, the furnace charge acting as the other electrode.

Direct Arc Heating. See **Direct Arc Furnace.**

Direct Current Arc Welding. **Arc Welding** (see), utilizing direct current at the arc.

Direct Fired Furnace. Furnace in which combustion takes place inside the heating chamber, so that the products of combustion are in contact with the material being heated.

Directional Property. Physical property of a metal (such as tensile strength, ductility, etc.) which differs with the crystal axis. In most metals, with crystals in random orientation, there are normally no directional properties, but in large single crystals or in drastically cold-worked metals, differences can readily be found. In deep drawing, for example, sheet metal having directional properties will show metal flowing more readily in one direction than in others.

Direct Process. (a) Process for the isolation of metal from ore by a single direct operation. (b) Process for making **Wrought Iron** (see) or steel directly from the ore, without first manufacturing pig iron.

Dirt. Any non-metallic material, present in massive metal, such as particles of slag or refractory. See **Sonim.**

Dirty Metal. Metal containing an excessive amount of non-metallic inclusions.

Dirty Steel. See **Dirty Metal.**

Discard. See **Crop.**

Disc Roll. See **Spool Roll.**

Disintegration. See **Pulverization.**

Disordered State. Random distribution of the atoms in an alloy system. See also **Order-Disorder.**

Dispersed Phase. In metals or alloys containing more than one phase, the phase which is distributed in isolated units, surrounded by a continuous background of the **Continuous Phase** (see).

Disperse System. Colloidal solution whose two phases are separated by relatively large surfaces.

Dispersion Hardening. See **Age Hardening.**

Displacement Series. See **Electromotive Series.**

Dissociation. Splitting of the molecules of a compound into two or more simpler units, such as smaller molecules, atoms, or electrically-charged ions.

Dissolved Carbon. Solution of carbon in molten or solid metal, usually iron or steel.

Distillation. Process of vaporizing a liquid and condensing the vapors by cooling.

Distillation Furnace. Furnace, most commonly of the reverberatory type, for heating closed vessels containing ore, in order to distill out volatilizable values, such as mercury, zinc, etc.

Distortion Test. Test on a specimen (usually tool or special steels) to determine behavior under heat treatment.

Distributor. Device for evenly distributing a furnace charge, as in a blast furnace.

Divalent. Having a chemical **Valence** (see) of two.

Dividers Method. Determining **Yield Point** (see) load by measuring the strain with a pair of dividers when the specimen is at the yield point.

Divorced Pearlite. See **Granular Pearlite.**

DM Steel. Series of low-alloy steels, used for oil still tubing, manufactured by Timken Steel & Tube Co.

"D" Nickel. Commercial grade nickel, with 4–5% manganese, manufactured by International Nickel Co.

Doghouse. Forechamber in a furnace, providing an incandescent wall surface for fuel ignition.

Doler-Brass. Silicon-containing yellow brass of about 34% zinc content, modified with up to 1.5% silicon; used for die casting (Doehler Die Casting Co.).

Doler-Zink. Zinc-base, die-casting alloy of about 4% aluminum and 3% copper content (Doehler Die Casting Co.).

Dolomite. Naturally-occurring double carbonate of calcium and magnesium, essentially $CaMg(CO_3)_2$. After calcining, it finds extensive use as a basic refractory.

Dolomitic Limestone. Limestone containing appreciable percentages of magnesium carbonate, making the mineral intermediate between **Limestone** (see) and **Dolomite** (see).

Dope. Special mixture, used as a single charge, to correct faulty **Blast Furnace** (see) operation.

Dopploy. High-alloy, chemically-resistant cast iron, with 17–20% nickel, about 2% chromium and 2% silicon (Sowers Mfg. Co.).

Doré (Bullion). Crude silver, containing fractional percentages of gold, usually in the form obtained after removal of lead in a cupelling furnace.

Doré Metal. See **Doré Bullion.**

Doré Silver. See Doré Bullion.

Dor Furnace. Regenerative furnace for the distillation of zinc, having heat-recuperating chambers at the end of the furnace.

Doublé. Sheet clad metal (see **Cladding**), made by rolling a base metal sheet together with a thinner sheet of a noble metal, such as gold.

Double-Acting Hammer. Forging hammer in which steam power is used on the downward pressure stroke and for lifting the hammer.

Double Annealing. Heating steel above the Ac_3 point on the iron diagram, then cooling to a point below the lower transformation range, followed by immediate reheating to about Ac_1, with subsequent slow cooling.

Double Base Box. See **Base Box.**

Double Belgian Mill. Combination of continuous rolling with looping rolling, using a series of stands of rolls, in parallel and tandem groups. See **Rolling Mill.**

Double Bell and Hopper. Arrangement, in a blast furnace system, of two **Bell and Hopper** (see) feeding devices, positioned one above the other.

Double Double (Iron). In pack mill operation, a pack of two sets of fours (see **Four**) for reheating and rolling.

Double Extra Strong Pipe. Pipe of great wall thickness, used for extremely high pressures.

Double Faggoted Iron. Wrought iron made by repeating the **Faggoting** (see) step.

Double Forging. Two **Forging** (see) operations, in sequence, with only one heating of the metal.

Double Furnace. Box **Annealing** (see) furnace with heating chamber adequate for two annealing boxes side by side; gas is usually admitted between the boxes.

Double Hammer. (a) See **Duplex Hammer.** (b) See **Double-Acting Hammer.**

Double Iron. See **Four.**

Double Lead Process. Lead **Patenting** (see) process in which two separate baths are used at different temperatures, one usually at a relatively high temperature for the heating, the other at a lower temperature for cooling.

Double Piercing (Process). Process of manufacturing seamless tubing (see **Piercing**), in which a solid billet is pierced to form a tube of relatively thick wall, followed by treatment, on a second piercing mill, to reduce wall thickness and simultaneously widen the tube.

Doubler. Power hammer, with long flat face, between a rolling mill and a sheet furnace, for flattening the curl in a pack caused by **Doubling** (see).

Double Refined Iron. See **Double Rolled Iron.**

Double Refining. See **Double Rolled Iron.**

Double Rolled Iron. Wrought iron, subjected to a second refining operation, after formation into **Merchant Bar** (see).

Doubles. See **Doubling** (a).

Double Shear. Shear stress, tending to fracture the material in two planes at right angles to a major axis.

Double-Shear Heat Steel. Steel, with about 1.00% carbon, made by the carburization of wrought iron.

Double Shear Steel. See **Shear Steel.**

Double Sintering. Process of roasting rich ores, in which the ore is first sintered at a rapid rate, for partially reducing the sulfur content, followed by crushing the resulting mass, and then sintering again, at a slower rate.

Double Treating. See **Toughening.**

Doubling. (a) In pack mill operations, folding metal sheets end to end, for doubling the number of sheets going through the mill. (b) See **Doubling Process.**

Doubling Floor. In the rolling of metal, the space between the heating furnace and the roll train.

Doubling Hammer. Hammer used in doubling operation to fold the metal sheet flat against itself. See **Doubling Process.**

Doubling Process. Process of purifying "singles" antimony, containing iron and sulfur as impurities, by heating, with or without additional iron, in a reverberatory furnace under an alkaline flux. The resulting metal, called "doubles," is further refined by "starring," a melting process using an alkaline sulfide-containing flux, melting lower than the antimony; on the final metal surface, when solidified, there is a fern-like structure, called "stars." The metal, in this form, is known as "star bowls."

Douglas Furnace. Cylindrical furnace, revolving around a horizontal axis.

Douglas Process. See **Hunt-Douglas Process.**

Dowmetal. Series of magnesium-base alloys, manufactured by Dow Chemical Co. See **Magnesium.**

Downcomer. See **Offtake.**

Downs Cell. See **Downs Process.**

Down Shearing. Specification, in ordering or selling steel, requiring that the shear direction of all four sides should be downward.

Downs Process. Method of producing sodium from fused sodium chloride, in which a central carbon electrode, rising through the bottom, releases chlorine, which is collected through a collecting dome, while side cathodes permit the collection of liquid sodium.

Downtake. See **Bulkhead** and **Offtake.**

Dow Process. Isolating magnesium by electrolysis of fused, dehydrated magnesium chloride.

Draft. (a) Amount of reduction in rolling or drawing procedure. Usually given in percentage representing the difference in cross-sectional area between the two passes. (b) See **Mold Taper.**

Drag. Lower half of a casting mold.

Dragout. Solution removed from a treating bath (such as electroplating, cyaniding, etc.), because of adherence to the work, when removed from the bath.

Drag-Over Mill. See **Pull-Over Mill.**

Draw. (a) Crack in a casting. (b) See **Tempering.**

Draw Bench. Frame holding apparatus for drawing bars, wire, or tubing.

Drawhole. See **Pipe (Cavity).**

Drawing. (a) See **Tempering.** (b) Reducing the cross-section of wire, tubing, etc., by passage through a die. The term also covers converting sheet metal into continuous regular shape, by passage through a die. (c) Removing solid material from a furnace. (d) Removing cast metal from the casting mold.

Drawing Back. See **Tempering.**

Drawing Machine. Equipment for removing solid material from a furnace or similar apparatus.

Drawing Quality (Steel Plate). Low-carbon steel plate ordered or sold on the basis of specifications governing suitability for deep-drawing.

Drawing Temper. See **Tempering.**

Draw Out. Device for pulling the end of a wire coil through a die.

Draw Plate. Plate, usually made of hardened steel, with successively smaller holes, through which wire is drawn in order to reduce its cross-section.

Draw Shrinkage. Undue amount of **Shrinkage** (see), resulting in stresses and brittleness; occurs in cast iron containing more than 0.3% phosphorus.

Dressing (of Rolls). Turning worn-down rolls to correct shape.

D Retort. Retort, consisting of a heavy cast iron pipe, shaped like a D, with the flat side horizontal, for heating mercury ores to recover the metal by condensation.

Drift Copper. Native copper found in gravel and clay, at a distance from the original ore body.

Drill Pipe. Steel pipe of exceptional strength, used in the drilling of wells by the rotary drilling system.

Drill Rod. Polished tool steel in accurately-rolled rod form, usually round in cross-section, used for making drills, punches, etc.

Drill Steel. See **Drill Rod.**

Drill Temper Steel. Tool Steel (see), with a carbon content of about 1.25%.

Drive Pipe. Steel pipe used in the drilling of wells.

Driver-Harris Alloy. Series of iron-nickel and nickel-base heat-resistant alloys, manufactured by Driver-Harris Co.

Driving-in Test. See **Penetration Test.**

Driving Spindle. See **Leading Spindle.**

Drop of Beam Method. Method of determining the **Yield Point** (see) stress in metals, in which the operator watches for the sudden drop of the scale beam.

Drop Forged Steel. See **Drop Forging.**

Drop Forging. Forging accomplished by means of dies, one in the falling weight, known as the tup or bob, the other in the stationary anvil below. Because of the high cost of the dies, drop forging can be used commercially only in large-scale production.

Drop Hammer. Hammer for forging, raised and then released to drop down to the anvil. See **Drop Forging.**

Drop Shot. Metal shot made by pouring molten lead into water.

Drop Test. (a) Method of determining impact strength of relatively large sections, such as rails, by dropping a heavy weight, usually one ton, from increasing heights until a pre-assigned deflection is reached. (b) Testing zinc coating by a drop of zinc-solvent chemical, optionally containing also other chemicals reacting with iron, to give deep coloration. The length of time needed for a drop to show color is an index of the quality of the zinc coating.

Drop Tin. Granulated tin made by pouring molten **Tin** (see) into water.

Drop Weight Test. See **Drop Test** (a).

Drop Zinc. Granular or globular zinc, usually made by pouring molten **Zinc** (see) into water.

Dross. (a) Metal oxides in molten metal, resulting from oxidation. Such metal may settle to the bottom or rise to the top. See **Hard Zinc.** (b) Material, usually of oxide nature, rising to the surface of impure molten metal.

Drossing. Removing oxides or undesired metals from molten metals, by means of a drossing spoon or agitation, causing the contaminant to float, so that it can be skimmed off the surface.

Drossing Spoon. See **Drossing.**

Dry Amalgamation. Treating gold or silver ores with hot, dry mercury.

Dry Assay. Determination of metal values in an ore by chemical methods, not involving use of solutions. See also **Assay.**

Dry Blast. Removing part of the moisture from air for blast furnace use, for better furnace operation. The process is not generally used, being unduly expensive.

Dry Casting. Casting (see) of metals into molds made of sand, dried after formation.

Dry Cell. See **Battery** (a).

Dry Cleaning. (a) Removing non-gaseous materials from blast furnace gas, e.g., by cyclone separation, bag filters, pad filters, Cottrell systems, etc. (b) Removing oil from tin, terne, or similar coated steel, after leaving the coating bath, by contact with bran or middlings.

Dry Copper. Copper containing oxygen exceeding the amounts in **Tough Pitch Copper** (see).

Dry Cyaniding. Hardening the surface of steel by heating in a mixture of ammonia and a carburizing gas; the surface absorbs carbon and nitrogen. See **Cyaniding** (a).

Dry Drawing. Drawing wire with lime (see **Lime Finish**) or sull (see **Sull Coated Wire**) as the drawing lubricant, together with soap or grease, held in the die container.

Dry Dross. See **Wet Dross.**

Dry Finishing. Process of cleaning terne and tin plate, in which a cleaner, or branner, consisting of a pair of cloth rolls, absorbs the oil on the plate and removes particles of bran and sawdust from previous cleaning steps. See **Terne** and **Tin Plate.**

Dry-Finish Plate. Terne or similar coated plate, leaving the coating bath practically free of oil. See **Terne.**

Dry-Finish Terne. See **Dry-Finish Plate.**

Drying Off. Evaporating mercury from a gold **Amalgam** (see), to leave gold, usually in film form.

Dry Method (of Analysis). Chemical analysis, carried out with dry reagents, such as **Blowpiping** [see **Blowpipe** (c)] in qualitative analysis, and **Assaying** (see) in quantitative analysis.

Dry Plate. Defective tin plate, with spots of dull surface and **Scruff** (see).

Dry Process. Heat treatment of ores, as opposed to **Wet Processes** (see), involving treatment by water solutions.

Dry Puddling Process. See **Puddling.**

Dry Quenching. See **Brunorizing.**

Dry Sand Mold. Mold, for casting metal, made from sand and dried, before pouring the metal into it.

Dry Streak. Defective condition in tin plate, due to spots of dull surface, contaminated with **Scruff** (see).

Dry Sweating. Process of purifying copper, in which impure blister copper is exposed, for a long period, to oxidizing heat below its fusion point.

Ductility. Characteristic of substances of being lengthened or flattened out, without losing continuity, when subjected to tensile stresses or rolling. Gold, silver, and platinum are the most ductile metals of normal usage. Copper, lead, and pure iron are also very ductile. Some metals, ordinarily nonductile, can be specially treated to make them amenable to drawing and rolling. Tungsten filaments, used in the lamp industry, are of this type, fine tungsten powder being compressed, heat-treated, and then swaged, after which the resultant tungsten bars are ductile enough to be drawn.

Ductillite. High-quality **Tin Plate** (see), cold-rolled from hot-rolled strip (Wheeling Steel Corp.).

Dull Finish. Non-glossy finish on cold-rolled strip or sheet metal, produced by roughened finishing rolls.

Dull Finish Auto Body Sheet. See **Deep-Drawing Dull Finish Auto Body Sheet.**

Dull Finish Roll. See **Dull Finish.**

Dull Finish Sheet. See **Dull Finish.**

Dull Iron. Iron, in **Cupola** (see) melting, which has not been heated to a high enough temperature for proper flowing.

Dull Red Heat. See **Low Red Heat.**

Dull Rolls. Rolls, used in the rolling of metal sheet, strip, or plate, with surface chemically or mechanically roughened.

Dumet. Alloy of iron-nickel (about 45% nickel), in wire form, coated with copper; used as lead-in wire in electric bulbs, radio tubes, etc.

Dumoulin Process. Process of obtaining electrolytic copper, in which the metal is deposited on a rotating mandrel, stripped off as a long strip, and then drawn directly into wire.

Duplex Hammer. Hammering apparatus for striking a metal mass, such as a bloom or a bar, simultaneously on opposite sides.

Duplexing. See **Duplex Metal** and **Duplex Process.**

Duplex Metal. See **Clad Metal.**

Duplex Process. Metallurgical process, in which the first operation is conducted in one type of equipment, and, after transfer of the metal, usually in molten form, a second operation is conducted in another type of equipment. In steel, the most common duplex process calls for preliminary treatment in an acid Bessemer furnace, followed by finishing in a basic open-hearth furnace. Another typical duplex process is represented by a basic open-hearth treatment, followed by finishing, under a reducing slag, in an electric furnace.

Duplex Steel. Steel made by a duplex process, pig iron being first partially purified by an acid Bessemer process, followed by final purification in a basic open-hearth furnace.

Dural. See **Duralumin.**

Duraloy. Series of high-chromium steels, with chromium up to 30%, manufactured by Duraloy Co. Nickel is optional, and may be as high as 15%. The alloys show the usual characteristics of the high-chrome steels, particularly corrosion and heat resistance.

Duralumin. Series of aluminum-base age-hardening alloys (see **Alcoa**), generally with about 4% copper and fractional percentages of magnesium and manganese.

Durana Metal. Yellow brass, with 30–40% zinc, and optional low percentages of aluminum, iron, lead, and tin.

Durbar Bronze. High-Lead Bronze (see), with 20–24% lead and 4–10% tin, used for bearings, etc. (Buffalo Die Castings Corp.).

Durco. Series of high-chromium steels (Duriron Co., Inc.), resistant to corrosion and heat, with 12–30% chromium, nickel up to 15%, about 1% silicon, and optional additions of 2–4% molybdenum.

Durex. Copper-tin alloy with graphite dispersed in the matrix, used as a bearing metal (General Motors Co.).

Durichlor. High-silicon cast iron, containing about 14% silicon and 3% molybdenum; resistant to chemical corrosion, including hydrochloric acid; used for chemical valves, fittings, etc. (Duriron Co.).

Durimet. Series of nickel-chromium steels, resistant to heat and corrosion, manufactured by Duriron Co., with 16–23% nickel, 11–22% chromium, and copper up to about 1.2%. Carbon is kept below 0.25%.

Duriron. High-silicon cast iron, with about 14% silicon, made by the Duriron Co. Despite its great brittleness, it is extensively used in chemical construction, due to its high resistance to many acids, particularly sulfuric and nitric acids.

Durodi Steel. Low-alloy steel (A. Finkl & Sons) with 1.5% nickel and fractional percentages of molybdenum and chromium, in addition to manganese and carbon.

Duro Gloss. Series of high-chromium steels with 12–29% chromium and low carbon content, made by Jessop Steel Co.

Duroline Pipe. Steel pipe, centrifugally coated with a low-lime cement, for use with corrosive solutions (National Tube Co.).

Durometer Method. Method of determining the surface hardness of materials, in which a ball is dropped on the polished specimen surface, the rebound being a measure of the hardness.

Durometer Test. See **Durometer Method.**

Duronze. Series of corrosion-resistant bronzes, containing about 2% tin and 1% silicon (Bridgeport Brass Co.).

Dust. In powder metallurgy, the material which passes through a 325-mesh screen.

Dust Bell. Seal near the bottom of a dust-collecting device, for periodic opening to remove flue dust or similar material.

Dust Catcher. Device, usually merely an enlargement of the flue, to permit deposition of dust from a gas, e.g., blast furnace gas.

Dust Chamber. See **Dust Catcher.**

Dust Gold. Gold (see) in the form of very fine powder.

Dust Plate. Iron plate, supporting the slag runner of a **Blast Furnace** (see).

Dust Precipitation. See **Cottrell (Electric Precipitation) Process.**

Dutch Leaf. See **Dutch Metal Leaf.**

Dutch Metal. Brass of approximately 20–24% zinc content, used for rolling imitation gold leaf.

Dutch Metal Leaf. Bronze Powder (see), polished by being revolved in drums with a lubricant, which results in leafing or filming.

Dutch Oven. See **Doghouse.**

Dutch White Metal. Britannia Metal (see) approximating 9% antimony and 10% copper, with tin as the remainder.

Dwight-Lloyd Process. Blast Roasting (see) in which air currents are drawn downward, through the ore mass.

Dwight-Lloyd Sintering. See **Dwight-Lloyd Process.**

Dwi-Manganese. See **Rhenium.**

Dy. Chemical symbol for **Dysprosium** (see).

Dynamic Indentation Test. See **Penetration Test.**

Dynamic Stress. Stress suddenly applied, or tending to produce motion in the part stressed. Both impact stress and fatigue stress are forms of dynamic stress.

Dynamo Grade Sheet. Silicon steel sheet, used in the manufacture of large motors, generators, and transformers. See **Electric Sheet.**

Dyo Steel. High-alloy steel (Carpenter Steel Co.), used for hot-die work and small tools. Composition approximates 14% tungsten, 4% chromium, 0.5% vanadium, and 0.30% carbon.

Dysprosium (chemical symbol Dy). Element No. 66 of the periodic system; atomic weight 162.46. One of the rare earth metals, it has never been isolated in pure form; consequently, no physical characteristics are known for the metal. It is trivalent chemically. Like all rare earth metals, it cannot be electrodeposited.

E

Ear. Wavy projection formed in the course of deep-drawing, because of differences in directional properties.

Earing. See **Ear.**

Earthy Fracture. Fracture resembling that of hard clay.

Earthy Manganese. See **Bog Manganese.**

Easy-Flo. Low-melting (1175°F) silver brazing alloy, manufactured by Handy & Harman. Its composition approximates 18% cadmium, 16.5% zinc, and 15.5% copper, with silver as the remainder; shows free-flowing characteristics at the melting point and is extensively used for both ferrous and non-ferrous metal joining.

Eberbach Hardness (Test). Method of determining the surface hardness of a material, using, as a microscopic indentor, a diamond, shaped into a 136° angle. See **Hardness** and **Microhardness.**

Eccentric Converter. Bessemer converter (see **Bessemer Steel**) with its mouth away from the center and to one side of the top, at an angle from a plane parallel to the bottom.

Economic Mineral. Mineral (see) of commercial value.

Economo Steel. Machine Steel (see), containing about 0.2% molybdenum (Wheelock, Lovejoy & Co.).

Eddy Current. Induced electric current, which lowers efficiency and increases temperature of rotating metallic objects in a magnetic field.

Eddy Current Loss. Loss of power, due to electric currents, self-induced in metal exposed to fluctuating magnetic flux.

Edge Finish. See **Number— Finish.**

Edge Rolls. See **Edging Rolls.**

Edging Pass. Pass (see) through which the metal is moved edgewise.

Edging Rolls. Rolls, in the rolling of metal sheets, which control the width of the sheet.

Edging Stand. See **Edging Rolls.**

Edison Gage. See **Metal Gage.**

Edison (Storage) Battery. See **Alkaline Storage Battery.**

Effervescing Steel. See **Rimmed Steel.**

Effusion. Characteristic of gases of passing through porous solids.

Ehrhard's Metal. Zinc-base type and bearing metal, with 3–4% copper, 4–6% tin, and 2–3% lead.

EHW Steel. Hot-die steel (Latrobe Electric Steel Co.), with about 9% tungsten, 3% chromium and 0.30% carbon.

Eichhorn-Liebig Furnace. Muffle Furnace (see) in which the charge is hand-agitated.

Eight. See **Double Double Iron.**

Eighteen-Eight Steel. Stainless Steel (see) of the common, austenitic composition, i.e., containing 18% chromium and 8% nickel.

Eisenbronze. See **Iron Bronze.**

E.I.S. (Steel). Series of low-alloy and high-alloy steels, manufactured by Heppenstall Co.

Eka-Cesium. See **Virginium.**

Eka-Iodine. See **Helvetium.**

Elastic Deformation. Change of dimensions, caused by external stress, the material being restored to substantially original dimensions after removal of the stress.

Elasticity. Property of all substances of elongating under stress, and of returning to original dimensions after removal of the stress. The characteristic of apparently complete elasticity holds for a limited range of stress only; beyond a certain stress value (different for each material, of course), complete elasticity no longer holds true, and strain increases at a rate greater than the applied stress.

Elastic Limit. Maximum stress in a material, before permanent deformation occurs, even after complete removal of the stress.

Elastic Ratio. Ratio of stress, at the **Yield Point** (see), to the **Tensile Strength** (see). For most steels, the elastic ratio is between 50 and 75%.

Elastic Strain. Strain within the elastic limit. See **Elasticity.**

Elastrodurometer. Method of determining the surface hardness of materials in which a ball is dropped on the polished specimen surface, the rebound being a measure of the hardness.

Elastuf (Steel). Series of machine steels of medium carbon content, manufactured by Horace T. Potts Co. and by Brown-Wales Co.

Elcomet. Series of modified stainless steels, manufactured by LaBour Co., with about 20% chromium, 10–20% nickel, silicon up to 1.25%, about 4% copper and 2–3% molybdenum. Carbon is kept below 0.15%, and manganese below 0.30%. The alloys show excellent chemical resistance to many corrosive agents, both acid and alkaline, and are, consequently, used for pumps and valves exposed to such corrosion.

Electric Accumulator. See **Storage Battery.**

Electrical Conductivity. Capacity to conduct electricity, which is particularly characteristic of metals. Conductivity is usually given in the form of its reciprocal, electrical resistivity, commonly stated as microhms per centimeter cube.

Electrical Dust Precipitation. See **Cottrell (Electric Precipitation) Process.**

Electrical Grade Sheet. Electric Sheet (see) of about 1% silicon content, used in the manufacture of small motors.

Electrical Potential. See **Voltage.**

Electrical Precipitation. See **Cottrell (Electric Precipitation) Process.**

Electrical Resistivity. Electrical resistance of any substance per unit length and unit cross-section area.

Electrical Sheet. See **Electric Sheet.**

Electrical Steel. See **Electric Sheet.**

Electric Arc Furnace. See **Arc Furnace.**

Electric Arc Welding. See **Welding.**

Electric Brazing. Brazing (see) process in which the work is electrically heated. See **Brazing.**

Electric Furnace. Furnace in which heat is generated electrically; may be of resistance, arc, or induction type.

Electric Furnace Steel. See **Electric Steel.**

Electric Grade Sheet. See **Electrical Grade Sheet.**

Electric Resistance. See **Electrical Resistivity.**

Electric Resistance Pyrometer. See **Pyrometer.**

Electric Sheet. Silicon Steel (see) rolled into sheet for use in laminated magnets.

Electric Steel. Steel made in any furnace where heat is generated electrically, almost always by arc. Because of relatively high cost, only tool steels and other high-value steels are made by the electric furnace process. See also **Electric Sheet.**

Electric-Welded Pipe. Pipe, usually of steel, made from strip or plate bent into shape and then welded along its edges, either by resistance heating or by fusion **Welding** (see).

Electric Welding. Welding (see) process in which the ends of the metals to be united are placed in contact, and a high-amperage electric current is passed through the metal junction, the high resistance of which raises that area to melting or welding temperatures.

Electrite (Tool) Steel. High-carbon alloy tool steels of varying composition, though most commonly with about 18% tungsten, manufactured by Latrobe Electric Steel Co.

Electroanalysis. Chemical analysis by means of electrodeposition.

Electrobronzing. Electroplating with bronze, most commonly from a cyanide bath. See **Plating.**

Electrocast Refractory. Refractory of high purity, molten in an electric furnace and cast into final form.

Electrochemical Cleaning. See **Electrolytic Cleaning** and **High Temperature Electrochemical Cleaning.**

Electrochemical Equivalent. Number of grams of an element set free or electrochemically dissolved by the passage of one coulomb of electricity (one ampere-second).

Electrochemistry. Science of the interrelationship between electricity and chemistry, including electroplating, electric cells, etc.

Electrocoppering. See **Copper Plating.**

Electrode. One of the two conductors leading electricity into an electrolytic process or equipment. See **Anode** and **Cathode.**

Electrodeposition. See **Electrolytic Deposition.**

Electrode Potential. Potential between a chemical element and a solution containing ions of that element in which it is immersed, under standard conditions.

Electroextraction. Recovery of metal, usually from ores, by dissolving out the metal and electrolysis of the solution.

Electrofacing. Coating a metal surface with a harder metal, for wear resistance, by means of **Electrolytic Deposition** (see).

Electroforming. Production or reproduction of articles by **Electrolytic Deposition** (see).

Electro-Galvanizing. Galvanizing (see) by electrodeposition of zinc on steel.

Electrolon. Silicon carbide product, manufactured by Simmons Saw & Steel Co., used for abrasive and refractory purposes.

Electrolysis. Process, in which chemical reaction produces electricity, or electrical currents create or affect chemical reactions, or cause decomposition. In metal technology, the term is frequently applied to the corrosion of metals under the influence of electric currents.

Electrolyte. Compound which, dissolved in an appropriate solvent like water, conducts electricity. All acids, bases (alkalies), and salts, if soluble, are electrolytes in water. The term is also used to designate the resulting conducting solution.

Electrolytic Cleaning. Cleaning, particularly degreasing, metal surfaces by electrolysis, using an alkaline bath and the metal as cathode; the hydrogen liberated by the electrolytic action mechanically removes oil and grease from the metal surface.

Electrolytic Condenser. Condenser utilizing aluminum electrodes, oxidized surfaces forming the insulating layer between pairs of electrodes.

Electrolytic Conduction. Conduction of electricity by solutions, the ions carrying the electric charge, the positively-charged ions (cations) travelling to the negatively-charged electrode (anode), and vice versa.

Electrolytic Copper. Copper that has been refined by electrolysis. The crude, impure copper is made anode in a bath containing copper sulfate, and is deposited on pure copper sheets (starting sheets) acting as cathodes. The resulting, purified metal assays 99.9% copper.

Electrolytic Deposition. Isolation of a metal on the cathode by electrolytic decomposition of a solution of a salt or compound of the metal. See **Plating.**

Electrolytic Iron. Iron isolated, as cathodic metal, from an electrolytic bath containing an iron salt. Either iron or inert anodes may be used in such electrolysis. Electrolytic iron is of high purity, except that a good deal of hydrogen is normally occluded by the metal, most of which can be removed by heating.

Electrolytic Lead. Lead isolated, as cathodic metal, from an electrolytic bath containing a lead salt. In practical operations, usually complex salts, such as the fluoborate, fluosilicate, and dithionate are used and not the simple lead salts, which do not deposit coherently (see **Lead Tree**). Like all electrolytic metals, electrolytic lead is of high purity.

Electrolytic Pickling. Cleaning metals, most commonly steel, by electrolysis, using an acid solution as electrolyte and the metal to be cleaned as cathode. In a modification of the process, for cleaning stainless steel, the latter is used as anode.

Electrolytic Plate. Thin coating of one metal on another, being applied by electrolysis. See **Plating.**

Electrolytic Process. See **Electrolysis.**

Electrolytic Refining. See **Electrorefining.**

Electrolytic Wirebar Copper. See **Electrolytic Copper.**

Electrolytic Zinc. Zinc isolated, as cathodic metal, from an electrolytic bath of a zinc compound. In practice, zinc ores are converted to the sulfate, and this is subjected to electrolysis in water solution, the metallic zinc being deposited on aluminum sheet cathodes.

Electromagnetic Test. See **Magnetic Test.**

Electromet. Alloys, particularly iron-base alloys, for use in steel-making and refining (Electro Metallurgical Co.).

Electrometallurgy. Science and technique of metallurgical operations conducted by electrical means, such as electrodeposition, electric furnaces, etc.

Electrometals Furnace. Two-phase electric **Arc Furnace** (see), with two electrodes through the roof and a neutral return in a buried hearth, through which relatively little current passes.

Electromotive Cell. See **Cell.**

Electromotive Force. Force (normally measured in volts) which causes an electric current to flow. See **Voltage.**

Electromotive Series. Chemical elements listed in sequence of descending electrode potentials. The more electropositive elements, the alkali metals, are at the top of the series; the electronegative non-metals are at the bottom. In general, a metal higher in the series will displace one lower in the series, when present in solution, releasing that metal in elementary form.

Electron. The ultimate indivisible negative charge, part of all atoms. Electricity is presumed to represent a flow of electrons from the negative pole to the positive one.

Electron Alloys. Series of magnesium-base alloys, developed by the German I. G. Farbenindustrie A.-G. For the most part, the magnesium content of these alloys is of the order of 90%.

Electron Microscope. Microscope using a beam of electrons instead of visible light. Because of their smaller size, electrons will yield greater magnification than is possible with ordinary light, or even X-rays.

Electroplating. See **Plating.**

Electrorefining. Process of purifying metals electrolytically; the impure metal acts as anode and highly pure metal deposits on the cathode. The impurities of the original metal remain behind as slimes (see **Anode Slime**), or accumulate, in dissolved form, in the electrolytic bath.

Electro Sheet Copper. Wide (up to 60 in.), thin copper sheet and foil (thickness from 0.0013 to 0.004 in.).

Electrostripper. Device for determining thickness of tin coating by removal of the coating electrolytically (Wilkens-Anderson Co.).

Electrothermics. Science and technique of electrical production of heat. Electric furnaces are most typical apparatus used for this purpose.

Electro-Tinning. Coating of steel with tin by electrodeposition. See **Tinning.**

Electrotype Metal. Lead-base type metal, with low contents of tin (3%) and antimony (4%).

Electrowinning. Process of obtaining a metal from its ores, involving the use of electrolytic methods. See **Electrolytic Deposition.**

Electrum. (a) Gold-base alloy, with 15–45% silver. (b) **Nickel Silver** (see) alloy, containing approximately 25% nickel and 25% zinc.

Electrunite (Tubing). Alloy steel tubing, made by welding by Republic Steel Corp.

Elektron. See **Electron Alloys.**

Element. Any of the ninety-two known chemical substances, which cannot be decomposed, by ordinary chemical methods, into simpler substances.

Elementary Analysis. See **Quantitative Analysis.**

Elephant Brand (Bronze). Series of copper-base alloys of varying composition, manufactured by Phosphor Bronze Smelting Co.

Elfur Iron. See **Cramp's Elfur Iron.**

Elinvar. Nickel-chromium steel (approximately 36% nickel and 12% chromium), optionally with fractional percentages of manganese and tungsten. Low temperature-expansion governs its uses.

Eliquation. See **Liquation.**

Elkonite. Tungsten composite, with copper or silver as the matrix, manufactured by powder metallurgy methods. It is used as resistance-electrode material, the copper or silver furnishing the major portion of the electrical conductivity (P. R. Mallory Co.).

Ellwood Bright Finish Tubing. Light steel tubing, of hard-drawn temper, subjected to a special surface treatment during cold-drawing.

Elmet Metal. Cemented composite of tungsten (or molybdenum) and copper (or silver), used for electrical contacts and similar purposes. Conductivity is high because of the copper or silver content, whereas the tungsten or molybdenum acts to prevent fusion when the metal is subjected to heat.

Elongation. Increase in length which occurs before a metal is fractured, when subjected to stress. This is usually expressed as a percentage of the original length and is a measure of the ductility of the metal.

Eloxal. Process, developed in Germany, for anodically oxidizing aluminum; an aluminum oxide coating results on the surface of the metal. See **Anodizing.**

Elverite. Wear-resistant chilled cast iron (Babcock & Wilcox Co.).

Ely Furnace. Rotating and oscillating furnace of oval-shaped internal cross-section, for the manufacture of **Wrought Iron** (see). The rotation and oscillation avoid the necessity for hand rabbling.

Ely Mechanical Puddler. See **Ely Furnace.**

Em Briquets. Briquets of ferro alloys of silicon, chromium, or manganese, manufactured by Electro-Metallurgical Co. The briquets are used as a convenient means to add these alloying elements to the cupola.

Emerald Copper. Non-metallic mineral, containing copper — dioptase, a hydrous silicate, approximating H_2CuSiO_4 in chemical composition.

Emerald Nickel. Non-metallic mineral containing nickel — zaralite, a hydrous carbonate, approximating $H_{12}Ni_3CO_{11}$ in composition.

Emetco. Series of bearing alloys, manufactured by Empire Metal Co.

E. M. F. See **Electromotive Force.**

Emissivity. Rate at which a body radiates heat, which is, under ideal conditions, proportional to the fourth power of the absolute temperature of the body. See **Thermal Radiation.**

Empire (Alloy). Series of low-alloy and high-alloy, ferrous-base, casting compositions, most commonly with chromium as the principal alloying agent (Empire Steel Castings, Inc.).

Ems Process. Process of condensing flue dusts by means of large flues filled with parallel rows of sheet iron baffles.

Emulsion. Mechanical mixture of two liquids which do not mix naturally.

Enamel. See **Vitreous Enamel.**

Enamel Ground Coat. See **Ground Coat.**

Enameling Steel. See **Enameling Stock.**

Enameling Stock. Black Plate (see) which has been hot-rolled, pickled, cold-reduced, annealed, and cold-rolled, for use in vitreous enameling.

End-Centered (Orthorhombic) Structure. Term used in describing internal crystal structures of materials, as determined by X-ray, in which the equivalent points are at the corners of the unit cell and at the centers of the faces perpendicular to the c-axis of the orthorhombic crystal.

End Flow. Flow of metal at the ends of steel rails, when in service, caused by impact and the pressure of railroad wheels.

End Hardening. See **Brunorizing.**

Endothermic Reaction. Chemical reaction which absorbs heat.

End-Quench Test. Standard test for the hardenability of steel, in which a stream of water cools one end of the piece of steel; the hardness is measured at various points from the hardened to the unhardened end.

End Sizing. Insuring accuracy of diameter on the ends of a large pipe, by die-sizing operation, in which either a mandrel is forced into the end of the pipe, or a die forced over the outside.

Endurance. Resistance of metal to **Fatigue** (see).

Endurance Limit. See **Fatigue Limit.**

Endurance Machine. See **Fatigue Test.**

Endurance Range. Maximum range of stress which a material will withstand in both directions under indefinitely continuous cycles of stress; the endurance range is twice the **Fatigue Limit** (see).

Endurance Ratio. See **Fatigue Ratio.**

Endurance Test. See **Fatigue Test.**

Enduro (Steel). Series of chrome- and nickel-chrome steels, manufactured by Republic Steel Corp. Chromium varies from 10% to 26%, whereas nickel, in the nickel-chrome series, reaches up to 21%; optional metals may be present in significant amounts (silicon, molybdenum, columbium, titanium). Both series are characterized by resistance to high temperatures, corrosive conditions, and abrasion.

Energizer. Material added to carburizing agents to accelerate the action or to improve carbon penetration. Barium and sodium carbonates are the most common energizers.

English Copper Process. Process of obtaining copper by reducing roasted copper ores in a reverberatory furnace.

English Cupellation. Refining silver in a reverberatory furnace with a movable bed; the bullion is charged gradually, and final refining is accomplished in the same furnace, after cupellation. See **Cupellation** (a).

English Furnace. Small furnace for zinc distillation; the zinc vapor passes downward and is then condensed.

English Lead Process. Smelting lead ores, in which a relatively large charge is roasted quickly at a high temperature in a furnace which is inclined toward the center of one of the sides, from which point the molten lead is periodically tapped.

English Method. See **English Lead Process** and **English Copper Process.**

English Process. See **English Lead Process** and **English Copper Process.**

English Zinc Furnace. See **English Furnace.**

Engorgement. Clogging of a furnace.

Engravers' Brass. Clock Brass (see) rolled into sheet, in hard temper.

"E" Nickel. Commercial-grade nickel, with about 2% manganese, manufactured by International Nickel Co.

Enlund Method. See **Enlund Test.**

Enlund Test. Determining the quantity of carbon and total alloying elements in steel by measurement of electric resistance characteristics.

Enriched Blast. Treating air for blast furnace use by adding oxygen (or by removing nitrogen), to increase heat and decrease heat losses. The process is not in commercial use as yet, because of undue cost.

Entering Guide. Device for supporting metal in the proper position and angle, when entering into a rolling mill.

Epstein Strip. Strip of metal, 3 x 50 centimeters in dimension, used as a sample in determining the magnetic characteristics of silicon-steel electric sheets. See **Epstein Test.**

Epstein Test. Method of determining total **Core Loss** (see) in electrical sheet steel, in which a series of standard strip samples (half taken in the direction of final rolling, and half against the direction of rolling) are used at a specified flux density, usually at 60 cycles per second.

Equicohesive Temperature. Temperature range in which the cohesive strength of the amorphous phase of a metal equals the strength of the crystalline phase, or in which mechanical failure is equally possible at grain boundaries and through the grain.

Equilibrium. In a reversible chemical reaction, the state where the rate of one reaction is equal to the rate of the opposing reaction, so that no measurable change occurs.

Equilibrium Diagram. See **Constitution Diagram.**

Equivalent Weight. Weight, in grams, of an element equivalent in its chemical combining power to 1 gram of hydrogen or 8 grams of oxygen.

Er. Chemical symbol for **Erbium** (see).

Eraydo. Series of zinc-base alloys of higher strength than commercial rolled zinc, manufactured by Illinois Zinc Co., with 1–3% of copper, and silver up to 0.1%. Used as sheet and ribbon.

Erbium (chemical symbol Er). Element No. 68 of the periodic system; atomic weight 167.2. One of the rare earth metals, isolated only in impure form by reduction of its anhydrous chloride with sodium; no data are available on physical constants. It is trivalent chemically. Like all rare earth metals, it cannot be electrodeposited.

Erichsen (Ductility) Test. Method for measuring the ductility of sheet metal, which involves determination of the width and depth of the impression made by a standardized plunger under the pressure required to fracture the metal.

Ermal. High-Test Cast Iron (see) (Erie Malleable Iron Co.) with about 2.3% total carbon, 1% silicon, and 0.8% manganese.

Ermalite. High-Test Cast Iron (see) (Erie Malleable Iron Co.) with about 2.5% total carbon, 2% silicon, 0.4% nickel.

Esco. Series of heat and corrosion-resistant steels of the stainless type (see **Stainless Steel**) with 12–30% chromium, and optionally up to 21% nickel (Electric Steel Foundry Co.).

Esco Alloy. Series of high-chromium and chromium-nickel corrosion-resistant steels (Electric Steel Foundry Co.).

Etchant. Solution used for **Etching** (see) purposes.

Etched Plate. Faulty **Tin Plate** (see) coating, caused by too long a pickling of the base metal, prior to the tinning operation.

Etch Figure. Marking caused by an **Etchant** (see) on a crystal surface.

Etching. In metallography, etching refers to the attack on a metal surface by means of reagents in order to reveal the structure, usually under the microscope. Deep etching, with relatively strong acids or alkalies, is used to reveal major effects, such as segregations, porosity, etc. Macro etching is usually a deep etching, which reveals the gross characteristics of the metal, such as mechanical rolling direction, etc. Most commonly, etching is of the micro etching or detail etching variety, bringing out greater detail, such as can be studied best, or only, under the microscope.

Etching Agent. See **Etchant.**

Etching Brass. Quarter-hard or half-hard sheet of yellow **Brass** (see).

Etching Pits. Deep pits, formed by any etching reagent and not related directly to revealing the structure of the metal.

Etching Reagent. See **Etchant.**

Etch Test. Deep-etching a ground and rough-polished metal sample with acid in order to examine it macroscopically for homogeneity.

Eu. Chemical symbol for **Europium** (see).

European Malleable Iron. See **White Heart (Malleable) Iron.**

Europium (chemical symbol Eu). Element No. 63 of the periodic system; atomic weight 152.0. One of the least frequently occurring among the rare earth metals, isolated by the electrolysis of its anhydrous trichloride in a bath of fused alkali chlorides, it shows a melting point of approximately 2100°F. Like most other rare earth metals, it is steel-gray in color. Europium is stable in air and in water. Chemically, it is primarily trivalent (europic), though divalent, europous, salts are also known. Like all rare earth metals, it cannot be electrodeposited.

Eutectic (Alloy). Composition in which liquidus curves or surfaces meet in a common point, this composition having the lowest melting point in that alloy area. Many alloy systems have only one eutectic, and consequently, this represents the lowest melting point for the system. During transition of an alloy from the molten to the solid state, the material forming a eutectic composition will remain liquid until the remainder of the metal has solidified.

Eutectic Change. Freezing, or melting, of an alloy of eutectic composition. See **Eutectic (Alloy).**

Eutectic Mixture. See **Eutectic (Alloy).**

Eutectic Point. See **Eutectic (Alloy).**

Eutectic Solder. Soft **Solder** (see) of lowest melting point in the lead-tin series, with about 63% tin and 37% lead.

Eutectic System. Alloy series characterized by the presence of a **Eutectic** (see).

Eutectoid. Composition in which solid solubility curves meet at a minimum temperature value.

Eutectoid Steel. Steel representing the eutectoid composition of the iron-carbon system, with about 0.83% carbon, the eutectoid temperature being about 1333°F. Such steel consists exclusively of pearlite. Steels with less than this quota of carbon are known as hypo-eutectoid and contain free ferrite in addition to the pearlite. When more carbon is present, the steel is known as hyper-eutectoid and contains free cementite. The presence of certain elements, such as nickel or chromium, lowers the eutectoid carbon point.

Eutectrol Process. Continuous gas **Carburizing** (see) process. The material moves continuously forward in a long muffle furnace, with a purging chamber at the admission end and a diffusion chamber at the discharge end; gas flow is controlled by means of valves and doors.

Evansteel. Abrasion-resistant, low-alloy steel (Chicago Steel Foundry Co.), with about 2% nickel and 1% chromium.

Evaporation. Change from the liquid to the gaseous state, with the absorption of heat.

Everbrite. Copper-base alloy series of about 30% nickel content, with low percentages of iron, silicon, and chromium optional, used for corrosion resistance and steam equipment.

Everdur. Copper-base alloys, with fractional percentages of silicon, up to 4%, and manganese up to 1.1%, a small amount of lead, 0.4%, being optional for easier machining. The alloys have high strength (akin to that of steel), good corrosion resistance, and high fatigue values (American Brass Co.).

Exolon. Fused aluminum oxide, manufactured by Exolon Co.; used for abrasive purposes.

Exothermic Ferrochrome. Mixture of ferrochrome, carbon, silicon, and a small proportion of an oxidizing agent, which reacts and generates heat when added to molten iron or steel, supplying molten ferrochrome to the metal. Usually, the material is equivalent to high-carbon ferrochrome in its effect.

Exothermic Ferrosilicon. Mixture of ferrosilicon and small proportions of an oxidizing agent, both usually finely ground, which reacts and generates heat when added to molten iron or steel, supplying molten ferrosilicon to the metal.

Exothermic Reaction. Chemical reaction which gives off heat.

Expanded Metal. Sheet, usually of steel, slit or punched, and then expanded, to form mesh; used for reenforcing concrete and plaster walls.

Expanding Metal. Alloys which expand on cooling; generally contain bismuth or antimony, frequently both. The expanding is important in type and similar castings requiring exactitude.

Expansive Metal. See **Expanding Metal.**

Extensometer. Mechanical device for indicating the elongation of metal while being subjected to tension.

Extra Best Best (Telephone & Telegraph Wire). Steel wire of lowest electrical resistance. See **Telephone & Telegraph Wire.**

Extra Clean Smooth Bright Wire. Steel wire, drawn dry, under conditions yielding an especially bright surface.

Extraction. (a) Science and technique of obtaining metals from their ores. (b) Proportion of metallic content in an ore obtained in the final metallurgical operation.

Extractor Box. See **Zinc Box.**

Extra Deep Drawing Dull Finish Auto Body Sheet. Automobile body steel sheet suitable for unusually severe drawing. See **Automobile (Body) Sheet.**

Extra Smooth (Hot-Rolled Annealed) Sheet. Hot-rolled, annealed steel sheet, given an additional cold-rolling to yield a smooth surface.

Extra Strong Pipe. Pipe with wall thickness greater than standard (see **Standard Pipe)** and suitable for greater pressures.

Extra-Terrestrial Metal. Metals which fall to earth in the form of meteorites. For the most part, iron and nickel are predominant in these, though smaller amounts of other metals are also frequently present.

Extrusion. Shaping a ductile metal into continuous form, such as tubing, by forcing it through a die of appropriate shape.

Extrusion Effect. Annular ring of defective material, found occasionally in extruded brass, consisting of oxides in a matrix of partially dezincified brass.

Eyelet Brass. Yellow **Brass** (see) of about 32% zinc.

EZ Stainless Steel. High-chromium (about 13%), corrosion-resistant steel, with 0.3–0.4% carbon and approximately 0.25% sulfur to aid machinability (Latrobe Electric Steel Co.). See **Stainless Steel.**

F

F. See **Fahrenheit.**

Faber du Faur Furnace. Tiltable crucible furnace of cubical shape, used in desilverizing zinc residues.

Face. See **Crystal Face.**

Face-Centered Cubic Space Lattice. See **Face-Centered (Cubic) Structure.**

Face-Centered (Cubic) Structure. Term used in describing internal crystal structures, as determined by X-ray, in which the equivalent points are at the corners of the unit cell and at the centers of the six faces of the cube.

Facing. Applying powdered coal or similar material to the surface of a mold, or mixing such material with the mold sand, to give a smooth surface to the cast metal. See **Casting.**

Facing Sand. Molding sand, in metal casting, for use in direct contact with the cast metal.

Factor of Safety. Maximum stress to which a structural unit can theoretically be subjected, without failure, divided by the maximum stress which it is likely to receive.

Faggoted Iron. See **Faggoting.**

Faggoting. Piling bars together, in the manufacture of **Wrought Iron** (see), prior to refining.

Fahlun (Brilliants) Metal. Alloy of approximately 60% tin and 40% lead, used for ornamental purposes.

Fahralloy. Nickel-chromium-iron heat and corrosion-resistant alloys of diverse composition (American Manganese Steel Co.).

Fahrenheit. Temperature scale in which the freezing point of water is taken as 32°, its boiling point as 212°. It is in common usage throughout the English-speaking countries.

Fahrite. Nickel-chromium-iron alloys of excellent heat and corrosion-resistance (Ohio Steel Foundry Co.) with 19–40% nickel, 18–32% chromium and optional low percentages of silicon and manganese.

Faint Red Heat. Temperature approximating 930°F, as judged visually.

Falding Furnace. Three-hearth **Muffle Furnace** (see), with mechanical means for raking, and the combustion flues located under the lowest hearth.

Falling Seam. In butt-welded pipe, a defect in the form of a depression along the weld, on the outside of the pipe.

Falling Weight Test. See **Drop Test** (a).

Falls Alloy. Series of master alloys, most commonly of copper-base and aluminum-base, used in foundry and related operations (Niagara Falls Smelting & Refining Corp.).

Fanning. Operation of blast furnace at low rate of output by lowering the air blast to 1–2 pounds inlet pressure. This is in contrast to banking, in which the blast is completely eliminated.

Fantail. Flue, fan-like in shape, leading from the **Slag Pocket** (see) in an open-hearth furnace to the top of the regenerator chamber.

Faraday. Unit representing 96,500 (absolute) coulombs of electricity, a quantity which will deposit one gram equivalent of a substance from its solution.

Faraday's Laws. (1) The amount of chemical change produced by electric current is proportional to the total quantity of electricity. (2) Quantities of each element which can be deposited or dissolved by a specific quantity of electricity are proportional to the chemical equivalent weight of the element.

Fastell Alloy. Series of silver-tungsten, silver-molybdenum, and copper-tungsten composites, made by powder metallurgy technique for heavy-duty electrical contacts (Fansteel Metallurgical Corp.).

Fatigue. Gradual deterioration of tensile and related properties of metals caused by repeated cycles of stress. It is believed to take place as a result of incipient microscopic cracks, broadening out under the influence of the stress cycles. Simultaneous corrosion and stress cycles are even more detrimental and will affect relatively resistant metals, such as stainless steel and Monel metal.

Fatigue Crack. See **Fracture.**

Fatigue Failure. See **Fatigue Fracture.**

Fatigue Fracture. See **Fracture.**

Fatigue Limit. Maximum stress which a given material will withstand in one direction, under indefinitely continuous cycles of stress. Usually such cycles are assumed to be complete reversals, equal in each direction from the normal. For ferrous metals, the fatigue limit is usually considered as about 40–45% of the ultimate tensile strength.

Fatigue Ratio. Ratio of the fatigue limit of a material to its tensile strength.

Fatigue Strength. See **Fatigue Limit.**

Fatigue Stress. See **Fatigue.**

Fatigue Test. Determining the fatigue limit of a metal, by subjecting a series of samples to rapid reversals of stresses of varying magnitude. The maximum stress range, which shows no failure by fatigue, represents the **Fatigue Limit** (see).

F.C.C. Steel. Series of high-carbon air-hardening and oil-hardening die and tool steels, manufactured by Forging and Casting Corp.

F.C.T. See **Ferro-Carbon-Titanium.**

Fe. Chemical symbol for **Iron** (see).

Feathered Tin. Granulated **Tin** (see), usually made by pouring molten tin into water.

Feather Shot. Granulated **Copper** (see), usually made by pouring molten copper into water.

Fecraloy. Electric-resistance iron-base alloy, approximating 15% chromium and 4% aluminum (Wilbur B. Driver Co.).

Federated Babbitt. Series of lead-base alloys, used for bearing purposes (Federated Metals Div.).

Federated Solder. Series of lead-base **Solders** (see), low in tin (6–30%), with 1.75% silver, 0.5–5.0% bismuth, and 0.5% antimony (Federated Metals Div.).

Feed. Raw materials supplied to a process or equipment, such as a furnace, etc.

Feeder. See **Riser.**

Feeling. Agitation of metal in a crucible [see **Crucible** (a)], by means of a rod, to determine whether fusion has been complete.

Fence Wire. See **Wire Fence.**

Fenton's Alloy. Bearing alloy with 80% zinc, 6–8% copper, and the remainder antimony or tin.

Feran. Aluminum-clad sheet iron. See **Cladding.**

Fernichrome. Low-expansion iron-base alloy, used for sealing metal wire into glass, with about 30% nickel, 25% cobalt, and 8% chromium (General Electric Co.).

Fernico. Low-expansion iron-base alloy, used for sealing metal wire into hard glass, with about 28% nickel and about 18% cobalt (General Electric Co.).

Fernite. Series of cast nickel-chromium-iron alloys, used for heat and corrosion-resistance, with chromium up to 28% and nickel up to 65% (Forging and Casting Corp.).

Ferric Furnace. Tall **Blast Furnace** (see) using bituminous coal, the conversion of the coal into coke occurring in the upper part of the furnace.

Ferriferous. Containing iron.

Ferrite. Substantially pure alpha or delta iron, containing only small amounts of other elements in solid solution. More generally, the term is used to designate solid solutions in which iron, in the alpha modification, is the solvent. It is soft, ductile but weak, and virtually non-hardenable. It is easily magnetized, and loses its magnetism just as easily.

Ferrite Divorcement. See **Abnormal Steel.**

Ferritic Alloy Steel. Alloy steel in which **Ferrite** (see) is the dominant phase.

Ferritic Steel. Steel in which ferrite is the dominant phase, representing pure alpha iron, or alpha iron as a solvent for alloying elements.

Ferritic Structure. See **Ferritic Steel.**

Ferro-. A prefix (derived from the Latin, ferrum) denoting iron base alloys, containing one or more other elements, except carbon, in sufficient quantity to warrant the use of the alloy as a vehicle for introducing such elements in steel•manufacture.

Ferroalloy. See **Ferro-.**

Ferro-Aluminum. Iron-base alloys, with 20–50% aluminum, used as master alloys for addition to steel or aluminum.

Ferro-Boron. Iron-base alloy, with 15–25% of boron, occasionally used as a deoxidizing and degasifying agent, and for adding boron to steel.

Ferrocarbo. Granular silicon carbide, manufactured by the Carborundum Co., used as a deoxidizing and graphitizing agent in iron and steel, being added to the cupola or ladle.

Ferro-Carbon-Titanium. Series of master alloys for addition of titanium to steel, for deoxidizing and scavenging. High-carbon ferro-carbon-titanium contains 15–18% titanium and 7–8% carbon, and is used for adding titanium to medium and high-carbon steels and cast irons. Medium-carbon ferro-carbon-titanium contains 18–21% titanium and 3.5–4.5% carbon. It is used for low-carbon steels. High-aluminum ferro-carbon-titanium contains 12–16% titanium, 18–22% aluminum, 2–4% silicon, and 3–5% carbon. It is used for making fine-grained steels by simultaneous additions of titanium and aluminum.

Ferrocerium. See **Misch Metal.**

Ferrochrome. See **Ferro-Chromium.**

Ferro-Chromium. Alloys of iron and chromium, with approximately 60–72% chromium, commonly used as a base for the manufacture of chromium-containing ferrous alloys. Carbon in the alloys will range from as low as 0.06% to about 8%, and the lower the carbon content the greater the cost. The process of manufacture involves the smelting of chromite ore (substantially $Fe(CrO_2)_2$) with coke and slagging materials for the high-carbon variety; low-carbon ferro-chromium is usually made by silicon reduction. Both low and high carbon ferro-chromium can also be obtained with high, 0.60–0.70%, nitrogen content.

Ferrocobalt. Alloy of cobalt with iron (commonly 50% of each), frequently used to add cobalt to iron-base alloys.

Ferro-Columbium. Iron-base alloys, commercially usually with approximately 50–60% columbium, made by smelting columbite ore (or oxide) with iron ore in an electric furnace. Its main use is as addition agent to stainless and chromium steels, to prevent intergranular corrosion.

Ferromagnetism. Property of certain metals, notably iron, of being attracted by a magnet. Other metals showing ferromagnetism are, in descending order of magnetic susceptibility: nickel, cobalt, manganese, chromium, titanium, palladium, platinum, and osmium. See also **Diamagnetism.**

Ferro-Manganese. Alloys of iron and manganese, usually high in manganese (approximately 80%), commonly used for additions of manganese to steel. The high-carbon variety contains about 6–7% of carbon; the medium, up to

1.5% carbon; the lower range, 0.10–0.75% carbon. Another grade, "Low-Iron," assays up to 2% iron, 3% silicon, and 7% carbon. See also **Spiegel.**

Ferro - Manganese - Silicon. Iron - base alloy, containing 20–25% manganese, 47–54% silicon and, at most, 0.6% carbon; used for adding manganese and silicon to cast iron in the ladle.

Ferro-Molybdenum. Alloys of iron and molybdenum, in which the molybdenum is commonly around 60%. Low-carbon ferro-molybdenum contains 0.5% maximum carbon, whereas the other common grade has a permissible carbon content of 2.0%. Ferro-molybdenum is commonly made by smelting molybdenite (MoS_2) or other molybdenum-containing ore or slag with iron ore and fluxing constituents.

Ferro-Nickel (Alloys). Alloys of iron with nickel, particularly those of about 36% nickel, whose thermal expansivity is extremely low. See **Invar.**

Ferro - Phosphorus. Iron - phosphorus alloys, commonly available with 17–25% phosphorus contents; the lower phosphorus content alloys being commonly made in the blast furnace, whereas the higher phosphorus-content alloys are products of the electric furnace; used for adding phosphorus to special-purpose steels, such as steel sheets, the phosphorus preventing these from sticking together during annealing.

Ferro-Selenium. Iron-selenium alloys, with about 50% selenium and fractional percentages of carbon, silicon, phosphorus, and sulfur. Used as a master alloy for addition to steels to achieve greater machinability.

Ferro-Silicon. Alloys of iron and silicon, with silicon varying from 15% to 90%. Ferro-silicon is commonly made in the electric furnace, by reducing iron ore and silica, and is extensively used as a steel deoxidizer.

Ferro-Silicon-Aluminum. Ferro-Silicon (see) with about 12–15% of aluminum; used most frequently as ladle addition and deoxidizer in the manufacture of steel.

Ferro-Silicon-Zirconium. See **Ferro-Zirconium.**

Ferro-Titanium. Commercially, alloys of titanium and iron, available in the low-carbon variety and also in high-carbon form called **Ferro-Carbon-Titanium** (see); used for special steel manufacture, and also to prevent intergranular corrosion in austenitic chrome-nickel steels. See also **Foundry Ferro-Titanium.**

Ferro-Tungsten. Iron-tungsten alloys, usually high in tungsten (70–85%), made by smelting the appropriate ores, together with slagging constituents, in an electric furnace; finds its main outlet in the manufacture of high-speed tool steels.

Ferro-Uranium. Master alloy of uranium with iron. Since uranium steels are no longer commercially used, there are no standard compositions for ferro-uranium.

Ferrous. Related to iron (derived from the Latin, ferrum). Ferrous alloys are, therefore, iron-base alloys.

Ferrous Alloy. See **Ferrous.**

Ferrous Metallurgy. Metallurgy of iron and steel.

Ferrous Sulfide. The compound FeS, which is formed by interaction of iron and the sulfur always present in iron-base alloys; it causes the objectionable characteristic of hot-shortness in sulfur-bearing steels.

Ferro - Vanadium. Iron - base alloys, with vanadium approximating 40%, silicon 1.5–8%, and carbon 0.2–3%; manufactured by electric furnace smelting, both carbon and silicon being reported as reducing agents; used as a master alloy for addition of vanadium to steel.

Ferroxyl (Indicator). Agar-agar jelly of potassium ferricyanide, phenolphthalein, and salt (or other corroding medium), used to test corrosion in iron.

Ferroxyl Test. See **Ferroxyl (Indicator).**

Ferro-Zirconium. Zirconium for addition to steel, commonly available as ternary zirconium-silicon-iron alloys, one grade approximating 14% zirconium, 40% silicon, and 46% iron; the higher grade runs 35–40% zirconium, 47–52% silicon, and the remainder iron;

used primarily as scavengers in steels, the zirconium content reacting with nitrogen and sulfur.

Ferruginous. Containing iron.

Ferrum. Latin name for iron, used in medicine and pharmacy.

Ferry Metal. Lead-base bearing metal with about 2% barium, 1% calcium, and fractional percentages of mercury.

Fery Pyrometer. Radiation **Pyrometer** (see) in which radiation from the source sighted on is concentrated on a minute thermocouple, the resulting millivoltage being read on a galvanometer.

Fettle. See **Fettling.**

Fettling. Refractory lining in open-hearth or reverberatory furnaces; usually formed from old furnace cinder, or low-silica ore. In the case of a puddling furnace, such lining may be of roll scale or of roasted cinder (also called bulldog).

Fiber (or Fibre). Direction in which metals have been caused to flow, as by rolling, with macroscopic evidence in the form of fibrous appearance in the direction of flow.

Fiber Stress. Unit stress which exists at any given point in a structural element subjected to load; given as load per unit area.

Fibrous Structure. See **Fracture.**

Field Grade Sheet. Low-silicon (about 0.25%) electric sheet, used in the manufacture of small motors.

Fiery Fracture. See **Crystalline Fracture.**

File Bronze. See **File Metal.**

File Hard. See **File Test.**

File Metal. Copper-base hardware alloy, with 18–31% tin, 7–9% lead, and optionally up to 7% zinc.

File Test. Rough, qualitative hardness test on metal in which the edge of a standard file is drawn across the surface under test. If the file does not take hold, or bite, the material is considered file hard.

Filing Board. Narrow plate to which a series of small files are attached, to remove burr from the edge of strip metal, after passing through slitting cutters.

Filler Metal. In **Welding** (see), the added metal which is molten by the arc or flame and acts to bind edges or fill empty zones.

Filler Sand. Coarse molding sand, in metal casting, used to afford good ventilation in the mold.

Film Test. Simple test for determining proper pouring temperature for molten metal. A small quantity of the metal is held in a spoon and the development of a visible film on the surface is timed.

Fin. Extra and, therefore, objectionable metal at the sides of rolled or forged metal, at the points where the rolls or hammer-and-anvil come together; also known as overfill, particularly when broad and not sharp. Further working of metal having overfills will cause laps.

Fine. See **Fineness.**

Fine Gold. Substantially pure **Gold,** (see).

Fine Grain. Appearance of metal with crystals of relatively small size.

Fine-Grained Fracture. See **Silky Fracture.**

Fine-Grained Steel. Steel which, on the basis of the A.S.T.M. austenitic standards, shows a grain size from #5 to #8.

Fine Metal. Crude copper matte, with about 60–80% copper content, obtained from the smelting of sulfide copper ores.

Fineness. (a) Size of particles in powdered material. (b) Proportion of gold or silver in alloys, containing either metal, expressed as parts per thousand.

Fine Round Wire. Round steel wire of #16 gage (Washburn & Moen steel wire gage) or finer.

Finery. Type of hearth, formerly used in the manufacture of **Wrought Iron** (see).

Fines. (a) Finely divided material, usually resulting from crushing or grinding. Such material frequently interferes with normal metallurgical operations. (b) Product passing through a screen or sieve, as opposed to **Shorts** (see).

Fine Silver. See **Silver.**

Fine Wire. Round steel wire of #20 wire gage (Washburn & Moen steel wire gage) or finer.

Fining. Manufacture of malleable iron from cast iron in a charcoal or similar hearth.

Finish. Surface appearance of metal, after final treatment, such as rolling, etc.

Finish Annealing. Annealing on steel, most commonly tubing, which leaves a blue oxide film.

Finished Steel. Steel ready for marketing.

Finishing. Operation giving metal its final surface characteristics, or **Finish** (see).

Finishing Mill. See **Rolling Mill.**

Finishing Pass. See **Finishing Rolls.**

Finishing Rolls. Rolls, in the rolling of metal strip or sheet, bringing the metal to exact thickness, and providing it with proper surface.

Finishing Stand. See **Finishing Rolls.**

Finishing Steel. Tool steel, maintaining good finishing cutting edges on tools when used at high speeds at relatively light cuts. In composition, such steels are essentially intermediate between **Carbon Steels** (see) and **High-Speed Tool Steels** (see).

Finishing Temperature. Temperature of final hot-working of a metal.

Finishing Tool Steel. See **Finishing Steel.**

Fink Anode. Insoluble anode consisting of a silicon-copper alloy (with 15–25% silicon), used in the electrolytic leaching of copper ores containing significant amounts of chlorides.

Firearmor. Nickel-base alloys, with about 12–20% chromium, 10% iron, and optionally low percentages of manganese and silicon; used for heat resistance purposes.

Fire Assay. See **Assay.**

Fireback. Rear wall of a furnace or fireplace.

Firebox. Combustion chamber in a furnace or hearth.

Fire-Box Plate. See **Boiler Plate.**

Fire-Box Steel. See **Boiler Steel.**

Fire Brick. Brick made from refractory clays or other refractories, used in furnace linings, flues, etc.

Fire Bridge. See **Bridge.**

Fireclay. Clay of refractory nature (because of low alkali content), used metallurgically to withstand high temperatures; usually, any clay with a softening point above 2900°F is considered a fireclay.

Firecoat. Film of oxide on metal, resulting from heating or, more frequently, direct flame.

Firecrack. Crack forming in metal during improper heating or annealing. See also **Fire-Cracking.**

Fire-Cracking. Surface-roughening of rolling-mill rolls, increasing with the use of the rolls. Certain alloying elements, notably carbon and chromium, increase this tendency, which seems to be due to overheating of the rolls and consequent surface cracking. Fire-cracked rolls tend to produce "orange peel" on the surface of the rolled metal.

Firedoor. Opening through which fuel is introduced into combustion equipment, such as a furnace.

Firedoor Terne. Terne Plate (see), ordered or sold on the basis of specifications governing use for terne-clad doors and shutters.

Fire Refining. Purifying crude metal, involving heating, by direct flame, with or without other processing.

First Forging. First of a sequence of **Forging** (see) operations.

First Law of Thermodynamics. Energy can neither be created nor destroyed.

Firthaloy. Cemented tungsten carbide, used for wire drawing purposes (Firth-Sterling Steel Co.).

Firth Hardness Test. See **Firth Hardometer.**

Firth Hardometer. Apparatus for measuring hardness, akin to the Vickers (see **Vickers Hardness Test**). In addition to a diamond indentor, small steel balls are also used.

Firthite. Cemented tungsten carbide, with cobalt as the cementing matrix, used for cutting tools (Firth-Sterling Steel Co.).

Firth-Sterling Steel. Series of high-alloy steels, including **Stainless Steel** (see), manufactured by Firth-Sterling Steel Co.

Fir Tree Crystal. See **Dendrite.**

Fir Tree Structure. See **Dendrite.**

Fish Scale. Tendency of faulty vitreous enamel to break away in small flakes.

Fissure. See **Flake** (a).

Five Point Star. See **Tin Flower.**

Fix. See **Fixing.**

Fixed Converter. Obsolete type of Bessemer converter (see **Bessemer Steel**) which cannot be rotated, so that raw materials must be charged from the top, and metal tapped out through a hole at the bottom.

Fixing. Process of **Fettling** (see) the hearth of a **Puddling** (see) furnace.

F.J.A.B. (Steel). Series of oil-hardening, carbon die and tool steels, manufactured by Milne & Co.

Flake. (a) Fine crack inside large-sized forgings or rolled metal, appearing, on fracture, as coarse, bright crystalline areas. (b) In powder metallurgy, flat particles, resembling fish-scales, of small thickness as compared to both other dimensions. (c) In galvanizing, see **Spangle.**

Flake Copper. Thin scales of metallic **Copper** (see), as found occasionally in nature.

Flame Descaling. Removing scale from steel by very rapid heating of the surface by means of a special oxy-acetylene torch. Differential expansion causes the scale to break off.

Flame Hardening. Surface-hardening by controlled oxy-acetylene torch heating, followed by cooling with water or cold air jets.

Flame Planing. Trimming and edging of steel plate by means of a gas torch.

Flame Priming. Removing loose scale, rust, and moisture from steel, preparatory to painting, by heating the surface by a gas torch.

Flame Reaction. Distinctive coloration imparted to a flame by certain elements, most particularly the alkali and alkaline earth metals, even in compound form, each element having specific characteristics as to color and intensity.

Flame Scarfing. Removing surface defects from steel, in ingot or semi-finished form, by means of a gas torch.

Flame Spectrum. Spectrum obtained by spectroscopically analyzing the flame resulting from volatilizing a substance in a non-luminous flame.

Flame Test. See **Flame Reaction.**

Flame Welding. See **Torch Welding.**

Flange Plate. See **Boiler Plate.**

Flange Steel. See **Boiler Steel.**

Flapping. Rabbling of molten copper to expose the metal to the atmosphere.

Flash. (a) Thin **Fin** (see). (b) Thin coating of electroplated metal, usually intended as foundation for another electroplated metal.

Flash Butt Welding. Resistance Butt Welding (see) in which the electricity is applied before contact, resulting in a series of arcs between the materials being welded.

Flash Coating. See **Flash** (b).

Flashlight Powder. See **Magnesium Powder.**

Flash Roasting. Process of **Roasting** (see) in which finely divided ore is blown, with air, into a combustion zone.

Flash Welding. Electric **Arc Welding** (see) in which the parts to be joined are clamped in position and an electric arc is established between them; when sufficiently heated, the parts are forced together under pressure, and held in close contact until cool.

Flask. (a) Frame made of metal or wood, used for containing the sand around the pattern, in mold making for metal casting. See **Cope** and **Drag.** (b) Iron bottle, of standard size, for holding 76½ pounds of mercury.

Flat. Narrow flat steel, up to one-quarter inch in thickness (including band, hoop, cotton tie, etc.), resulting from hot rolling.

Flat Bar. Bar (see) metal of any rectangular cross-section, except square, more than one-quarter inch thick, and up to six inches wide.

Flat-and-Edging Method (of Rolling). (a) Rolling steel into angles; the metal is first rolled to the proper thickness and then bent to form a right angle. (b) Method of controlling the width of flat sections, in rolling mill operations, in which the metal is rolled on edge in deep grooves in the rolls.

Flat Hot-Rolled Steel Products. Class of steel products of relatively flat nature, including hot-rolled sheet and strip, slabs, plates, bars, etc.

Flat Rod. Rod, usually of steel, rolled to flattened shape in a **Rod Mill** (see).

Flat Rolling. Rolling, in which one or more passes (see **Pass**) form the metal section, by spreading in one direction and compression in another.

Flatting Mill. (a) **Rolling Mill** (see) for producing sheet or strip metal. (b) Rolling mill for reducing brittle metal grains to metal dust.

Flatting Pass. Pass (see) in which the metal is passed through flatwise.

Flat Wire. Product of the cold-rolling of drawn round wire, in flattened form.

Flecto Iron. Malleable cast iron, manufactured by Ohio Brass Co.

Flexo Steel. Silicon-manganese steel, containing approximately 2% silicon, 0.8% manganese, and 0.60% carbon; used for the manufacture of springs (Carpenter Steel Co.).

Flexure Stress. Stress resulting from bending.

Flintshire Furnace. Reverberatory furnace, with a depression in the side of the hearth, for the roast-reaction treatment of lead ores (see **Roast-Reaction Process**).

Float Copper. (a) Extremely fine copper powder, most commonly made by stamping; because of its fineness, the powder does not settle readily in water. (b) Copper, found native, which has been carried away, by natural influences, from its original rock.

Float Gold. Extremely fine gold powder; because of its fineness, the powder does not settle readily in water, and can easily be carried off in a water stream. See **Gold.**

Florentium. European name for **Illinium** (see).

Floss. Molten **Cinder** (see) floating on the surface in a **Puddling** (see) furnace.

Floss Hole. Opening, in the uptake of a hand **Puddling** (see) furnace to enable removal of cinder.

Flotation. Concentrating mineral values from ores by agitation of the ground ore with water, oil, and flotation chemicals. The mineral values, normally, are wetted by the oil and are lifted to the surface by air bubbles, where they are floated off.

Flour Copper. See **Float Copper** (a).

Floured Mercury. Mercury, in finely granulated state, as occurs in amalgamation and other processes; the contaminated metal globules stay isolated and resist union into compact fluid metal.

Flouring. See **Floured Mercury.**

Flour Gold. See **Float Gold.**

Flow. (a) In powder metallurgy, standard test of the rate at which a given metal powder flows through a standard orifice. (b) See **Creep.**

Flower. See **Spangle.**

Flow Figures. See **Lüder's Lines.**

Flowing Furnace. Reverberatory furnace, with an inclined hearth, used for heating roasted lead ores.

Flue Bridge. See **Bridge.**

Flue Dust. Solid material, in powder form, carried over from any furnace or process, and tending to deposit in the flues or to be carried off with the gases produced by the furnace or operation.

Fluid. Material in the liquid or gaseous state.

Fluid Deformation. Deformation in an amorphous material, in which material flows freely under stress.

Fluidity. Ease of flowing (particularly with reference to a material in the molten state).

Fluorescent Method. Method of detecting surface flaws in metal castings or forgings; the metal is suspended in the vapors of a fluorescent compound; tiny crystals tend to fill the flaw, and ultraviolet light will reveal their presence.

Fluorescent Screen. Activated sheet or board which becomes visible at any point exposed to X-rays.

Fluorspar. Natural calcium fluoride ore, corresponding to the formula CaF_2. It is the principal neutral flux, and is used commonly in basic open-hearth practice on steel, aiding substantially in making the slag more fluid.

Flushing. Removal of cinder from a **Blast Furnace** (see).

Fluting. Kinking or breakage due to curving of metal strip on a radius so small, with relation to thickness, as to stretch the outer surface above its elastic limit.

Flux. Material which, by its chemical and physical action on non-metallic constituents, aids in the separation of a metal from these non-metallics. Fluxes of the solid acid oxide or basic oxide type are extensively used in all smelting operations; they are also required in many welding, brazing, and soldering operations, as well as in the hot-coating of steel.

Fluxing Ore. Ore which contains, in addition to metal values, sufficient fluxing agents (see **Flux**).

Fluxing Stone. **Limestone** (see) or **Dolomite** (see), used in blast furnace operations.

Flux Process. Process for manufacturing tin or terne plate, in which molten zinc chloride, with or without ammonium chloride, is used as flux on the surface of the fused metal pot. See **Terne** and **Tin Plate.**

Flux Spoon. Small hand ladle for removing a sample of molten metal from a container, such as a furnace.

Flux Spot. Flaw on zinc or similar coated steel, caused by low acid or high iron content in the pickling bath, used prior to coating.

Flying Micrometer. Device for determining the thickness of continuous metal, such as sheet, plate, etc., while in movement during working and processing.

Flying Shears. Apparatus for cutting billets or bars while in motion; the shears are synchronized with the rate of speed of travel of the metal.

Foeppl (Hardness) Test. Method of testing hardness, in which two pieces of the material under test are pressed together under constant load.

Foeppl-Pertz Damping Tester. Apparatus for determining the **Damping Capacity** (see) of a material, in which a specimen is held vertically, and caused to vibrate; the time required for cessation of the vibration is a measure of the damping capacity of the material.

Foil. Metal in any width, with thickness not greater than about .005 inches.

Follow Board. See **Bottom Board.**

Fontainmoreau Bronze. Not a bronze, nor a brass. Zinc-base alloys, with up to 8% copper and, optionally, up to 1% iron and up to 1% lead.

Fool's Gold. See **Pyrites.**

Forechamber. See **Doghouse.**

Forehearth. Settling reservoir, independently heated, for holding furnace residues, to permit gravity separation of metal and slag.

Fore Plate. (a) Iron plate, used as the bottom of the hearth in a wrought iron **Puddling** (see) furnace. (b) Large metal plate, on entering side of a **Rolling Mill** (see), for holding metal, prior to insertion into the rolls.

Fore-Span Plate. See **Bloomery.**

Forge. Open hearth or fireplace, with air blast, particularly one used in the manufacture of **Wrought Iron** (see) directly from iron ore.

Forge Cinder. Slag or dross, resulting from **Forge** (see) operations.

Forged Steel. Steel product made by hammering or pressing (usually hot). See **Forging.**

Forge Heating. Heating metal, usually steel, to a temperature suitable for **Forging** (see) operations. Frequently, forge heating follows a preliminary heating.

Forge Pig Iron. **Pig Iron** (see) suitable for use in the manufacture of wrought iron.

Forge Roll. **Rolling Mill** (see) for the conversion of puddled steel blooms or slabs into bars.

Forge Scale. Coating of oxide, formed on iron or steel during forging operations.

Forge Train. See **Forge Roll.**

Forge Welding. **Welding** (see) in which the parts are heated in a furnace, and then pressed or hammered together.

Forging. Hammering or pressing of metal, usually at elevated temperatures. The term is also used to describe the metal produced by such hammering or pressing.

Forging Brass. Yellow brass (approximately 60% copper), used in forging

operations; lead, iron, and tin, in low percentages, are common additions.
Forging Bursting. See **Bursting** (b).
Forging Ingot. Metal **Ingot** (see) intended for forging operations.
Forging Press. Hydraulic press for pressing metal (as opposed to hammering).
Forging Quality Billet. See **Forging Quality Steel.**
Forging Quality Bloom. See **Forging Quality Steel.**
Forging Quality Hot-Rolled Strip Steel. Hot-Rolled Strip Steel (see), ordered or sold on the basis of specifications insuring uniformity, soundness, and freedom from surface imperfections.
Forging Quality Slab. See **Forging Quality Steel.**
Forging Quality Steel. Hot-rolled steel, in the form of billets, blooms, and slabs, substantially free from piping and suitable for forging into final form.
Forging Quality (Steel Plate). Steel plate, ordered or sold on the basis of specifications governing suitability for forging, heat-treating, and other uses where uniformity and homogeneity are essential.
Forging Round. Semi-finished stage in the conversion of ingot metal into a finished forging.
Forging Steel. Steel, usually made in basic open-hearth or electric furnaces, adaptable to hot hammering or pressing. Medium-carbon and alloy steels are generally used for this purpose.
Forging Strain. Strain created in metal by forging itself or by subsequent cooling.
Former. In a closed pass type of rolling (see **Closed Pass**), the portion of one roll which indents into the cross-section of the other roll.
Forming Properties. Properties of metal which permit forming and distortion from the original shape.
Forward Creep. Tendency of the speed of the metal being rolled to be slightly increased on the delivery side of the rolls, so that the speed of the metal is slightly greater than the roll peripheral speed.
Forward Slip. See **Forward Creep.**
Founding. See **Casting** (a).
Foundry. Plant for manufacturing metal castings.
Foundry Alloy. See **Master Alloys.**
Foundry Coke. Coke (see) larger than 2.5 in. in any dimension.
Foundry Ferro-Titanium. Master Alloy (see) for additions of titanium to cast iron, containing about 17–21% titanium, 17–21% silicon, with carbon and aluminum below 0.5% each.
Foundry Iron. See **Foundry Pig (Iron).**
Foundry Pig (Iron). Pig Iron (see), with phosphorus in the 0.50–1.00% range, used for iron castings.
Foundry Sand. Sand used for making metal-casting molds; should have good binding properties, be refractory, and be of proper grain-size.
Four. In pack mill operation, a pack of two sets of **Pairs** (see) for reheating and rolling.
Fourdrinier Alloy. Alloy used for Fourdrinier wire in paper-making screens, with about 80% copper, 20% zinc, and fractional percentages of tin.
Four-High Mill. Rolling mill consisting of four rolls in vertical position; the middle two are known as the working rolls, and the metal is passed through these. The top and bottom rolls are known as the reinforcing rolls or backing-up rolls.
Four-Pass Stove. Stove in **Blast Furnace** (see), in which the gas from the combustion chamber passes through three regenerative flues.
Fractional Crystallization. Process wherein one solute is crystallized from a solution of two or more solutes; used for the purification of crystalline substances.
Fracture. Surface appearance of metals when broken. Typical among forms of fracture are: Conchoidal fracture represents a surface, either convex or concave, resembling a shell. Columnar fracture represents a surface with rectangular, finger-like projections. Silky fracture represents a brilliant and lustrous appearance, akin to silk. Usually, metals of silky fracture are tough and strong. Crystalline fracture represents the appearance of crystals at the broken surface. Crystalline fracture

usually reflects brittleness and low tensile strength. Fibrous fracture represents a surface of long stringy nature. Metals of fibrous fracture are usually tough and ductile. Granular fracture represents a metal surface with a rough appearance, akin to sandstone. Woody fracture represents a surface in which slag particles, as found in dirty steel, give rise to an appearance akin to wood. Irregular fracture, also called ragged fracture and broken stick fracture, represents a break in which there is no regularity, and which follows no specific plane; frequently, long pieces break away from the test specimen, at the zone of fracture. Cup and cone fracture represents a break in which one part of the broken specimen has the appearance of a pointed cup, whereas the other, mate to it, is cone-like in shape. Half-cup fracture represents a break in which part of the cross-section is cup-like in shape. Star fracture, also called rosette fracture, represents a broken surface resembling a star or rosette in appearance. In steel, it is commonly associated with **Sorbite** (see). Shear fracture, also called angular fracture, is a straight-plane break at about 45° to the axis of the specimen; it is found occasionally in non-ferrous metals of moderate tenacity and ductility. Fatigue fractures or cracks, also known as progressive fractures or as detail fractures, start from a nucleus where there is an abnormal concentration of stress, and propagate through the metal. The surface is smooth, and frequently shows concentric markings, with the nucleus as the center.

Fracture Standards. See **Penetration-Fracture Test.**

Fracture Test. Nicking and breaking a billet by means of sudden impact, to enable macroscopic study of the fractured surface.

Frame. See **Wire-Drawing Frame.**

Francisci Furnace. Furnace, for heating roasted **Zinc** (see) ores and reducing them to metal, consisting of muffles superimposed over one another and all connected to a common condensing chamber.

Frankite. High-chromium (12–39%) cast iron, with carbon ranging from 0.70 to 2.00% (Frank Foundries Corp.).

Frary Metal. Lead-base alloy series, hardened with about 1.5% calcium and 1.5% barium; excellent for low-pressure bearings at moderate temperatures.

Freckle. Flaw on the surface of **Tin Plate** (see), in the form of a spot of dull appearance.

Free. Chemically uncombined; refers usually to metal found, in elementary form, in nature, e.g., free gold.

Free Carbon. See **Total Carbon.**

Free Cementite. Cementite (see) present in iron-carbon alloys as such, and not in pearlitic form. See **Pearlite.**

Freecut (Steel). High-sulfur, free-cutting steel, manufactured by Republic Steel Corp.

Free-Cutting. Metal to which one or more alloying constituents have been added, usually in small amounts, to prevent tearing while being machined, and to cut freely under the machining tools. Free-cutting steel is usually made by the open-hearth or Bessemer process; the free-cutting property is achieved by fractional percentages of sulfur above the normal value of 0.08%, up to 0.40%.

Free-Cutting Brass. See **Leaded Yellow Brass.**

Free-Cutting Commercial Bronze. Commercial Bronze (see) with about 2% lead, added for easy machining.

Free-Cutting Nickel Silver. See **Leaded Nickel Silver.**

Free-Cutting Phosphor Bronze. Phosphor Bronze (see) with up to 4% lead. A common composition contains about 4% each of lead, tin, and zinc.

Free-Cutting Steel. See **Free-Cutting.**

Free Energy. Energy, accompanying a chemical reaction, which is available for external work.

Free Ferrite. Ferrite resulting from the decomposition of austenite when cooled, prior to the appearance of carbide.

Free Gold. See **Free.**

Free Silver. (a) See **Free.** (b) Fiscal policy of free minting of silver into coin, usually at a fixed value ratio to gold. See **Bimetallism.**

Free-Turning. See **Free-Cutting.**

Freezing Point. See **Melting Point.**

Freezing Point Depression. Lowering of freezing point of liquid, due to the presence of substance dissolved in it. Within limits, the depression is proportional to the concentration of the dissolved substance.

Freezing Range. See **Solidification Range.**

Freiberg Amalgamation. See **Pan-Amalgamation Process.**

Fremont Impact Test. Method of measuring resistance of a material to impact, in which a falling hammer is used, with heights of drop varied, to cause bending or fracture.

Freudenberg Plate. One of a group of iron plates, suspended in a dust chamber, to aid in condensing and settling furnace fume and dust.

Frismuth Aluminum Solder. Series of tin-base solders, for aluminum, of variable composition. Lead may be present up to 29%, zinc up to 48%, copper up to 6%, aluminum up to 10%, and silver up to 6%.

Fritting. Formation of slag, or similar material, by incipient or partial fusion, without complete melting.

Frontier Alloy. Series of copper-base alloys of the **Brass** (see) and **Bronze** (see) types (Frontier Bronze Co.).

Fulguration. See **Blick.**

Full Annealing. See **Annealing.**

Full Blow. Continuing the blow in a Bessemer converter (see **Bessemer Steel**) till the drop of the flame has become pronounced.

Full Cherry Red Heat. Temperature approximating 1300° F, as judged visually.

Full Finish Black Plate. See **Full Finish Plate.**

Full Finish Plate. Steel plate, reduced either hot or cold, cleaned, annealed, and then cold-rolled to a bright finish.

Full Mill. **Rolling Mill** (see) after dressing of rolls, before the mill becomes hollow.

Full Pickling. **Black Pickling** (see) of steel sheet in which the metal is subjected to the pickling operation in the bar or breakdown stage and again after completion of the rolling.

Full Quenching. Normal type of quenching, using a single quenching medium, as opposed to **Interrupted Quenching** (see).

Full Size Test. See **Specimen Test.**

Full Weight Coil. Coil of sheet or strip metal within 12½% of coil weight as specified when ordered.

Full Yellow Heat. See **Lemon Color Heat.**

Fulminating Gold. Not a metal, but an explosive compound made by addition of ammonia to gold chloride.

Fulminating Mercury. Not a metal, but an explosive, mercuric fulminate, frequently used in percussion caps. The compound is obtained by the action of nitric acid and alcohol on metallic mercury.

Fulminating Silver. Not a metal, but an explosive compound obtained by addition of ammonia to silver oxide.

Fume. Solids carried over, or formed in, the gases of furnaces or similar apparatus, and tending to disperse colloidally in the gas.

Funnel. See **Gate.**

Furnace. Apparatus for the processing of material by use of high temperatures.

Furnace Bridge. **Bridge** (see) at the end of a furnace, acting to constrict the cross-section of the hot gas flow, and thus speed up the draft.

Furnace Cadmia. See **Furnace Cadmium.**

Furnace Cadmium. Mixture of zinc and cadmium oxides, accumulating in the flues of **Zinc** (see) smelting furnaces.

Furnace Charger. Apparatus for **Charging** (see) a furnace.

Furnace Cooling. Slow cooling of heated metal in a hot furnace in which the source of heat has been cut off.

Furnace Holding the Iron. See **Holding the Iron.**

Furnace Lining. Non-metallic **Refractory** (see) in contact with the molten

charge, capable of withstanding heat and the chemical action of the charge.
Furnace Losing the Iron. See **Losing the Iron.**
Furnace Refining. Purification of metal by treatment, while molten, in a reverberatory furnace. The term is most commonly used in connection with the refining of copper.
Fused Bath Electrolysis. Extraction of metals, such as aluminum, sodium, etc., by electrolytic decomposition of their fused salts, or of electrolytically-decomposable compounds, dissolved in materials inert under the conditions of electrolysis.
Fused Electrolyte. Molten compound capable of passing and being decomposed by electricity. Many metals are manufactured by electrolysis of such fused electrolytes, important among these being aluminum, magnesium, sodium, and calcium.
Fused Refractory. See **Electrocast Refractory.**
Fusibility. Ease or difficulty of melting a material.
Fusible Alloys. Alloys having melting points below that of tin (449°F); usually alloys of bismuth, lead, and tin, frequently also containing cadmium. Mercury may also be used as an ingredient. These alloys are frequently, but not always, known by the names of their inventors. Anatomical Alloy. Melting point 140°F. Composition: bismuth 54%, lead 17%, tin 19%, mercury 11%. Bismuth Solder. Melting point 235°F. Composition: bismuth 40%, lead 40%, tin 20%. Dalton's Alloy. Melting point 198°F. Composition: bismuth 60%, lead 25%, tin 15%. Darcet's Metal. Melting point 199°F. Composition: bismuth 50%, lead 25%, tin 25%. Guthries Alloy. Composition: bismuth 47%, lead 19%, tin 20%, cadmium 13%. Lichtenberg's Metal. Melting point 198°F. Composition: bismuth 50%, lead 30%, tin 20%. Lipowitz's Metal. Melting point 154°F. Composition: bismuth 50%, lead 26.7%, tin 13.3%, cadmium 10%. Mellotte's Alloy. Melting point 211°F. Composition: bismuth 50%, lead 19%, tin 31%. Newton's Alloy. Melting point 201°F. Composition: bismuth 50%, lead 31.2%, tin 18.8%. Onion's Alloy. See Lichtenberg's Metal. Rose's Alloy. Melting point 212°F. Composition: bismuth 50%, lead 28%, tin 22%. Wood's Metal. Melting point 154°F. Composition: bismuth 50%, lead 25%, tin 12.5%, cadmium 12.5%.
Fusible Metals. See **Fusible Alloys.**
Fusion Point. See **Melting Point.**
Fusion Welded Pipe. Pipe, usually of steel, made from strip or plate bent into shape, the edges bevelled, and these joined together by **Fusion Welding** (see).
Fusion Welding. Welding (see) in which the metals are joined, by means of molten metal, most commonly an added filler metal.

G

Ga. Chemical symbol for **Gallium** (see).
Gad. (a) Wedge of steel. (b) Ingot or bar of metal, most commonly steel.
Gadolinium (chemical symbol Gd). Element No. 64 of the periodic system; atomic weight 156.9. Rare earth metal, isolated by reduction of the anhydrous trichloride with sodium, but as yet not sufficiently pure to determine physical constants. Chemically, it is trivalent. Like all rare earth metals, it cannot be electrodeposited.
Gage. See **Metal Gage.**
Gage Length. Portion of specimen subjected to test which is measured to determine changes in dimensions.
Gage Number. See **Metal Gage.**
Gagger. In metal casting technique, steel unit for reinforcing a sand mold and holding it together, after removal of the **Cope** (see).
Galena. Chief ore of lead, chemically approximating PbS.
Galenite. See **Galena.**
Gallery Furnace. Furnace for heating retorts, used in the distillation of **Mercury** (see).
Galling. Damaging of a surface by friction against another metal.
Gallium (chemical symbol Ga). Element No. 31 of the periodic system; atomic weight 69.72. Gray, lustrous

solid, soft in texture, melting just above room temperature (86°F); boiling point around 4172°F; specific gravity 5.91 (solid), 6.09 (liquid). Chemically, it is primarily trivalent (gallic); in some compounds it is divalent (gallous); may be obtained by electrolysis of an alkaline solution of a gallium salt (theoretically, an alkali gallate solution); the salt is prepared from residues of zinc distillation. Not extensively used.

Galvanic Cell. See **Cell.**

Galvanized Iron. See **Galvanizing.**

Galvanized Rope. Steel rope made of galvanized wire. See **Galvanizing.**

Galvanized Sheet. See **Galvanizing.**

Galvanized Sheet Gage. Gage, based on the United States Steel **Gage** (see), but with each number 2.5 ounces per square foot heavier than the same United States Steel Gage number.

Galvanized Wire. See **Galvanizing.**

Galvanizing. Coating steel with zinc, principally for rustproofing purposes. In the older, standard process, sheet steel is dipped in molten zinc after preliminary cleaning, yielding a spangled surface appearance. A more recent, but less used, process involves electrodeposition of the zinc. Because of the higher position of zinc in the electromotive series, it protects the iron from oxidation.

Galvanizing Embrittlement. Slight embrittlement, due to **Aging** (see) of steel galvanized after cold forming.

Galvanizing Flux. Flux used in the **Hot Galvanizing** (see) of steel; most commonly it consists of ammonium chloride and a frothing agent.

Galvanizing Machine. Equipment for passing steel sheet through molten zinc in the **Hot Galvanizing** (see) process.

Galvanizing Pot. Vessel for holding molten zinc in **Hot Galvanizing** (see) steel. Usually, the term is understood to include the furnace for heating the metal and the **Galvanizing Machine** (see).

Galvannealing. Heating galvanized sheets, taken directly from the galvanizing pot, in a furnace at a temperature near or above the melting point of zinc, forming a complete surface coating of iron-zinc alloy.

Galvanoluminescence. Emission of feeble glow from anode in certain electrolytic cells.

Galvanoplasty. See **Electroforming.**

Galv-Weld. Lead-base **Solder** (see) of low tin content (Galv-Weld, Inc.).

Gamma Antimony. Common, metallic form of **Antimony** (see).

Gamma Arsenic. Common, metallic form of **Arsenic** (see).

Gamma Brass. Form of brass which contains 58.5–67.5% zinc; it corresponds to the compound Cu_5Zn_8; with 50.0–58.5% zinc, a mixture of beta and gamma brasses is in equilibrium.

Gamma-Forming Element. Element, manganese and nickel being the most common, which, as an alloying constituent with iron, tends to suppress the alpha phase in iron and steel and to form the gamma phase.

Gamma Iron. Allotropic modification of iron, stable in the 1670–2552°F range. It is of face-centered cubic crystal structure.

Gamma Loop. In the iron transformation diagram, a restricted area, in which the gamma form is stable; the restriction results from the presence, in solid solution, of any of the following elements: aluminum, antimony, arsenic, beryllium, chromium, columbium, germanium, molybdenum, phosphorus, silicon, tantalum, tin, titanium, tungsten, vanadium. Outside the range of compositions of this loop, there are no allotropic changes, and the alloys outside the loop are, therefore, not subject to heat treatment.

Gamma Phase. See **Gamma Iron** and **Gamma Brass.**

Gamma Rays. Extremely short waves in the ether, emitted by radium and other radioactive substances.

Gamma Ray Test. Test of the soundness of metal by photography, using **Gamma Rays** (see).

Gangue. Portion of an ore which is commercially undesirable and, therefore, must be removed before further processing.

Garrett Mill. Combination of billet mill with groups of rod mills, to yield a

Looping Mill (see) arrangement of greater-than-normal capacity.

Garrett Rod Mill. See **Garrett Mill.**

Gas. One of the three fundamental states of matter; in the gaseous state, a material has neither constant volume nor definite shape, but fills any confining vessel.

Gas Brazing. Brazing in which a gas flame is the source of heat. See also **Gas Welding.**

Gas Carburizing. Carburizing process, in which the carbon is supplied to the steel in the form of a gas introduced as such into the retort. Carbon monoxide and the methane-series hydrocarbons are the most common gas **Carburizing** (see) agents.

Gas Carburizing Machine. Equipment for gas **Carburizing** (see), including a chamber for holding the metal and a heating furnace.

Gaseous State. See **Gas.**

Gasflux. Process and equipment (Automatic Gasflux Co.) for the continuous automatic fluxing of areas being welded or brazed; the fluxing material, in vaporizable liquid form, is entrained directly in the gas used in the welding torch and thus the area being heated is always automatically under flux protection.

Gas Furnace. Furnace (see) using gas as fuel.

Gas Port. Inlet for gas in gas-fired furnace, particularly of the reverberatory type.

Gas Producer. Equipment for converting coal into gas, the incandescent coal being reacted with steam and air to yield a mixture of carbon monoxide and hydrogen, together with large amounts of nitrogen and smaller quantities of carbon dioxide and hydrocarbons.

Gassing. (a) Absorption of gas by a metal. (b) Evolution of gas from a metal, during melting operations, or on solidification. (c) See **Gassing of Copper.**

Gassing of Copper. Conversion of ductile copper into brittle metal by exposure to reducing gases. The deterioration of the metal is due to reduction of the copper oxide content of the original metal, with the resulting gases (H_2O vapor and CO_2) penetrating along the crystal boundaries, causing breakdown of cohesion between crystals.

Gas Spectrum. Absorption spectrum obtained as a result of spectroscopically analyzing the light which is passing through a gas or vapor.

Gas Welding. Welding (see) process in which a gas flame is the source of heat. Most frequently, oxygen is used in conjunction with acetylene or hydrogen; however, other fuels may be used as well, and, for economy, air is occasionally substituted for oxygen.

Gate. Inlet through which metal flows into a casting mold.

Gating. See **Gate.**

Gathered Metal. See **Laded Metal.**

Gathering. Defective condition in the rolling of steel sheet; small portions of the sheet, in contact with the roll, tend to stick to the roll, giving it an uneven surface.

Gateing. See **Gate.**

Gauge. See **Metal Gage.**

Gayley Process. Improving the air blast, in a **Blast Furnace** (see), by lowering the temperature of the air to remove the moisture in the form of ice.

G Bronze. Gun Metal (see), approximating 88% copper, 10% tin, and 2% zinc.

Gd. Chemical symbol for **Gadolinium** (see).

Ge. Chemical symbol for **Germanium** (see).

Gear Bronze. Bronze, commonly phosphor-bronze, used for casting gears, wheels, etc.

Genalloy. Series of corrosion and heat-resistant iron-nickel-chromium alloys, with 31–40% nickel, 12–21% chromium, and up to about 0.50% carbon (General Alloys Co.).

Genelite. Graphited **Bearing Metal** (see), with a matrix metal of bronze approximating 13% tin, 10% lead, and the metallic balance copper; the composite is made by powder metallurgy technique. Graphite represents about 5% of the total (General Electric Co.).

Genesee (Steel). Series of heat and

corrosion-resistant chromium and nickel-chromium steels, with chromium up to 25%, and nickel up to 65% (Symington-Gould Corp.).

Genuine Babbitt. See **Babbitt (Metal).**

Genuine Wrought Iron. **Ingot Iron** (see), with or without low-alloy additions of nickel and copper, made by A. M. Byers Co.

German Cupellation. Process of cupellation [see **Cupellation** (a)], using a large reverberatory furnace, with a removable roof; the charge is not refined in the same furnace.

Germanium (chemical symbol Ge). Element No. 32 of the periodic system; atomic weight 72.60. Gray crystalline metal, it is brittle in nature, akin to its congener, silicon, and is stable to atmospheric influences; melting point 1756°F; boiling point about 4900°F; specific gravity 5.35. Chemically, it is normally tetravalent (germanic), though it is divalent in some compounds (germanous). Germanium is isolated by reduction of the dioxide with hydrogen or carbon; commercially not used; cannot be electrodeposited.

German Process. See **Continental Process.**

German Reduction Process. Process of obtaining copper in which the ore is roasted to yield a matte of 30–40% copper content, called coarse metal, followed by another roast to 60–70% copper content, called fine metal, and a final roasting to copper oxide.

German Silver. See **Nickel Silver.**

German Silver Solder. Series of alloys, with relatively low silver content (10–20%) and varying amounts of copper, zinc, cadmium and tin; used for the usual purposes of **Silver Solder** (see).

German Steel. Steel made from charcoal iron reduced from **Bog Iron Ore** (see).

Germination. See **Stead's Brittleness.**

Gerstenhofer Furnace. Shaft furnace, for roasting ores, containing shelves, the ore falling from one shelf to another as **Roasting** (see) proceeds.

Getter. Metal or alloy, such as barium and/or magnesium, used in a closed space, most usually in radio tubes, to increase vacuum by chemical reaction and physical adsorption of gases. The getter material is usually flashed, i.e., converted to vapor, by induction heating. In this form, it reacts readily with residual gases.

Ghost. (a) In spectroscopy, light line or band of a metal constituent, relatively difficult to discern because of the predominance of other, darker lines. (b) Line or band structure appearing in hot-rolled or hot-worked metal, as a result of segregates and blow-holes extended in the direction of the rolling or working.

Ghost Band. See **Ghost** (b).

Ghost Line. See **Ghost** (b).

Ghost Structure. See **Ghost** (b).

Gilding. Coating with gold or an alloy having the appearance of gold; most commonly laying thin layers of the gold or alloy on other materials.

Gilding Metal. High-copper red **Brass** (see), with 90–97% copper and zinc as the remainder, used for jewelry and cartridge cap fabrication.

Girod Furnace. **Direct Arc Furnace** (see) in which an electrode is brought through the roof of the furnace, whereas steel electrodes are set into the heart to make contact with the furnace charge.

Gl. Chemical symbol for glucinum (see **Beryllium**).

Glance Cobalt. Non-metallic mineral, cobaltite, a cobalt sulfarsenide, approximating CoAsS in chemical composition.

Glance Copper. Non-metallic mineral, chalcocite, approximating Cu_2S in chemical composition.

Globular Carbide. See **Granular Pearlite.**

Globular Cementite. See **Spheroidal Cementite.**

Globular Pearlite. See **Granular Pearlite.**

Glucinum. See **Beryllium.**

Glyco. Lead-base alloys, with up to 22% antimony and up to 8% tin (Joseph T. Ryerson & Sons).

Godfrey Furnace. **Roasting** (see) furnace, having an annular hearth.

Gohi Iron. Copper-bearing iron (Newport Rolling Mill Co.), very low in impurities and in carbon (.02% maximum), containing about 0.25% copper.

Gold (chemical symbol Au). Element No. 79 of the periodic system; atomic weight 197.2. It is of distinctive yellow color, soft, extraordinarily malleable and ductile, and immune to most corrosive influences and chemicals; melting point 1945°F; boiling point about 4910°F; specific gravity 19.32. Chemically, gold is principally monovalent (aurous) and trivalent (auric). It occurs native, and is extracted from ores by the cyanide process or by amalgamation. Used as a monetary standard and for jewelry. Gold electroplating is commercially accomplished by use of either a double alkali cyanide or an auric chloride-hydrochloric acid bath.

Gold Amalgam. Alloy of gold and mercury. See **Amalgam.**

Gold-Beaters' Mold. Pack of partially beaten gold foils with layers of goldbeaters' skin, made from cow caecum, between each layer of **Foil** (see), for beating into gold leaf.

Gold Beating. Process of converting gold sheet into fine gold foil by a hammering process. Commonly, the gold sheets are separated by vellum or goldbeaters' skin during the hammering operations. See also **Gold-Beaters' Mold.**

Gold Bronze. Bronze (see) of goldlike color; a typical composition contains about 7% tin and 3% zinc.

Gold Dust. Gold, in the form of fine particles, particularly as obtained by placer mining.

Gold-Filled. Relatively heavy gold plating or cladding on base metal. It is extensively used in jewelry.

Gold Foil. See **Foil.**

Gold Latten. Gold, in very thin foil form, or thin base metal coated with gold.

Gold Leaf. Gold beaten or rolled extremely thin, used for gilding purposes. The term is frequently used to describe imitation gold leaf, commonly made of brass.

Gold Paint. Paint pigmented with "gold powder," really a **Bronze Powder** (see).

Gold Plate. Product of **Gold Plating** (see).

Gold Plating. Coating metal objects with gold; the object to be coated is made cathode (negative electrode) in an electrolytic bath, containing a decomposable gold salt, usually a double alkali cyanide. Either gold metal, or insoluble carbon, is used as the positive electrode (anode).

Gold Powder. See **Bronze Powder.** It does not contain gold.

Goldschmidt Process. (a) Process for removing the tin coating from scrap tin plate (see **Detinning**) by use of chlorine, resulting in the formation of liquid tin tetrachloride, while the iron remains unattacked. (b) See **Thermit.**

Gold Shell. See **Abyssinian Gold.**

Gold Solder. Gold-base alloys, used for joining gold, usually containing substantial proportions of silver and copper. Zinc is optional, usually in lesser quantities. The quality of the solder, as a rule, is made compatible with the carat rating of the metal which is being soldered, in order to achieve color matching.

Gong Metal. Copper-tin alloy, for the fabrication of gongs, commonly containing about 22% tin.

Good Roasting. Complete **Roasting** (see) of an ore, to the limit of oxidizability.

Gothic. Square-like cross-section of metal bars, with convex sides and highly-rounded corners.

Gothic Pass. Rolling mill, in which the grooves in the rolls are curved or rounded, resulting in the rolled metal being circular or elliptical in shape.

Government Bronze. See **Gun Metal.**

Grac. Lead-base anti-friction alloy, hardened with about 10% tin, and small quantities of graphite, copper, antimony, arsenic, and either nickel or cadmium (Graphitized Alloys Corp.).

Grade. Referring to steel, grade indicates divisions within a **Type** (see), based on chemical or physical differences.

Grade "A" Nickel. See **"A" Nickel.**
Grade "D" Nickel. See **"D" Nickel.**
Grade "E" Nickel. See **"E" Nickel.**
Grade #1 Tin. See **Chempure Tin.**
Grade #2 Tin. Commercially available second-grade tin of about 99.9% purity; used for type metals, tin coatings, etc.
Grade #3 Tin. Commercially available lowest grade virgin tin of about 99% purity; used principally for alloying, in solders, etc.
Grain. Imperfect metallic crystals. See also **Allotrimorphic Crystal.**
Grainal. Series of **Process Alloys** (see), manufactured by Vanadium Corp. of America, for addition to steel; contain 15–20% titanium, 10–20% aluminum, and either 13–25% vanadium or 6% zirconium. A few pounds of Grainal, added per ton of steel, increase hardness, tensile and impact strength, and ductility.
Grain Boundary. Bounding surface between crystals. When alloys yield new phases (as in cooling), grain boundaries are the preferred location for the appearance of the new phase. Certain deteriorations, such as season cracking and caustic embrittlement, occur almost exclusively at grain boundaries.
Grain Boundary Network. See **Network Structure.**
Grain Gold. Gold (see), in granular form, resulting from heating operations.
Grain Growth. Increase in metallic crystal size as annealing temperature is raised; growth occurs by invasion of crystal areas by other crystals. A common error is to regard fatigue failure as being caused by crystal growth.
Graining. Pulverizing metal by rapid stirring while freezing.
Grain Refinement. Heating hardenable steel to the lower critical range, Ac_1, at which temperature pearlite grains are transformed into small-size crystals of austenite. Increase in temperature from this point will cause growth of the grain size. More generally, grain refinement can be achieved on cast metals by hot-working.
Grain Rolls. Cast iron rolls, for rolling metal, cast in sand molds.
Grain Size. Average diameter of grains in the metal under consideration, or, alternatively, the number of grains per unit area. Since increase in grain size is paralleled by lower ductility and impact resistance, the question of general grain size is of great significance. The addition of certain metals affects grain size, for example vanadium and aluminum tend to give steel a fine grain. The ASTM (American Society for Testing Materials) has set up a grain size standard for steels, and the **McQuaid-Ehn Test** (see) has been developed as a method of measurement.
Grain Size Test. Determination of grain size in metal, especially after a special treatment for the purpose. See also **McQuaid-Ehn Test.**
Grain Structure. See **Grain.**
Grain Tin. (a) Non-metallic tin oxide ore in the form of grains or crystals. (b) Highest-grade of metallic **Tin** (see) as obtained by charcoal reduction.
Gramix. Graphited (Bearing) Metal (see), with a matrix metal of copper-tin; the composite is made by powder metallurgy technique (United States Graphite Co.).
Gram-Mol. See **Mol.**
Grampus. Tongs for manipulation of billets in **Bloomery** (see) operations.
Granite City (High Tensile Steel). Low-alloy, high-strength steel, with carbon about 0.15%, silicon about 0.20%, manganese about 0.80%, chromium about 0.15%, and copper about 0.25% (Granite City Steel Co.).
Granodized Steel. See **Granodizing.**
Granodizing. Forming a rust-resistant coating on iron or steel by immersion in a hot solution of acid zinc phosphate. The process can also be used on zinc or cadmium.
Granular Fracture. See **Fracture.**
Granular Pearlite. Structure, resulting from ordinary pearlite in steel when it is annealed for long periods below the lower critical point. The cementite forms granules or spheroids within the ferrite, breaking up the normal lamel-

lar structure. The structure is also known as Spheroidite.

Granulated Aluminum. Aluminum Powder (see), which has not been coated with a lubricant. This term is also used for relatively large masses of aluminum, such as are used for addition of aluminum in alloying with other metals.

Granulated Steel. Steel made from pig iron by a process involving preliminary granulation of the pig iron.

Granulating Machine. Device for transforming liquid metal into granular or fine-size particles. The most common form utilizes a rapidly revolving disc which scatters the liquid metal centrifugally in tiny particle form.

Granulating Pit. Large pit, partially filled with water, into which hot slag, by-product of blast furnace operations, is dumped, usually with a stream of water against the stream of slag.

Granulation. (a) Production of coarse metal particles in roughly-outlined shape, by any method, such as pouring molten metal through a screen, by strong agitation of a fused metal during solidification, etc. (b) Formation of crystals, during the solidification of an ingot, after dendritic formation has ceased.

Graphalloy. Composition (Graphite Metallizing Corp.), consisting of graphite, impregnated at high pressure with a metal, such as copper, lead, silver, or cadmium. It is self-lubricating, and, therefore, used for bearings and similar purposes.

Graphic Gold. Crystals of naturally-occurring sylvanite ore, a mixed gold-silver telluride, occurring in regularity so as to give the appearance of written symbols; also known as graphic tellurium.

Graphic Tellurium. See **Graphic Gold.**

Graphite. Soft, black, allotropic form of carbon, found in many parts of the world and also made artificially from coke in the electric furnace. When elementary carbon separates out from iron, as in certain cast irons, its normal form is graphite.

Graphited (Bearing) Metal. Bearing metal composition in which graphite has been incorporated. The composition is sometimes manufactured by pressure impregnation, or electrolytically.

Graphitic Carbon. Graphite, in the form of tiny flakes, distributed through the metal, which forms in gray cast iron (see **Pig Iron**) during cooling, as a result of the precipitation of some of the dissolved carbon.

Graphitizing. Process of annealing which converts the combined carbon of white cast iron into free carbon, in nodular form. With the carbon in this form, the metal is no longer brittle, and is known as **Malleable Cast Iron** (see).

Grapho Babbitt Metal. Babbitt (see) metal impregnated with graphite, for use in bearings subject to high pressures or velocities (Lehigh Babbitt Co.).

Gravimetric Analysis. Quantitative Analysis (see) in which the desired constituent is determined by weight, usually in the form of one of its compounds.

Gravity. See **Specific Gravity.**

Gravity Die-Casting. See **Die-Casting.**

Gray Antimony. Not a metal; the ore, stibnite, approximating the formula Sb_2S_3.

Gray Cast Iron. See **Cast Iron** and **Pig Iron.**

Gray Coating. Defect in galvanized steel sheet, in which areas have a dull gray color and are devoid of spangles or have defective spangling. See **Galvanizing.**

Gray Cobalt. Non-metallic ore of cobalt, smaltite, chemically approximating cobalt diarsenide, $CoAs_2$.

Gray Copper. Non-metallic ore of copper and antimony, tetrahedrite, chemically approximating Cu_3SbS_3.

Gray Cut Cobalt Steel. High-speed alloy tool steel (Vanadium Alloys Steel Co.), with about 20% tungsten, 12% cobalt, 4% chromium, 1.5% vanadium, and 0.6% molybdenum.

Gray Iron. (a) See **Gray Cast Iron.** (b) See **Gray Coating.**

Gray Manganese. Non-metallic ore of manganese, manganite, hydrated manganese oxide, chemically approximating $Mn_2O_3.H_2O$.

Gray Metal. Non-metallic shale of gray color.

Gray Mill. Special type of **Rolling Mill** (see) used in rolling beams and H sections.

Gray Pig Iron. See **Cast Iron** and **Pig Iron.**

Gray Sheet. See **Gray Coating.**

Gray Slag. Mixture of lead, lead compounds, lime, and cinders, by-product of the reverberatory smelting of lead. Lead is normally recovered from the slag by crushing the material and then subjecting it to smelting in a blast furnace.

Gray Top. Narrow band on the edge of **Tin Plate** (see), of dull color, caused by an inadequate thickness of tin coating.

Grease Mark. Defect in the appearance of **Tin Plate** (see), due to excess of palm oil, in streaks, remaining on the plate after withdrawal from the molten tin and oil.

Grease Pot. Pot or tank in **Tin Plating** (see), holding palm oil for fluxing the metal plate.

Grease Spot. Area on black plate on which an accidental drop of oil or grease, carbonized during annealing heating, prevents proper coating in galvanizing or tin coating.

Greasy Gold. See **Fine Gold.**

Greasy Streaky Finish. Special finish on terne plate made by the combination process (see **Combination Process**), created by grooved exit rolls in the second coating machine.

Greaves-Etchells Furnace. Direct **Arc Furnace** (see), utilizing three-phase current, one of the phases being buried in the hearth of the furnace.

Green Casting. See **Green Sand Casting.**

Greene Furnace. Three-phase direct arc electric furnace, generally analogous to the **Heroult Furnace** (see).

Green Gold. Gold-base alloys with about 25–35% silver and small percentages of other elements for the lower gold-content alloys.

Green Hole. Improperly made furnace **Tap Hole** (see), resulting in metal leakage.

Green Plate. Defective tin plate, with spots of dull surface and **Scruff** (see).

Green Sand Casting. Metal cast in sand mold which has not been subjected to any baking or drying.

Green Sand Core. In metal casting technique, a sand core which has not been subjected to any drying. See **Core** (b).

Green Sand Mold. See **Green Sand Casting.**

Grid. Perforated or grating-like lead plate in a **Storage Battery** (see), for holding chemically active electrolytic material.

Grilio Furnace. Muffle Furnace (see), continually fed by mechanical means.

Grinders' Asthma. Lung disease (phthisis), resulting from prolonged inhaling of metal and grinding dust in grinding metals.

Grinders' Rot. See **Grinders' Asthma.**

Grinding Crack. Crack in metal caused by faulty grinding operations, or by grinding improperly heat-treated metal.

Grip. Clamp, on tension testing machine, for holding the specimen.

Gronwall-Dixon Furnace. See **Electrometals Furnace.**

Gross Calorific Value. Heat produced on complete combustion of a gram of fuel under specified conditions, any water resulting from the combustion being in the liquid state (in contradistinction to net calorific value).

Gross Heating Value. See **Gross Calorific Value.**

Ground Coat. In processes which use more than one coat of material, such as enameling, the coat in direct contact with the metal.

Growth. (a) In powder metallurgy, any increase in dimensions occurring during sintering. (b) In massive metal, any increase in dimensions due to physical or chemical changes; such growth is usually conducive to distortion.

Grunter. Curved rod, used in the handling of crucibles.

G.S.G. See **Galvanized Sheet Gage.**

Guaranteed Case Carburizing Quality (Steel Plate). Steel plate ordered or sold on the basis of specifications governing suitability for carburizing and heat treatment, or of specified grain structure after such carburizing.

Guaranteed Hardenability. Specification, in the ordering or selling of steel, requiring specified hardness after specified heat treatment.

Guaranteed Segregation Limits. Method of taking samples for **Check Analysis** (see) which is more restrictive than the routine method for taking such samples.

Guard Plate. Plate covering the **Tap Hole** (see) of a furnace.

Guide. Device for holding the metal in the proper position, during rolling.

Guide Cage. Steel frame, bolted to the mill housing, which holds the **Guide** (see) on a rolling mill.

Guide Mark. Flaw on rolled metal in the form of a line, caused by scratching against a guide, as a result of incorrect rolling.

Guide Mill. Small rolling mill, with device for supporting the metal in the proper position and at the proper angle while being rolled.

Guide Round. Round (see), formed by preliminary reduction to a round-cornered square, and then formed into an oval, and finally to a round form by appropriate passes.

Guillery Cup Test. Form of **Cupping Test** (see).

Guillery Hardness Test. Test in which a standard-sized ball is impressed into the surface of the specimen, under spring pressure, and the size of the resulting indentation measured to indicate **Hardness** (see).

Guillery Impact Test. Impact Test (see) in which the loss of energy of a rotating hammer is used as the measure of impact strength of the sample being tested.

Guillotine. Machine for breaking metal by means of a falling weight.

Guillotine Shears. Metal shears, in which a cutting blade drops down and shears the metal off.

Guinea Gold. (a) Gold of 22 **Carat** (see) value, used for the coining of British Guineas. (b) Red **Brass** (see) with about 12% zinc, used for cheap jewelry.

Gun Iron. Fine-grained cast iron of relatively low total carbon and sulfur content.

Gunite. High-Test (see) cast iron (Gunite Foundries Corp.), with about 3% total carbon and 2% silicon.

Gun Metal. Copper-base alloy, containing either 10% tin and 2% zinc or 8% tin and 4% zinc. Fine-grained, of good casting qualities, excellent strength, toughness, and chemical resistance to mildly corrosive influences, such as sea water.

Guthries Alloy (or Metal). See **Fusible Alloys.**

Gutzkow's Process. Process for parting **Bullion** (see), when large quantities of copper are present; the metal is treated with sulfuric acid, the resulting silver sulfate being reduced with charcoal or ferrous sulfate.

H

H. Chemical symbol for **Hydrogen** (see).

Haas Furnace. Multiple-muffle furnace, resembling the **MacDougall Furnace** (see), with the hearths separated by flues.

Hadfield (Manganese) Steel. See **Manganese Steel.**

Haenisch and Schroeder Process. Process of recovering sulfur dioxide from waste gas, in which the sulfur dioxide is obtained in liquefied form.

Hafnium (chemical symbol Hf). Element No. 72 of the periodic system; atomic weight 178.6. It is a steel-gray metal and may be assumed to be ductile when pure, akin to its congener, zirconium; melting and boiling points are believed to be about 4040°F and 5800°F, respectively; specific gravity 13.3. Chemically, it is tetravalent. Pure hafnium can be isolated by decomposition of the anhydrous tetraiodide on a heated tungsten filament, the iodide being formed from oxide, obtained by repeated fractional crystallizations from related zirconium compounds. It probably cannot be electrodeposited, as is true of zirconium.

Hähner Furnace. Shaft furnace, operating continuously, for roasting mercury ores, into which ore and charcoal are charged in alternate layers.

Hair Crack. Very narrow crack.

Hairline Crack. Small crack occurring inside large forgings or rolled metal or on the surface of smaller units after quenching or grinding.

Hair Plate. See **Bloomery.**

Hair Seam. Very fine **Seam** (see).

Halcomb (Steel). Series of low-alloy and high-alloy steels, including **Stainless Steel** (see), manufactured by Halcomb Steel Co.

Half-and-Half (Solder). Soft **Solder** (see), composed of approximately equal parts of lead and tin.

Half Bloom. Mass of puddled iron, prior to squeezing. See **Puddling.**

Half-Cup Fracture. See **Fracture.**

Half Flask. Unit which, together with its mate, functions as a complete flask [see **Flask** (a)].

Half-Hard Temper. In strip or sheet, cold-rolled metal suitable for bending, especially across the grain.

Half-Life. Period of time in which the radioactivity of a substance drops to half its original value.

Hall Effect. Potential difference created between two edges of a metal strip carrying a longitudinally flowing electric current, when the plane of the strip is placed perpendicularly across a magnetic field.

Hall Furnace. Modification of the **Wethey Furnace** (see).

Hall Process. (a) Standard method for manufacturing **Aluminum** (see); the purified oxide is dissolved in fused cryolite and then electrolyzed. (b) See **Pig Boiling Process.**

Halo. Term used, in X-ray pattern terminology, to designate an indistinct, circular band of indefinite and diffuse limits. The presence of a halo indicates the absence of orientation.

Hammered Bloom Iron. See **Busheled Iron.**

Hammer Forging. Forging by steam or pneumatic hammers, with deformation accomplished by impact.

Hammer Hardening. Hardening of metal, through cold-working, by mechanical hammering.

Hammering. See **Hammer Forging.**

Hammer Refining. Hammering or otherwise mechanically working a metal, particularly iron and steel, previously heated to a temperature conducive to the growth of large crystals, in order to obtain small grain size; frequently used in welding, since certain welding zones show an intense growth of grain.

Hammer Scale. Iron oxide formed on the surface of steel during hot-working operations.

Hammer Test. See **Drop Test** (a).

Hammer Welding. See **Forge Welding.**

Hampden Steel. High-carbon chrome tool steel (Carpenter Steel Co.), with about 12% chromium, about 2.1% carbon, and fractional percentages of nickel and manganese.

Hand Mill. Rolling mill in which all manipulation of the ingots, bars, or sheets is done by hand.

Hand Puddling. See **Puddling.**

Hand Rolling. See **Hot-Rolling.**

Hand Round. Round (see) formed by preliminary reduction in diamond-shaped passes and subsequent repeated hand-feeding into an oval or round pass until round form is achieved.

Handy Silver Solder. Series of **Silver Solders** (see) most commonly containing zinc, copper, and cadmium in addition to silver, manufactured by Handy & Harman.

Hanging. Faulty operation of a blast furnace whereby a pasty mass of the raw materials adheres to the wall of the furnace, causing blocking of inflowing gas and downflowing raw materials.

Hanging Guide. Guide (see) held against the underside of the bottom roll, to aid in removal of metal from a rolling mill.

Hanging Test. Tensile test applied to wire. See **Tensile Strength.**

Hanover White Metal. Britannia Metal (see) approximating 8% antimony and 5% copper, with tin as the remainder.

Hansgirg Process. Process for manufacturing magnesium; magnesium oxide, mixed with carbon, is heated in an electric furnace; the resulting mixed vapors of magnesium metal and carbon monoxide are cooled rapidly by means of refrigerated hydrogen, or natural gas.

Hanson - Van Winkle - Munning Electrolytic Pickling Process. Process of **Electrolytic Pickling** (see); steel is first treated, as cathode, in 10–20% sulfuric acid, followed by treatment, as anode, in a 50% sulfuric acid solution, to remove smut; current density is about 100 amperes per square foot in both steps.

Hard Bearing Bronze. See **Hard Gear Bronze.**

Hard Bronze. Bronze used for tough, dense castings, usually containing about 88% copper, with the residue tin and zinc or lead.

Hard Chill. See **Chilled Rolls.**

Hard Chromium (Plating). See **Chromium Plating.**

Hard-Drawing. Drawing metal wire through a **Draw Plate** (see), to reduce cross-section and to increase tensile strength.

Hard-Drawn. Wire or tubing drawn to high tensile strength by a high degree of cold work.

Hard-Drawn Finish. Finish on tubing or similar drawn material, resulting from the final cold-drawing.

Hard-Drawn Spring Wire. Steel wire, like **Music Wire** (see), but of lower grade, being usually made of basic open-hearth steel.

Hard - Drawn Tubing. See **Hard-Drawn Finish.**

Hardenability. Relative ease of fully **Hardening** (see) a metal, usually steel, by appropriate heat treatment.

Hardenability Curve. Curve demonstrating the change in hardness of steel, after appropriate heat treatment, from the outside surface to the interior.

Hardenability Test. Test to determine hardening characteristics of metal, particularly steel, when subjected to appropriate heat treatment.

Hardened Steel. Steel which has been subjected to a **Hardening** (see) process.

Hardening. Sequence of temperature-time treatment of metal which results in its greater hardness.

Hardenite. Martensite (see) containing about 0.9% carbon.

Hard-Facing. Surfacing a metal, usually by welding, with a layer of a hard, abrasion-resistant metal or alloy.

Hard Gear Bronze. Bronze used for casting gears, wheels, etc., with 13–15% tin, and low percentages of zinc, nickel, and phosphorus.

Hard Head. Tin-iron alloy, as obtained in the smelting of slags from the first treatment of tin concentrates.

Hardite. Series of nickel-chromium-iron heat-resistant alloys, manufactured by Hardite Metals, Inc. Nickel may vary from as little as 6% to as high as 93%, with 12–25% chromium, 2–5% silicon, and normally about 2% manganese.

Hard Lead. See **Antimonial Lead.**

Hard Metal. (a) Alloy of about two parts copper and one part tin. (b) Impure copper matte, containing relatively large quantities of tin.

Hardness. Resistance of a substance to cutting, abrasion, or indentation. The hardnesses of metals will differ somewhat with the specific apparatus and technique of measuring. For details concerning the various types of apparatus used in measuring hardness, see **Brinnel Hardness, Rockwell Hardness, Vickers Hardness,** and **Scleroscope Hardness.**

Hardness Penetration. See **Penetration Hardness.**

Hardness Scale. See **Moh's Scale.**

Hardness Test. See **Hardness.**

Hard Pale Solder. Soft **Solder** (see), containing about two parts of tin and one part of lead.

Hard Plating. See **Hard Chromium (Plating).**

Hard Solder. See **Solder.**

Hard Steel. See **Hardened Steel.**

Hard Temper. In strip or sheet, stiff and springy, cold-rolled metal, not suitable for bending in any direction.

Hardtem (Steel). Low-alloy, oil-hardening die steel, with fractional percentages of chromium, nickel, vanadium, and molybdenum (Heppenstall Co.).

Hardware Bronze. Lead-bearing brass (not bronze) containing about 13%

zinc and 1.8% lead; used in hardware manufacture.

Hard Zinc. Iron-zinc alloy which collects as a dross in **Galvanizing** (see) pots, after long operation.

Hargus (Steel). Non-Deforming Steel (see), with about 1.00% manganese and 0.30% chromium; carbon approximates. 1.00% (Ziv Steel & Wire Co.).

Harris Process. Process of removing arsenic, antimony, tin, and zinc from virgin or secondary lead; the metal is contacted with molten caustic soda and salt; the impurities oxidize and the resultant oxides are dissolved in the caustic.

Hartmann Lines. See **Lüder's Lines.**

Harveyizing. Hardening the surface of flat steel by carburization at high temperature, followed by rapid chilling with a jet of water, in order to obtain a very hard case.

Harvey Process. (a) Manufacturing magnesium by electrolysis of a suspension of magnesium in a fused bath of mixed fluorides of magnesium, barium, and sodium. (b) See **Harveyizing.**

Hascrome. High-chromium iron alloy containing about 4% manganese, manufactured by Haynes Stellite Co.; primarily used for hard-facing welding rods, the material being self-hardening after deposition by welding torch or rod.

Hasenclever-Helbig Furnace. Roasting furnace, with a sharply inclined hearth, the ore rolling down the hearth during **Roasting** (see).

Hastelloy. Series of nickel-base alloys, made by Haynes Stellite Co., with varying amounts of molybdenum, chromium, manganese, copper, silicon, and iron; remarkably resistant to chemicals, including hydrochloric acid, and to high temperatures; one grade, Hastelloy C, is even resistant to wet chlorine.

Haupt Furnace. Furnace of the Silesian type (see **Silesian Furnace**), heated by gas, the air being preheated by means of recuperative chambers.

Hauzeur Furnace. Zinc distillation furnace, of dual type, in which waste heat from one group of retorts is used for heating the second group.

Hayden Process. Electrolytic refining of copper, by the **Series System** (see), in which a single anode is located at one end of the tank, and a single cathode at the other, the raw unrefined copper, in plate or slab form, being placed parallel to the electrodes.

Haynes Metal. Series of cobalt-chromium-iron and cobalt-tungsten-chromium alloys; the former group contains 5–25% cobalt and 20–30% chromium; the latter group contains 45–62% cobalt, and 28–40% tungsten, with chromium as the remainder.

Haynes Stellite. See **Stellite.**

Haystellite. Cast tungsten carbide, made by Haynes Stellite Co., used for cutting tools and similar purposes requiring great hardness and wear-resistance.

H Beam. Steel beam of H cross-sectional shape, rolled to standard dimensions.

Head. (a) See **Riser.** (b) See **Whirl.**

Header. (a) See **Riser.** (b) Mass of charcoal, sand, or cinder, at end of zinc pot in the galvanizing of wire, used to remove excess zinc from the wire.

Heading. In lead annealing, a mass of sand or crushed coal, located at the end of the lead bath, into which the wire is led directly if galvanizing or other finishing is not intended.

Heap Matte. Matte resulting from **Heap Roasting** (see).

Heap Roasting. Roasting of sulfide ores, under oxidizing conditions, with the ore heaped on the ground in masses shaped like pyramids cut off at the top. Ignition of the ore is generally started by a layer of wood at the bottom of the heap.

Heap Sand. Sand used in casting of metal to fill in behind the **Facing Sand** (see).

Hearth. Zone in which a metallurgical reaction is conducted, at elevated temperature, frequently with the metal in the molten state. See also **Blast Furnace** and **Open-Hearth Process.**

Hearth Chill. See **Chill** (c).

Hearth Ends. Lead ore, in unreduced form, blown out of a roasting or reducing furnace.

Hearth Roasting. See **Reverberatory Roasting.**

Heat. Cycle of melting operations in a furnace, starting with the charging of raw materials, and ending with the tapping of molten metal.

Heat of Adsorption. Quantity of heat evolved in adsorption of a definite quantity of gas on a bare surface.

Heat Blocking. See **Pigging Up.**

Heat of Combustion. Amount of heat evolved when one gram-molecule of a substance is burned in oxygen at constant volume.

Heat Conductivity. See **Thermal Conductivity.**

Heat Convection. See **Thermal Convection.**

Heat of Formation. Quantity of heat evolved or absorbed in the formation of one mol of a chemical compound. No matter what the sequence of reactions involved, the heat of formation is always identical for any compound at any given temperature and pressure (Hess' Law).

Heat of Fusion. See **Latent Heat.**

Heating Back. Zone behind a forge, where the air blast is preheated before entering the forge hearth.

Heating Curve. Time-temperature curve for slow heating of a metal sample. See also **Cooling Curve.**

Heating Lag. Tendency for physical changes to lag behind temperature changes during heating.

Heating Value. See **Gross Calorific Value** and **Net Calorific Value.**

Heating Zone. Zone in any **Continuous Furnace** (see) for heat-treatment, in which the metal is brought to, and held at, the desired temperature.

Heat Radiation. See **Thermal Radiation.**

Heat of Reaction. Quantity of heat evolved or absorbed during a chemical reaction.

Heat Resistant Alloy. Nickel-chromium alloy, usually with iron as a third constituent, used for high-temperature equipment and also for acid resistance. High-chromium iron alloys (about 30% chromium) are in the same category.

Heat Resistant Steel. See **Heat Resistant Alloy.**

Heat Tinting. Surface-coloring metal, by means of surface oxidation in air. Heat tinting can occasionally be used as a method for studying the detailed micro-structure of metal, after polishing, since different constituents oxidize at different rates.

Heat of Transformation. Quantity of heat evolved or absorbed during a metallurgical **Transformation** (see).

Heat Treating. See **Heat Treatment.**

Heat Treatment. Altering the properties of a metal by subjecting it to a sequence of temperature changes, time of retention at specific temperatures being as important as the temperature itself. Heat treatment usually markedly affects strength, hardness, ductility, malleability, and similar properties of both metals and their alloys.

Heat of Vaporization. See **Latent Heat.**

Heat Working. Operating open-hearth furnace, including additions of raw materials, tapping, etc.

Heavy Burden. See **Burden.**

Heavy Case. Case (see) on metal with a total penetration exceeding 0.06 inches; usually accomplished by pack carburizing.

Heavy Cold-Rolled Sheet. Cold-rolled steel sheet of 0.0625 inch thickness or more.

Heavy Gold. Gold, as found in nature, in relatively large particle size.

Heavy Goods. Standard steel products of relatively large size and weight, such as structural steel, etc.

Heavy Iron. Defective spangle on galvanized steel (see **Galvanizing**) caused by the presence of excess iron dissolved in the molten zinc bath.

Heavy Metal. Crystals of iron-tin alloy suspended in the molten tin bath, used in tin plating (see **Tin Plate**). See also **Heavy Metals.**

Heavy Metals. Chemical term designating metals which can be precipitated analytically, in acid solution, by means of hydrogen sulfide. The more common among the metals of this group are: lead, silver, gold, mercury, bismuth, copper, cadmium, arsenic, antimony, and tin.

Heavy Products. See **Heavy Goods.**

Hegeler Furnace. Gas-fired variation of the **MacDougall Furnace** (see) in which there are two long compartments, with seven or more hearths, and raking is accomplished intermittently by semi-mechanical means; has been extensively used in the smelting of zinc blende ores.

Heliarc Welding. Electric **Arc Welding** (see) in an atmosphere of inert helium gas.

Heller's Steel. Series of low-alloy and high-alloy tool steels, manufactured by Heller Bros. Co.

Helvetium. Element No. 85 of the periodic system. Its presence has been proven among the ultimate decomposition products of radium, but nothing is known of its characteristics. Theoretically, it is a non-metal, resembling iodine.

Hematite. Most important ore of iron, essentially Fe_2O_3. Most American iron and steel production is from hematite.

Hematite Pig (Iron). Pig Iron (see) obtained from hematite iron ore.

Henderson Process.. Chloridizing Roasting (see) copper sulfide ores, forming copper chlorides, which are subsequently leached from the mass and recovered.

Heppenstall (Steel). Series of low-alloy steels, manufactured by Heppenstall Co.

Heptavalent. Having a chemical **Valence** (see) of seven.

Herbert Pendulum (Hardness) Test. Method of measuring hardness in which a mass of metal, in arch shape, is balanced on a metal ball. By adding or adjusting weights on the arched unit, the center of gravity can be raised or lowered above the ball. Various types of hardness can be read from the amplitudes and times of swing under varying conditions.

Herculoy. Series of corrosion-resistant silicon bronzes, of about 3.3% silicon and about 0.5% tin content; extensively used for hardware, tanks, and similar structural purposes (Revere Copper & Brass Co.).

Heroult Furnace. Direct Arc Furnace (see), commonly of the three-phase current type with the electrodes spaced sufficiently far apart to prevent arcking between them; extensively used in the manufacture of electric furnace steel.

Heroult Process. See **Hall Process.**

Herrenschmidt Process. Treatment of antimony ore with coal in a cupola or blast furnace; the roasting is continued to the point of oxide formation, and the volatile trioxide, after leaving the furnace, condensed in a series of condensing chambers.

Herreshoff Furnace. Modern type of MacDougall vertical cylindrical roasting furnace, mechanically rabbled. See **MacDougall Furnace.**

Hessite. Natural mineral silver telluride, approximating Ag_2Te in composition.

Hess' Law. See **Heat of Formation.**

Heterogeneous. Having varying characteristics; non-uniform.

Heusler Alloys. Alloys containing 4–15% aluminum, 54–76% copper, and 16–30% manganese. The alloy series is remarkable for being ferromagnetic, i.e., attracted by magnet.

Hexagonal (Crystal System). Crystal form having three equal axes, lying equidistant in a plane and intersecting a fourth axis perpendicular to that plane.

Hexavalent. Having a chemical **Valence** (see) of six.

Heyn's Reagent. Solution of ammonium copper chloride, used in ferrous metallography to reveal ferrite.

Hf. Chemical symbol for **Hafnium** (see).

Hg. Chemical symbol for **Mercury** (see).

Hidden Maximum System. See **Peritectic System.**

Hiduminium. Series of British aluminum-base alloys of varying composition.

High. Polished, as applied to surface condition of metal sheet or strip.

High-Alloy Roll. Roll, for rolling metal, composed of cast iron containing relatively large quantities of alloying elements.

High-Alloy Steel. See **Alloy Steel.**

High-Aluminum Ferro-Carbon-Titanium. See **Ferro-Carbon-Titanium.**

High Brass. See **Brass.**

High-Carbon Cast Steel. Cast steel of more than 0.35% carbon content. See **Cast Steel** (a).

High-Carbon Ferro-Carbon-Titanium. See **Ferro-Carbon-Titanium.**

High-Carbon Ferro-Chromium. See **Ferro-Chromium.**

High-Carbon Ferro-Manganese. See **Ferro-Manganese.**

High-Carbon High-Chromium Die Steel. Die Steel (see) with carbon approximately 1.2%, and chromium exceeding 12%.

High-Carbon Low-Tungsten Steel. Tool Steel (q.v.) used for small hand and finishing tools, with carbon in the 0.90–1.30% range, and tungsten ranging from 1.0 to 2.5%. Fractional percentages of chromium and vanadium are optional, in addition to normal content of about 0.30% each of manganese and silicon.

High-Carbon Steel. See **Hypereutectoid Steel.**

High-Carbon Strip. Strip Steel (see), of high-carbon composition, furnished either in the as-rolled condition or after heat treatment.

High-Carbon Wire. Steel wire of moderately high carbon composition, such as rope wire, piano wire, etc.

High-Conductivity Copper. See **Conductivity Copper.**

Highflex. Bimetal (see) used in thermal equipment (H. A. Wilson Co.).

High-Frequency Furnace. See **Coreless (Induction) Furnace.**

High-Frequency Induction Furnace. see **Coreless (Induction) Furnace.**

High-Grade Zinc. Metallic zinc with total impurities not over 0.10%. See **Zinc.**

High-Lead Bronze. Tin bronzes containing relatively high percentages (up to 30%) of lead dispersed in the bronze matrix; are used primarily for bearings and bushings. Frequently, additional metals, such as nickel and arsenic, are added in small amounts to aid in maintaining the suspension of the lead.

High-Manganese Screw Stock. Open-hearth or Bessemer steel, high in sulfur (up to 0.40%), with manganese approximating 1.5%.

High-Nitrogen Ferro-Chromium. See **Ferro-Chromium.**

High-Phosphorus Pig (Iron). Pig Iron (see) of higher-than-normal phosphorus content.

High-Quality Steel. See **Alloy Steel.**

High Shoulder. See **Round.**

High-Silicon Pig (Iron). See **Silicon Pig (Iron).**

High-Speed Brass. See **Free-Cutting Brass.**

High-Speed (Tool) Steel. Steel which does not soften even at red heat, and therefore can be used at elevated temperatures for drills, cutters, and other machine tools; the most common alloying element is about 12–22% tungsten; chromium, vanadium, and molybdenum and other elements may also be present. More recently, molybdenum high-speed steels (with molybdenum up to 9% and tungsten 1–5%) have become more common.

High Steel. See **Hypereutectoid Steel.**

High-Strength Brass. See **High-Tensile Brass.**

High-Strength (Structural) Steel. See **Low-Alloy High-Strength Structural Steel.**

High-Temperature Electrochemical Cleaning. Electrolytic cleaning of steel by immersion into molten sodium hydroxide, with current passing at a rate of 100–150 amperes per square foot, the steel acting as cathode.

High-Temperature Steel. Steel, usually of intermediate or high-alloy composition, suitable for high-temperature service. See **Corrosion-Resistant Alloy.**

High-Temperature Test. Test of tensile strength and **Creep** (see) of metal when subjected to stress at high temperatures. Ordinary tensile tests are known as short-time tests; creep tests, as long-time tests.

High-Tensile Brass. Copper-base alloy, of excellent tensile strength, with 22% zinc and 2% aluminum.

High-Tensile (Structural) Steel. See **Low-Alloy High-Strength Structural Steel.**

High-Test Cast Iron. Cast iron with

low percentages of alloying elements, such as nickel, chromium, molybdenum, etc., insufficient to classify the material as an alloy cast iron, but adequate to increase materially tensile strength; frequently made from steel scrap and much finer in grain than ordinary cast iron.

High-Test Iron. See **Semi-Steel.**

Hi-Gloss. Stainless steel series of 18% chromium 8% nickel, or 12% chromium 12% nickel content (Jessop Steel Co.).

Hills-McCanna Alloy. Series of copper-base alloys, of **Brass** (see) and **Bronze** (see) type of composition (Hills-McCanna Co.).

Hioloy (Steel). Series of carbon and low-alloy corrosion- and wear-resistant steels, manufactured by Ohio Steel Foundry Co.

Hipernick. Alloy of equal parts of nickel and iron, used as **Permeability Alloy** (see).

Hi-Steel. Low-alloy, high-strength steel, with carbon usually below 0.12%, containing about 1% copper, 0.5% nickel, 0.6% manganese, and silicon not over 0.3% (Inland Steel Co.).

Hitenso. Series of **Cadmium Bronzes** (see), manufactured by American Brass Co.

Ho. Chemical symbol for **Holmium** (see).

Hob. Table on to which tinned or terne plates are delivered from the first tinning or terne bath in the **Combination Process** (see), before being passed into the second bath.

Hobson's (Steel). Series of carbon and alloy tool steels, manufactured by Hobson, Houghton & Co.

Hoepfner Process. Recovering copper from its ores by leaching with cupric chloride and salt solution, and electrolyzing the resulting cuprous chloride liquor; the electrolytic by-product, cupric chloride, is returned for further leaching.

Holding the Iron. Faulty condition of **Blast Furnace** (see) operation, in which a low amount of iron is obtained despite normal feeding of raw materials.

Hole. See **Soaking.**

Holloway Process. Treating copper or iron sulfides by fusion and blowing with air, to burn out the sulfur content.

Hollow Roll. Rolling Mill (see), after continued use has worn the center so that rolling is not adequately flat.

Holmium (chemical symbol Ho). Element No. 67 of the periodic system; atomic weight 164.94. One of the rarest of the rare earth metals, it has never been isolated; consequently, no data are available regarding its physical constants. Chemically, it is trivalent. Like all rare earth metals, it cannot be electrodeposited.

Holtite Brazing Solder. See **Spelter Solder.**

Homogeneity Test. Test, particularly on steel, to determine areas of non-homogeneity. Such tests include fracture tests, macroetching, sulfur prints, etc.

Homogeneous. Having identical characteristics throughout.

Homogenizing. Heating a metal or alloy to some temperature, below melting, at which diffusion is adequate to create a more homogeneous structure.

Hood and Fender Stock. Flat automobile body steel sheet, of high polish and good drawing qualities; used for hoods, fenders, etc. See **Automobile Body Sheet.**

Hooker Brass. See **Clock Brass.**

Hooke's Law. Up to the elastic limit, **Strain** (see) is proportional to **Stress** (see).

Hoop. See **Hoop Iron.**

Hoopes Aluminum. See **Hoopes Process.**

Hoopes Process. Purifying commercial aluminum to 99.99% aluminum content; the aluminum is alloyed with pure copper and then used as anode in a fused fluoride bath, heavier than aluminum; pure aluminum collects at the cathode, above the fused bath.

Hoop Iron. Strip steel of low carbon content, usually made by the basic open-hearth or Bessemer process, suitable for steel hoops. More broadly, the term is used to cover all relatively thin material up to about 8 inches in width; in this sense, hoop steel is identical with strip steel.

Hoop Steel. See **Hoop Iron.**

Höpfner Process. Treating copper sulfide ores with solutions of cupric and sodium or calcium chlorides, the resulting solution being electrolyzed in cells, with impure copper anodes protected by diaphragms.

Horizontal Mill. (a) Rolling mill in which the material is held in a horizontal position during rolling. (b) **Rolling Mill** (see) in which the axes of the rolls are horizontally positioned.

Horn. Projection, causing an end, shaped like a fishtail, on rolled metal, when a rolling mill is hollow, the sides of the metal being more elongated than the edges.

Horn Quicksilver. Mercurous chloride, or calomel (not a metal), chemically HgCl.

Horn Silver. Non-metallic mineral, cerargyrite, corresponding approximately to silver chloride, AgCl, in composition.

Hoskins Alloy. Series of nickel-base alloys, with low percentages of alloying elements, such as manganese and silicon, or high percentages of nickel (up to 35%) and iron (Hoskins Mfg. Co.).

Hot Bed. Large area, commonly composed of rails near one another, for holding hot, partially-rolled metal.

Hot-Blast. Air blast in **Blast Furnace** (see), after preheating.

Hot-Blast Stove. See **Blast Furnace.**

Hot-Blast Valve. Valve in **Blast Furnace** (see), usually water-cooled because of the high temperature of the gases; controls the blast from the hot blast stove.

Hot-Die Steel. Alloy steels resistant to shock, particularly at high temperatures. Chromium steels, or, preferably, tungsten steels, are most frequently used for hot-die purposes.

Hot-Dip Process. See **Hot Galvanizing.**

Hot-Drawing. Drawing (see) metal at elevated temperatures.

Hot-Drawn. See **Hot-Drawing.**

Hot Finish. Surface finish on metal, obtained after hot-working; frequently there is a bluish-black scale, adhering to the metal, when so finished.

Hot Finishing Mill. Final **Stand** (see) or stands (see **Rolling Mill**) in the hot-rolling of metal sheet.

Hot Galvanizing. Dipping sheet or strip steel in molten zinc. See **Galvanizing.**

Hot-Heading. See **Upset Forging.**

Hot Junction. See **Pyrometer.**

Hot Leveller. Roller Levelling (see) machine for levelling metal sheet, after the final pass through the finishing rolls.

Hot Melt Process. Process of melting metal in which the raw charge consists of molten metal. See **Duplex Process** and **Triplex Process.**

Hot Metal. Molten metal, particularly prior to casting into pig form.

Hot Metal Mixer. Rotating or tilting vessel, usually made of steel and lined with brick, acting as a reservoir for molten pig iron in steel manufacture. The mixer maintains the heat of the metal and, in addition, serves to make the composition uniform.

Hot Metal Process. (a) See **Hot Metal Refining.** (b) See **Hot Melt Process.**

Hot Metal Refining. Process of refining molten pig iron by treatment, in an electric arc furnace, first under an oxidizing slag and then under a reducing slag.

Hot Mill. Finishing mill (see **Rolling Mill**) in pack mill operations.

Hot Mill Oxide. See **Mill Scale.**

Hot Pack Mill. Rolling Mill (see) for **Pack Rolling** (see).

Hot Pack Rolling. See **Hot-Rolling.**

Hot-Pressing. In powder metallurgy, forming and heating compacts simultaneously.

Hot-Pressing Plate. See **Hot-Pressing Quality (Steel Plate).**

Hot-Pressing Quality (Steel Plate). Steel plate ordered or sold on the basis of specifications governing suitability for hot-pressing, flanging, or bending work, as opposed to cold-forming or deep-drawing.

Hot-Pressing Steel. See **Hot-Pressing Quality (Steel Plate).**

Hot-Reduction. Reduction of metal size, particularly thickness, accomplished while the metal is hot.

Hot-Rolled Annealed Plate. Hot-rolled steel plate, subjected to box annealing, with blued edges, and without surface polish.

Hot-Rolled Annealed Sheet. Steel sheet of light gage (about 0.05 inch or lighter), hot-rolled, and then either cold-rolled and box annealed, or box annealed and cold-rolled, with subsequent flattening.

Hot-Rolled Bar. Bar, particularly of steel, hot-rolled to a thickness of one-quarter inch or more, with width up to six inches.

Hot-Rolled Commercial Quality Steel Sheet. Ordinary steel sheet, carrying mill scale on the surface, obtained directly after hot-rolling, without particular specifications.

Hot-Rolled and Normalized Sheet. Steel sheet of relatively heavy gage, hot-rolled and subjected to **Normalizing** (see).

Hot-Rolled Physical Quality Steel Sheet. Hot-rolled steel sheet, ordered or sold on the basis of special specifications as to physical properties, such as tensile strength, drawing quality, stretcher strains, etc.

Hot-Rolled Pickled and Annealed Sheet. Hot-Rolled Annealed Sheet (see) subjected to **Pickling** (see) to remove scale after hot-rolling.

Hot-Rolled Pickled Commercial Quality Steel Sheet. Ordinary steel sheet (see **Hot-Rolled Commercial Quality Sheet Steel**), subjected to final pickling, to remove scale.

Hot-Rolled Pickled Drawing Quality Steel Sheet. Hot-rolled steel sheet, subjected to final pickling, to remove scale; ordered or sold on the basis of specifications governing deep-drawing quality.

Hot-Rolled Pickled Physical Quality Steel Sheet. Hot-Rolled Physical Quality Steel Sheet (see), subjected to final pickling, to remove scale.

Hot-Rolled Pickled Sheet. Steel sheet, usually 0.016 inch or heavier in thickness, made by hot-rolling and then subjected to **Pickling** (see); annealing and flattening are optional additional steps.

Hot-Rolled Pickled Strip. See **Hot-Rolled Pickled Sheet.**

Hot-Rolled Products. Steel sheet, strip, and similar flat stock, rolled at temperatures exceeding 750°F.

Hot-Rolled Sheet. Steel sheet, 0.016 inch or heavier in thickness and up to four feet in width, hot-rolled, without subsequent cleaning. Optional additional treatments include annealing, normalizing, flattening, or a single cold-rolling pass.

Hot-Rolled Special Surface Quality Steel Sheet. Hot-rolled steel sheet, ordered or sold on the basis of specifications governing any special surface requirements.

Hot-Rolled Strip. See **Hot-Rolled Strip (Steel).**

Hot-Rolled Strip (Steel). Hot-rolled flat steel of dimensions up to eight or nine feet wide, and up to one-quarter inch thick. Narrow hot-rolled strip includes widths up to 26 inches; wide hot-rolled strip from 26 to 48 inches; broad hot-rolled strip more than 46 inches finished width.

Hot-Rolling. Diminishing the thickness of metal by rolling at elevated temperatures. The more modern hot-rolling mills are continuous in operation.

Hot Saw. Rotary saw used for cutting hot metal in bar or billet form.

Hot Short. Metal unsuitable for hot-working because of brittleness at the working temperatures is called hot short. Sulfur is frequently the cause of hot shortness in steel.

Hot Spot. Area or zone, in heated equipment, overheated because of faulty operating conditions.

Hot-Stamping Brass. See **Forging Brass.**

Hot Strip Mill. Rolling Mill (see) for the hot-rolling of **Strip Steel** (see).

Hot-Tinning. Dipping sheet or strip steel in molten tin. See **Tinning.**

Hot Top. See **Sinkhead.**

Hot-Top Big-End-Up Mold. See **Big-End-Up Mold.**

Hot-Work Die Steel. Steel suitable for the fabrication of steel dies for the working or drawing of metals at high temperatures. Chromium approximating 3.5%, chromium 1.3–4.5% with 9–19% tungsten, and chromium and tungsten approximating 6% each, are the most common types.

Hot-Working. Plastic deformation, such as rolling, hammering, etc., at a temperature sufficiently high not to create strain hardening (work hardening), i.e., above the recrystallization temperature.

Hot-Working Steel. See **Hot-Die Steel.**

Housing Window. See **Window.**

Howell Furnace. Roasting (see) furnace of the rotating type.

H. S. Steel. See **High-Speed (Tool) Steel.**

Hubbelite. Magnesium oxychloride cement, in which metallic copper powder has been dispersed, to achieve greater strength and germicidal characteristics.

Huey Nitric Acid Test. Corrosion test on **Stainless Steel** (see); a sample is heated in boiling 60% nitric acid for five repeated periods of 48 hours each and the loss in weight of each period recorded; normally, the average loss from the last three readings is taken as indicative of the quality of the steel.

Hull Cell. Experimental electrolytic cell, used for determining operating current densities; the slanting arrangement of the cathode causes a variation of cathode current density, which, at each point of the cathode, is a logarithmic function of the distance from the high current density end of the cathode plate.

Hull (X-Ray) Method. See **Debye-Scheerer (X-Ray) Method.**

Hunt-Douglas Process. Recovering copper from its ores by leaching with ferrous chloride, the dissolved copper being precipitated on scrap iron. An alternative form is leaching with dilute sulfuric acid, then treatment with sulfur dioxide and calcium chloride, followed by reaction with hydrated lime and reduction to metal.

Huntington-Heberlein Roasting Process. Process of **Roasting** (see) lead sulfide ores, in which burnt lime is added to prevent premature melting of the sulfide.

Huron Aluminum Alloy. Series of aluminum-base alloys, with fractional or low percentages of copper, manganese, magnesium, nickel, cobalt, chromium or cadmium.

Huron Steel. High-carbon chrome steel series (Ludlum Steel Co.) with chromium about 12%, vanadium about 1%, and fractional percentages of manganese and silicon. Carbon is about 2.1%.

Husmann's Alloy (or Metal). Tin-base bearing metal, with about 11% lead, 11% antimony, and 4% copper.

Hybinette Process. Recovering nickel by treating a **Matte** (see) of nickel-copper-sulfur with the copper-containing electrolyte from the nickel-plating tanks, resulting in deposition of the copper and solution of the nickel; this solution is sent to electrolytic tanks for nickel deposition.

Hyblum. Aluminum-base alloy, with about 1% chromium and 1% nickel.

Hybnickel. Nickel-chromium-iron alloys (Hybnickel Alloys Co.) of excellent heat and corrosion resistance, containing 5–35% nickel and 18–35% chromium.

HYCC Steel. High-chromium (about 12%) high-carbon (about 2.2%) abrasion-resistant steel, with fractional percentages of vanadium (Crucible Steel Co.).

Hydrargyrum. Latin name for mercury, used in medicine and pharmacy.

Hydraulic Bronze. Bronzes, usually of the copper-zinc-tin-lead type, used for casting of pumps, valves, and similar hydraulic fittings.

Hydride. Compound of hydrogen with any metal (those with non-metals are not called hydrides). Such hydrogen is abnormal since it is acting electronegatively instead of electropositively.

Hydrizing. Heat treating metals in an atmosphere of hydrogen.

Hydrogen (chemical symbol H). Element No. 1 of the periodic system; atomic weight 1.0080. Colorless gas, liquefying at —423°F and solidifying at —434°F. It is chemically monovalent and reacts primarily with non-metals. However, it also unites with the alkali and alkali earth metals. It is strongly adsorbed by many metals, such as tantalum and titanium, forming compounds dissociable in heat and vacuum. Hydrogen is frequently "plated out" together with metals, such as chromium and iron, during electroplating, and is one cause of brittleness. Made commercially by electrolysis of water, it finds considerable use in oxy-hydrogen gas torches, and for reduction of metals from their oxides.

Hydrogen Annealing. Annealing (see) in hydrogen atmosphere.

Hydrogen Diffusion Method. Determining critical points of iron and steel; one side of a thin specimen is exposed to hydrogen, with the other under high vacuum; hydrogen diffuses through the metal, the rate increasing constantly with temperature, though discontinuities are to be found at the A_0, A_2, and A_3 points.

Hydrogen Electrode. Noble metal, coated with an adherent layer of finely divided noble metal, to increase its surface and thus its gas-absorbing capacity, with a stream of hydrogen gas passing over the surface.

Hydrogen Embrittlement. Brittleness of metal, resulting from the occlusion of hydrogen (usually as a by-product of pickling or by co-deposition in electroplating). See **Hydrogen.**

Hydrogen Evolution Test. Test of the continuity of tin or terne plating, in which the metal is immersed in dilute acid, under standard conditions, and the rate of hydrogen evolution determined.

Hydrogen Ion (Concentration). See **pH.**

Hydrogenite. Dry mixture of ferrosilicon and sodium hydroxide, liberating hydrogen upon the addition of water.

Hydrometallurgy. Science of metal recovery by any process involving solution in an aqueous medium, such as acid or cyanide solution, etc.

Hydrone. Sodium-lead alloy, with about 30% sodium; manufactured by E. I. du Pont de Nemours Co. for generating hydrogen by addition to water.

Hydropyrometer. See **Water Pyrometer.**

Hydroxyl Ion (Concentration). See **pH.**

Hy-Glo Steel. High-chromium steel (Latrobe Electric Steel Co.), highly resistant to corrosion, containing approximately 18% chromium, and about 0.65% carbon. Used primarily as tool steel and for cutlery.

Hylastic Steel. Medium **Manganese Steel** (see), with about 1.6% manganese and 0.10% vanadium (American Steel Foundries).

Hypereutectoid Steel. Steel, containing more than about 0.85% carbon, the eutectoid carbon value; made by the open-hearth or the crucible process, or in the electric furnace.

Hypoeutectoid Steel. Steel containing less than about 0.85% carbon, the eutectoid carbon value.

Hysteresis. (a) See **Cooling Lag.** (b) See **Hysteresis Loop.**

Hysteresis Loop. Loop formed when magnetic flux density is plotted against magnetizing force in a magnetization-demagnetization sequence on a magnetizable metal. Both power and time are required for magnetization and demagnetization; this phenomenon is known as hysteresis.

Hysteresis Loss. Power and time lost in magnetizable metals as a consequence of repeated magnetization and demagnetization. See **Hysteresis Loop.**

Hytemco. Nickel-base electric resistance alloy, with about 30% iron (Driver-Harris Co.).

Hytensl Bronze. Series of bronzes, manufactured by American Manganese Bronze Co., with varying amounts of zinc, aluminum, manganese, and iron.

Hy-Ten (Steel). Series of low-alloy and high-alloy steels, manufactured by Wheelock, Lovejoy & Co.

I

I Beam. Metal beam, of I cross-sectional shape, rolled to standard dimensions.

Idiomorphic Crystal. Perfect crystal, usually large; forms only under especially favorable conditions, and not normally, because of viscosity, rapidity of cooling, agitation of the liquid or melt, etc. The normal (imperfect, or allotrimorphic) crystals are commonly known as grains in metallography.

Idle Loop. See **Controlled Loop.**

Idria Furnace. See **Leopoldi Furnace.**

Igewsky's Reagent. Carbon steel etching solution, 5% picric acid in absolute alcohol.

Ihrigizing. See **Siliconizing.**

Il. Chemical symbol for **Illinium** (see).

Illinium (chemical symbol Il). Element No. 61 of the periodic system; estimated atomic weight 146. An extremely rare member of the rare earth metal series. Neither the metal itself nor any of its compounds has been isolated in pure condition. By analogy, it may be assumed to be trivalent chemically. Like all rare earth metals, it probably cannot be electrodeposited.

Illium. Series of nickel-base alloys, containing 18–24% chromium, 3–7% molybdenum, 4–8% iron, 1–8% copper, 0.4–1.5% manganese, and fractional percentages of silicon. Primary use is for laboratory apparatus, especially equipment subjected to pressure, such as calorimeter bombs (Burgess-Parr Co.).

Ilmenite. See **Titanic Iron Ore.**

Impact Failure. Instantaneous rupture without plastic distortion, under instantaneously applied stress.

Impact Fatigue. Resistance of a material to repetitive stress by impact.

Impact Hardness. See **Impact Fatigue.**

Impact Machine. See **Impact Test.**

Impact Strength. See **Impact Test.**

Impact Stress. Stress involving a sudden blow on the part under consideration.

Impact Test. Test designed to determine the resistance of metal to breakage by impact, usually by concentrating the applied stress in a notch. See **Izod (Impact) Test** and **Charpy (Impact) Test.**

Impact Value. See **Impact Test.**

Impoverishment. Loss of any valuable constituent, as by oxidation, liquation, or volatilization.

Impregnation. (a) Generally, diffusion of a second metal (or carbon) into the surface of a metal. For example, carbon, aluminum, zinc, silicon, or chromium may be diffused into iron-base metals. (b) In powder metallurgy, filling the pores of a compact, usually after sintering, with any other substance. (c) In aluminum technology, filling blow-holes in castings with water-glass, bakelite, or similar filler.

Improving. See **Softening.**

In. Chemical symbol for **Indium** (see).

Inclusion. Non-metallic impurity, such as oxide, sulfide, etc., entrapped in metal during manufacture or formed by reaction within the metal. (Carbides are not viewed as inclusions when they represent essential phases of the metal structure.) Normally, inclusions will segregate within the inter-space between the crystals of the metal.

Inconel. Series of nickel-base alloys, manufactured by International Nickel Co., with about 13% chromium, 6% iron, and small amounts of manganese, silicon, and copper. Because of high resistance to heat and oxidation, used in food technology and for automotive parts.

Inconel-Clad Steel. Clad metal (see **Cladding**) in which a heavier layer of steel is coated with a lighter coating of **Inconel** (see).

Indentation Test. Hardness (see) test, by pressing a pointed or rounded testing element into the surface, under a substantially static load.

Indiana Furnace. Modification of the **Belgian Furnace** (see) for zinc distillation, using gas for combustion, the gas chamber being under the lowest group of zinc retorts.

India Steel. See **Wootz.**

Indices of Crystal Face. Reciprocals of intercepts of a crystal plane on reference axes based on a chosen system of coordinates.

Indigo Copper. Covellite; non-metallic mineral approximating CuS in chemical composition.

Indirect Arc Furnace. Arc Furnace (see) in which the current flows between two electrodes, the furnace charge being heated indirectly, principally by radiation.

Indirect Arc Heating. See **Indirect Arc Furnace.**

Indirect Fired Furnace. Furnace in which combustion takes place in a chamber separated from the material being heated.

Indirect Heating. See **Indirect Arc Furnace** and **Indirect Fired Furnace.**

Indium (chemical symbol In). Element No. 49 of the periodic system; atomic weight 114.76. An extremely soft, silvery metal, malleable and ductile, not affected by atmosphere or water; melting point 312°F; boiling point above 2650°F; specific gravity 7.28. Chemically usually trivalent, though in some compounds it is monovalent or divalent. Indium, quite rare in occurrence, is normally obtained from its concentrate in the flue dust of zinc smelters, by chemical purification, conversion to an electrolyzable salt such as the sulfate, followed by electrodeposition. The metal has recently become of industrial importance, particularly in plated form and for special low-melting alloys. Available from the Indium Corp. of America.

Induction Brazing. Brazing (see) in which the material is heated by electric induction.

Induction Furnace. Electric furnace, usable at both low and extremely high temperatures, in which electric current is induced within the metal charge by coils carrying alternating current. Low frequencies require the use of an iron core; with higher frequencies, the charge acts as the core.

Induction Hardening. Hardening of metal by heat treatment by means of high-frequency induction. Usually, steel, when so hardened, is heated to above the critical range and then quenched with water sprays.

Induction Heating. See **Induction Furnace.**

Industrial (Steel). Series of corrosion-resistant high-chromium and nickel-chromium steels (see **Stainless Steel**), manufactured by Industrial Steels, Inc.

Inert. Material, in a system, which does not enter into chemical reaction or does not substantially affect the physical properties of the system.

Inert Element. See **Inert Gas.**

Inert Gas. Gaseous element of the group O of the periodic system, such as helium, argon, etc., which is substantially non-reactive under any condition.

Ingaclad (Stainless-Clad) Steel. Clad metal (see **Cladding**) in which a thin layer (about 20%) of stainless steel has been bonded to a thicker base of simple steel (Ingersoll Steel & Disc Division).

Ingate. See **Gate.**

Ingersoll DBL (High Speed Steel). High-speed steel, containing tungsten, molybdenum, and vanadium (Ingersoll Steel & Disc Division), used primarily for hack-saw blades.

Ingot. Solidified mass of metal, usually of round or rectangular cross-section, resulting from the pouring of molten metal into an appropriately shaped mold. The ingot stage is usually intermediate in rolling or other mechanical shaping operations.

Ingot Iron. Highly refined steel made by the basic open-hearth process, with a maximum of 0.15% of total impurities. Because of its high purity, it has excellent ductility and resistance to rusting.

Ingotism. Structure of coarse columnar and equiaxed crystals, formed during the solidification of ingots.

Ingot Mold. Mold into which metal is poured, for the casting of **Ingots** (see).

Ingot Stripper. Device for removing mold from metal ingots after solidification, usually in the form of two lifting jaws acting on the mold, while a plunger operates between them, pushing the ingot down. See **Ingot.**

Ingot Structure. Crystal structure sequence within a normal ingot, with chill crystals at the surface, graduating into more normal structures in the center. See **Ingot.**

Ingot Topping. See **Topping.**

Inherent Hardenability Variation. Variation in the hardenability of steels of the same composition, due to austenitic grain size differences in the steels.

Inhibitor. Material which, when added in small quantities to an acid pickling bath, minimizes attack on the metal, but does not materially affect chemical reaction with the scale. See also **Rust Inhibitor.**

Initial Creep. See **Creep.**

Initial Creep Stress Test. See **Time Yield Stress.**

Inland Hi-Steel. Low-alloy steel made by Inland Steel Co., containing about 1% copper and 0.5% nickel; shows relatively high strength and resistance to atmospheric corrosion.

Insoluble Anode. Anode (see) in electrodeposition or electrometallurgy which does not go into solution during the electrolytic process.

Insulated Cooling. Cooling of heated metal, in which the rate of cooling is controlled by covering the metal, directly or inside an underground pit, with an inert insulating material, such as sand or lime.

Insuluminum. Steel impregnated with aluminum on its surface (see **Calorizing**), resistant to oxidization at high temperatures (General Electric Co.).

Intensity of Force. See **Fiber Stress.**

Intensity of Stress. See **Fiber Stress.**

Intercommunicating Porosity. In powder metallurgy, porosity allowing a continuous flow of a fluid from one end to another.

Intercrystalline Cracking. Cracks which form in metals, following the lines of crystal boundaries; due to either corrosive conditions or prolonged stress, or, more frequently, both of these acting simultaneously.

Intercrystalline Failure. See **Intercrystalline Cracking.**

Intercrystalline Fracture. See **Intercrystalline Cracking.**

Interface. Surface between two different **Phases** (see).

Intergranular Attack. See **Intergranular Corrosion.**

Intergranular Corrosion. Corrosion which tends to localize at crystal boundaries. See **Intercrystalline Cracking.**

Inter-Lamination Resistance Test. Test of resistance of the insulation of oxide or other non-metallic coatings on silicon steel sheet.

Intermediate Constituent. See **Intermetallic Compound.**

Intermediate Cooler. See **Cinder Cooler.**

Intermediate Manganese Steel. See **Medium Manganese Steel.**

Intermediate Structure. Microstructure, in steel, intermediate between normal and abnormal. See **Normal Steel** and **Abnormal Steel.**

Intermediate Zinc. Metallic zinc with total impurities not over 0.50%; lead maximum is 0.20%, iron 0.03%, and cadmium 0.50%, with no aluminum allowed.

Intermetallic Compound. Compound of two or more metals, with characteristics sufficiently defined to indicate chemical individuality, as opposed to mere alloying. Like all compounds, the constituents are always present in definite proportions, though they frequently do not follow normal valence rules.

Internal Crack. Crack which occurs under the surface of metal, usually in the form of a tiny discontinuity.

Internal Fissure. See **Internal Crack.**

Internal Stress. Stress within a solid object or system, due to conditions existing during cooling or working, or to the mechanical structure of the metal (for example, a coiled spring); it must always be kept within the elastic limit of the metal.

Interrupted Quenching. Quenching in one medium, as in water or in oil, for a limited time, followed by air-cooling or, after holding in air for some given time, reimmersion in the original quenching medium; at times, will help avoid objectionably high internal stresses which may result from full quenching.

Interstitial Compound. Semi-metallic, usually hard, compound formed by penetration of atoms of small diameter between the atoms of the transition metals.

Intertype Metal. Type Metal (see) with about 12% antimony and 4% tin.

Invar. Iron-base alloy of high nickel content (about 36%), which has an extremely low coefficient of expansion at ordinary temperatures; consequently, it is used in pendulums, thermostat elements, etc. It is made by Driver-Harris Co.

Invaro (Steel). Non-Deforming Steel (see), with about 1.1% manganese, 0.5% tungsten, 0.5% chromium, and 0.3% vanadium; carbon approximates 0.90% (Firth-Sterling Steel Co.).

Inverse Rate Method. Variation of the time-temperature curve method for determining critical temperatures in metal (see **Critical Point Test**), in which the time required for a specimen of the metal to rise or fall a preassigned temperature interval is plotted against the temperature change.

Inverse Segregation. Segregation, occurring occasionally in various copper and aluminum alloy systems, in which certain alloy constituents tend to concentrate in the first portions solidifying. Inverse segregation can be largely eliminated by pouring into hot molds or in other ways reducing the rate of cooling of the casting.

Inwall. See **Blast Furnace.**

Iola Furnace. Gas-fired furnace, containing a series of hearths, used for the distillation of zinc.

Ion. Electrically charged atom, or group of atoms, in a solution.

Ir. Chemical symbol for **Iridium** (see)

Iridium (chemical symbol Ir). Element No. 77 of the periodic system; atomic weight 193.1. A yellowish-white member of the platinum metals family; it is the most noble of metals, hard, brittle, and difficult to work, resistant to practically all chemical reagents and to all corrosive influences; melting point 4449°F; boiling point is assumed to be about 9550°F; specific gravity 22.42. Chemically, it is most frequently trivalent or tetravalent; forms compounds also with other valences. Found native, in combination with other members of the platinum family, it is separated by chemical means, usually to the final form of the double ammonium chloride, and then reduced to metal by simple ignition. It is normally used as an alloy with platinum or osmium, to which it imparts hardness. Iridium can be electrodeposited but there has been no commercial operation along these lines.

Iridosmine. See **Osmiridium.**

Iron (chemical symbol Fe). Element No. 26 of the periodic system; atomic weight 55.85. The most abundant and most common of metals, used extensively in all the arts; unfortunately is subject to atmospheric rusting. In pure form, it is a silvery white, soft, extremely ductile metal of high magnetic characteristics; its properties are drastically affected by relatively small percentages of carbon (see **Steel**); melting point 2786°F; boiling point about 5430°F; specific gravity 7.86. Chemically, it shows both divalence (ferrous) and trivalence (ferric). Found normally in nature as an oxide; the commercial metal is usually manufactured by carbon reduction (see **Blast Furnace**). Pure metal can best be obtained by hydrogen reduction of the oxide followed by vacuum fusion. Iron can be electrodeposited, using either a sulfate or chloride bath, but this process is little commercialized.

Iron Black. Not iron. Finely divided black antimony, prepared by reduction of an antimony salt solution by zinc or a similar metal.

Iron Bronze. Copper-base alloy series, with 4–40% zinc, 1–9% tin, 1–4% iron, and optional fractional percentages of other metals such as lead and aluminum.

Iron-by-Hydrogen. See **Reduced Iron.**

Iron Carbide. See **Cementite.**

Iron Chamber. Zone, in a **Puddling** (see) furnace, in which the charge is heated.

Iron Foundry. See **Foundry.**

Ironless Induction Furnace. See **Coreless (Induction) Furnace.**

Iron Loss. Sum of **Hysteresis Loss** (see) and **Eddy Current Loss** (see).

Iron Notch. See **Blast Furnace.**

Iron Plate. See **Bloomery.**

Iron Plating. Coating metal objects with iron, or depositing iron for building up worn areas, in which the object is made cathode (negative electrode) in an electrolytic bath containing a decomposible iron salt, usually the chloride or sulfate. Iron metal is commonly used as anode (positive electrode).

Iron Powder. Iron, in finely divided form, most commonly made by hydrogen reduction of finely ground oxide. See **Reduced Iron.**

Iron Pyrites. See **Pyrites.**

Iron Reduction Process. See **Precipitation Process.**

Iron Runner. Spout through which molten iron flows from a **Blast Furnace** (see) tapping hole.

Iron Saw. Rotary saw used for cutting iron or steel, most commonly when in bar or billet form.

Iron Scale. Scale which forms on hot iron and steel during working; consists principally of a mixture of oxides of iron.

Iron Shears. Equipment for cutting iron in bar, plate, or sheet form.

Iron Shot. Iron in the form of globules or irregular jagged pieces, manufactured by pouring molten iron directly into water; used as an abrasive.

Iron Sponge. Metallic iron in spongy form, obtained by low-temperature reduction of iron oxide by gaseous reducing agents.

Iron Steel. Composite, consisting of soft iron coated on or clad with steel (see **Cladding**).

Irregular Fracture. See **Fracture.**

Irregular Sketch Sheet. Sheet metal sheared into shapes of more than four straight sides.

Isometric (Crystal System). See **Cubic (Crystal System).**

Isotope. See **Atomic Weight.**

Isotropy. The quality of having the same properties in all directions, as opposed to **Anisotropy** (see).

Izod (Impact) Test. Method of testing resistance to impact; a test piece, 1 cm. square, notched to a depth of 2 mm., is held in a vise which forms part of the testing machine, and a pendulum hammer is permitted to strike from a standard distance; the force required to break the metal is determined and expressed in foot-pounds of energy.

J

Jack. See **Mock Lead.**

Jacona Metal. Lead-base bearing metal, with about 20% antimony and 10% tin.

Jail and Safe Steel. Composite, consisting of two or more layers of steel of different compositions, difficult to cut through with tools or torches.

Jalcase (Steel). Series of low-alloy steels, with manganese up to about 1.65% as the principal alloying element; manufactured by Jones & Laughlin Steel Co.

Jal-Ten (Steel). Low-alloy, high-strength steel, with carbon not over 0.35%, 1.25–1.75% manganese and not over 0.40% copper (Jones & Laughlin Steel Corp.).

Jam Welding. Welding (see) performed between heated square-butted edges.

Japan. Varnish which requires baking to complete oxidation; frequently used to protect steel articles.

Jelliff Alloy. Series of nickel-chromium and nickel-copper electric-resistance alloys (C. O. Jelliff Mfg. Corp.).

Jessop Steel. Series of steels, including low-alloy and high-alloy tool steels and corrosion-resistant steels; manufactured by Wm. Jessop & Sons, Inc.

Jewelers' Metal. Red **Brass** (see), with 6–17% zinc and tin optional up to 2%.

Jewell Alloy. Series of heat and corrosion-resistant, low-alloy, cast irons, with fractional or low percentages of silicon, manganese, nickel and chromium (Jewell Alloy & Malleable Co.).

Jewelry Sweep. Gold, silver, and platinum, in finely divided form, collected from the benches and floors of jewelry workers.

J Metal. Cast form of **Stellite** (see).

Joachimstal Process. Process of treating silver sulfide ores, in which the metal values are converted to silver chloride, subsequently leached by sodium hyposulfite, and the silver precipitated with sodium sulfide.

Jobbing Mill. Rolling Mill (see) for rolling sheets of about two to six feet width, and 0.06 to 0.12 inch thickness.

Johnson Bronze. Copper-base alloy series (Johnson Bronze Co.), with 70–87% copper, 5–12% tin, lead optional up to 19%, and zinc optional up to 5%; used primarily for bearings.

Johnson McKay Retort. Retort consisting of a six-foot length of cast iron, about twelve inches in diameter, for recovering mercury from its ores by direct heating and condensation of the resulting metallic vapors.

Johnston's Apparent Elastic Limit. See **Johnston's Limit.**

Johnston's Limit. Point, on the stress-strain curve, at which the deformation rate is 50% greater than at the beginning.

Joint Board. See **Bottom Board.**

Jominy Hardenability Curve. Curve illustrating changes in the hardness of steel along a test bar, heated to a predetermined temperature, followed by quenching of one end of the bar in a special apparatus.

Josephinite. Naturally-occurring alloy of nickel and iron, approximating Fe_2Ni_5 in chemical composition.

Josephtown Condenser. Apparatus for condensing metal vapor, diluted with non-condensible gases, directly to coherent metal, in which the incoming gases are bubbled through a bath of molten metal.

J-S Steel. Tough, oil-hardening tool steel, with about 2.2% tungsten and 1.4% chromium; carbon is approximately 0.50% (Firth-Sterling Steel Co.).

Jump Mill. See **Roughing Stand.**

Jump Welding. See **Butt Welding.**

Jupiter Process. See **Jupiter Steel.**

Jupiter Steel. Cast steel of relatively high strength and ductility, with about 3% aluminum, 2% ferrosilicon, and 0.5% ferromanganese, added while molten.

K

K. Chemical symbol for **Potassium** (see).

K. Kelvin temperature. See **Absolute Temperature.**

Kalium. Latin name for potassium, used in medicine and pharmacy.

Kämmerling Furnace. Zinc smelting furnace, of the Belgian type (see **Belgian Furnace**), having two combustion chambers, separated by a longitudinal wall.

KA-2-Mo Steel. Stainless Steel (see), of the 18% chromium, 8% nickel, low-carbon range, modified with about 3% molybdenum.

Kaolin. Relatively pure form of clay.

Karat. See **Carat.**

KA-2 Steel. Stainless Steel (see) of the 18% chromium, 8% nickel, low-carbon range; originally developed in Germany, but now sold by various American firms.

Kast Furnace. Small shaft furnace of circular cross-section, for smelting lead ores.

Katadyn Process. Sterilizing water by small traces of silver or other metals.

Kathion. See **Cation.**

Kation. See **Cation.**

Keen Edge Steel. Steel of high-carbon content capable of maintaining a keen cutting edge. Alloying elements may be present.

Keil Furnace. Zinc distillation furnace, heated by gas, employing vertical retorts.

Keith Process. Electrolytic refining of lead, using lead acetate or chloride as electrolyte, anodes of impure lead, encased in muslin bags; the cathode deposit, consisting of crystalline lead, falls aways from the plate electrode and is collected on the bottom of the tank.

Kelcaloy. High-chromium steel series, with 11–15% chromium, manufactured by M. W. Kellogg Co.

Keller Furnace. Roasting (see) furnace of multiple-hearth construction, used for roasting sulfide ores.

Keller's Etch. Etchant for macrostructural study of aluminum, consisting of 2.5% nitric acid, 1.5% hydrochloric acid, and 0.5% hydrofluoric acid, in water.

Kelvin Scale. See **Absolute Temperature.**

Kelvin Temperature. See **Absolute Temperature.**
Kennametal. Tungsten-titanium carbide of the formula $WTiC_2$, manufactured by McKenna Metals Co. Used as a sintered carbide for high speed cutting and similar tools.
Kernel Roasting. Heating sulfide ores of copper, to concentrate the metal inside the lumps of heated ore.
Kettle Dross. Mixture of lead and lead oxide, obtained as skimmings from **Desilverization** (see) of lead.
Keyhole Specimen. Specimen for use in connection with the Charpy impact test, in which a keyhole-like notch is cut into the metal.
Kicker. Molten steel of relatively high carbon content, particularly as used in **Duplex Process** (see) for mixing with other steel compositions.
Killed Steel. Steel deoxidized by aluminum, ferrosilicon, etc., to the point at which there is no gas evolution during solidification of the ingot in the mold, causing the top surface of the ingot to freeze, with the consequent formation of a central pipe.
Killed Wire. Wire given a permanent circular set so as to conform to the circle of the drawing block, thus preventing its uncoiling off the block; this is accomplished by setting the draw plate at less than a right angle to the emerging wire.
Killing. (a) See **Killed Steel.** (b) Operating pickling bath, when iron has reached allowable limits, so as to spend it completely; no new acid is added, and the time of immersion is increased until it becomes too long for commercial operation.
Killing Off. Adjusting the composition of over-reduced steel, by adding ore and **Washed Metal** (see) prior to casting.
Kiln. Furnace for **Calcining** (see) coarse ore.
Kind. Referring to steel, "kind" indicates the process of manufacture.
King Pot. Large, central crucible, or pot, in brass founding operations.
King's Silver. Silver of high purity.
Kinite. Strong, wear-resistant die steel with about 14% chromium, 1.1% molybdenum, and fractional percentages of cobalt, nickel, manganese, and silicon; carbon is about 1.5% (Kinite Corp.).
Kink. Defect in metal wire, bar, or sheet, showing sideway waviness.
Kish. (a) Graphite separating from molten, high-carbon cast iron and floating on the surface, or floating loosely in the air. (b) Dross floating on the surface of molten lead.
Kiski Steel. Non-Deforming Steel (see), with about 1.5% manganese, 0.60% tungsten, 0.50% chromium, and 0.20% vanadium (Braeburn Alloy Steel Co.).
Kiss Process. Modification of the **Patera Process** (see), in which calcium hyposulfite is used for ore leaching and the silver is precipitated by means of calcium polysulfide.
Kitchen. (a) See **Arsenic Kitchen.** (b) See **Laboratory.**
Kjellin Furnace. Induction Furnace (see) in which a core of laminated sheet iron is used. A coil of copper wire or water-cooled copper tubing acts as primary, and an annular channel, filled with metal, as secondary.
Kleeman Condenser. Clay pipe of rectangular cross-section, used for condensing zinc vapor in zinc smelting.
K Monel. Age-hardening modified **Monel Metal** (see), containing approximately 3% aluminum. It possesses the corrosion resistance of standard Monel metal, and greater strength and hardness. Used in marine service, springs, etc.
Knobbled Iron. See **Finery.**
Knobbling. See **Spellerizing.**
Knobbling Fire. See **Finery.**
Knobbling Rolls. See **Spellerizing.**
Knoop Hardness Test. See **Microhardness.**
Knox and Osborne Furnace. Wood-fired shaft furnace, operating continuously, for roasting mercury ores, the furnace fireplace being located on one side.
Koehler Furnace. Revolving muffle furnace of cylindrical design.
Konel. Nickel-base alloy, manufactured by Westinghouse Electric & Mfg. Co., with about 17% cobalt, 9%

titanium, and fractional percentages of other metals; because of excellent heat resistance, extensively used in radio tube work and for certain automotive parts.

Konik (Steel). Low-alloy, high-strength steel, with 0.10–0.20% carbon, 0.10–0.30% copper, 0.30–0.50% nickel, and up to 0.30% chromium (Continental Steel Corp.).

Kovar. Low-expansion, iron-base alloy used for sealing metal wire into hard glass, with 28–30% nickel, 15–18% cobalt, and fractional percentages of manganese (Westinghouse Electric & Mfg. Co.).

Kramer Alloy. Series of copper-base alloys, manufactured by H. Kramer & Co.

Krohnke Process. Process of treating silver ores with a solution of cupric chloride, prior to further treatment by **Amalgamation** (see).

Krupp Process. (a) Process for refining pig iron, in which the molten iron is stirred with molten iron oxide and manganese oxide, to oxidize most of the silicon and phosphorus. (b) Process of hardening the surface of steel, such as armor plate, by cementation (see **Metallic Cementation**).

Krupp Washing Process. See **Krupp Process** (a).

K Series. Series of X-ray spectrum frequencies of an element, thought to be due to change of electrons from various higher quantum states to the state whose principal quantum number is 1.

K. S. Magnet Steel. Magnet steel, containing 30–40% cobalt, up to 9% tungsten, 3% chromium, and 0.40–0.80% carbon.

Kunheim Metal. Pyrophoric Alloy (see), made by heating **Misch Metal** (see), or alloys of misch metal with magnesium and aluminum, in a stream of hydrogen, forming hydrides of the rare earth metals.

Kutch. Package of parchment interspersed with gold sheets for the initial steps of **Gold Beating** (see).

K. W. Steel. Low-alloy steel (Carpenter Steel Co.), used for dies, small tools, and points where wear-resistance is of value; contains 3.5% tungsten in addition to about 1.3% carbon.

L

La. Chemical symbol for **Lanthanum** (see).

La Belle (Steel). Series of low-alloy and high-alloy tool steels, manufactured by Crucible Steel Co.

Laboratory. Zone between the fire bridge and flue bridge in a reverberatory furnace, comprising the **Hearth** (see).

La Bour (Alloy). Series of complex nickel-chromium alloys, highly resistant to corrosion (La Bour Co.), with 8–63% nickel, approximately 25% chromium, 6% copper, 2–4% molybdenum, and 1–54% iron.

Lacquer. Coating material, consisting of cellulose derivatives, or similar substances, dissolved in a readily volatile solvent.

Laded Metal. Non-metallic mass of glass, taken from a glass-melting pot for casting.

Ladle. Pot designed for holding molten metal. The metal may be poured over the lip by tilting the ladle, or it may be tapped from the bottom (bottom-pouring).

Ladle Analysis. Chemical composition of molten metal, before it is poured into ingot form; usually, a sample is taken from the molten metal while being poured into the molds.

Lafond's Bearing Metal. See **Lafond's Bronze.**

Lafond's Bronze. Series of copper-base alloys, used for bearings and pump parts, with 2–18% tin, and about 2% zinc.

Lag. See **Cooling Lag** and **Heating Lag.**

Lake Copper. Copper smelted from Lake Superior ores. Because of high natural purity, it is sold without being refined electrolytically, despite the fact that it does contain low fractional percentages of both silver and arsenic.

Lamellar. Composed of thin plates or layers.

Lamellar Pearlite. See **Pearlite.**

Laminating. Forming into thin sheets, as by beating or rolling.
Laminating Machine. Rolling Mill (see) designed for making thin sheets, particularly of precious metals.
Laminating Roller. Adjustable roll for regulating the thickness of rolled metal (see **Rolling Mill**).
Lancaster Process. Process for manufacturing **Wrought Iron** (see), in which a second hearth is superimposed above the first **Finery** (see) to complete the process.
Lanthanum (chemical symbol La). Element No. 57 of the periodic system; atomic weight 138.92. One of the more common of the rare earth metals; a silvery white metal, quite malleable but not ductile; melting point 1494°F; boiling point about 3270°F; specific gravity 6.15. Chemically, it is always trivalent. Lanthanum can be isolated by electrolysis of the anhydrous trichloride in a fused bath of alkali and alkaline earth chlorides. In common with the other rare earth metals, it cannot be electrodeposited and is not used commercially.
Lanyon Shield. Curtain, made of sheet iron, hanging in front of a zinc furnace, to protect the operator from the heat of the furnace.
Lap. Fold of metal which has been rolled or hammered down against the surface of rolled metal, but has not welded into sound metal.
Lap Weld. Weld formed by overlapping the edges of the parts to be welded, followed by hammering or pressure at the appropriate temperature.
Lap Welded Pipe. Pipe made by beveling the edges of the skelp, overlapping the edges after forming the pipe, followed by welding.
Lap Welding Process. See **Lap Welded Pipe.**
Large Wire. Steel wire of greater diameter than common wire (see **Coarse Wire**).
Latent Heat. Heat absorbed or given up when a substance changes its state from solid to liquid (latent heat of fusion), or from liquid to vapor (latent heat of vaporization).
Lateral Spreading. Spreading of width of metal when subjected to rolling.
Latrobe (Steel). Series of carbon tool steels (carbon about 1.0%), manufactured by Vanadium Alloys Steel Co.
Latten. Metal, most commonly brass, in thin sheet form.
Latten Brass. See **Latten.**
Lattice. See **Space Lattice.**
Lattice Constant. Distance between successive planes of a specified plane-family in a crystal.
Laue (X-ray) Method. Method of X-ray examination of crystals which uses a small crystal held in a fixed position and refracting "white" X-rays (X-rays of many wave-lengths).
Lautal. Age-hardening aluminum alloy containing about 5% copper, with small amounts of silicon and manganese. It is extensively used in Germany for aeronautical work.
Lauth Mill. Three-High rolling **Mill** (see) in which large top and bottom rolls of equal size are power driven, whereas the middle roll, driven by friction, has a diameter about two-thirds of that of the others.
Laxal Process. Process of forming rust-resisting and paint-adherent coating on steel, by treatment with hot oxalic acid.
Laying Reel. Reel for reeling steel rod, in which the delivery pipe ends in an arc, causing the rod to coil by a final circular twist.
Leaching. Extracting metallic values of an ore by water or a chemical solution, such as dilute acid. The solution is later treated to recover the metal.
Lead (chemical symbol Pb). Element No. 82 of the periodic system; atomic weight 207.21. Bluish-gray metal of bright luster when untarnished, it is extremely soft and malleable, quite ductile but of low tenacity; highly resistant to ordinary corrosion and to many chemical agents, including acids; melting point 622°F; boiling point 3191°F; specific gravity 11.35. Chemically, it is most frequently divalent (plumbous), with some tendency to tetravalence (plumbic). Usually found in nature as the sulfide, galena, and the metal is obtained by smelting. It is ex-

tensively used for chemical work, in storage batteries, as a base for alloys of the solder, bearing and type metal varieties, for water pipes, etc. It can be electrodeposited, using either a fluoborate or a fluosilicate electrolyte, but this process is not extensively used.

Lead Annealed Wire. Steel wire annealed (see **Annealing**) by passage, usually in single strand form, through a bath of molten lead, followed by cooling in air.

Lead Annealing. See **Annealing.**

Lead-Antimony-Tin Alloy. See **Type Metal.**

Lead Ash. Slag, resulting from the processing of lead ores or lead metal.

Lead Bath. (a) Mass of molten lead, for heat treating of metal, as in **Patenting** (see). (b) Furnace for melting gold or silver ores with lead.

Lead Burning. Joining lead sheet or pipe by means of lead; usually a hydrogen or acetylene flame is employed as a source of heat.

Lead Coating. Process of coating metal, usually iron or steel, by lead or a high-lead alloy. Hot-dipping processes are difficult to carry out; electrolytic methods, though effective, are too expensive for general use. See also **Terne Plate.**

Leaded Brass. Brass (see), either high or low in zinc, containing about 2% lead as an added ingredient. The lead is present as dispersed particles and increases machinability.

Leaded Bronze. Copper-base alloys with approximately 2% lead, added to increase machinability. Malleability and ductility are lowered by the presence of the lead, interfering with severe hot or cold working.

Leaded Commercial Bronze. Commercial Bronze (see) with approximately 1.5% lead, added for easier machinability.

Leaded Copper. Commercially pure copper, with approximately 1% lead, added for easier machinability.

Leaded Gun Metal. Gun Metal (see), with up to 1% each of lead and nickel.

Leaded High Brass. See **Leaded Yellow Brass.**

Leaded Nickel Silver. Nickel Silver (see), with 1–2.8% lead, added for easier machinability.

Leaded Phosphor Bronze. Phosphor Bronze (see), to which approximately 1% of lead has been added for easier machinability. A special grade of leaded phosphor bronze contains about 4% of lead and 4% zinc, yielding a metal about twice as readily machinable.

Leaded Red Brass. Brass (see) of approximately 20% zinc content, with about 1.5% lead, added for easier machinability.

Leaded Steel. Free-machining steel, containing, in fine dispersion, about 0.25% lead.

Leaded Yellow Brass. Yellow **Brass** (see), containing about 33% zinc, with up to 3% lead, added for easier machinability.

Leader Pass. First forming pass (see **Rolling Pass**) after all roughing passes have been completed.

Lead Foil. Lead in **Foil** (see) form.

Lead Fume. Fume (see) from lead processing.

Lead Improving. See **Softening.**

Leading Spindle. Shaft through which power is transferred from the prime mover to a rolling mill.

Lead Patenting. Patenting (see), using a bath of molten lead for heating and quenching. The heating bath is normally kept above the critical point (1450–1900° F), whereas for quenching the 930–1110° F range is most common.

Lead Plating. Coating metal objects with lead; the object to be coated is made cathode (negative electrode) in an electrolytic bath, containing a decomposable lead salt. Because the simple lead salts yield "treeing" (see **Lead Tree**), it is necessary to use the more complex fluosilicate or fluoborate. Lead metal is commonly used as anode (positive electrode).

Lead Quench. In the heat treating of steel, the step of immersing the hot metal in molten lead in order to gain quenched and tempered properties in one step and to avoid the martensitic condition. Hypoeutectoid steels, in small sections, obtain a fine pearlitic structure by this procedure.

Lead Sealed. Lead-coated sheet steel (Continental Steel Corp.).

Lead Shot Metal. Lead containing 0.3–0.8% arsenic, which improves the fluidity of the lead and increases its tendency to form perfect spheres when the metal is dropped through a lead shot tower.

Lead Softening. See **Softening.**

Lead Storage Battery. Electric storage battery in which the positive pole consists of lead peroxide held in lead plates, the negative pole is metallic lead, and a solution of sulfuric acid acts as electrolyte. Its normal voltage is slightly above 2 per cell.

Lead-Thallium. Series of alloys of lead and thallium, with thallium up to 65%; chemically more resistant than lead.

Lead Tree. Tree-like forms which lead assumes when it is electrodeposited from simple salts, such as the nitrate or acetate. To avoid tree formation, lead must be electroplated from baths of more complex salts, such as the fluoborate, fluosilicate, or perchlorate. Occasionally, the phenomenon of treeing is manifested also by other metals.

Lead Wool. Shredded lead, in fine-diameter threads, used as a packing material.

Lead Works. Plant for extraction of lead from its ores.

Leaf. See **Metal Leaf.**

Leafing. Formation of a continuous, brilliant film by flakes of metallic powder.

Leantin. Lead-base bearing metal (Lumen Bearing Co.) low in tin content.

'Lectromelt Furnace. Direct Arc Furnace (see), capable of quick melting. It is of three-phase-current construction, and is extensively used for iron and steel foundry metal melting (Pittsburg Electric Furnace Corp.).

Leddel Alloy. Zinc-base bearing and die-casting series of alloys, with about 5–6% copper, and 5–6% aluminum.

Ledeburite. Eutectic of austenite and cementite, with 4.3% carbon. It forms at 2066°F, from liquid metal, and may be found in cast iron and some of the high-alloy steels.

Ledloy. Leaded Steel (see), manufactured by Inland Steel Co.

Ledrite Brass. Free-Cutting Yellow **Brass** (see) with about 3% lead (Bridgeport Brass Co.).

Lee Process. See **Extrusion.**

Leg Pipe. See **Tuyere Stock.**

Lemon Color Heat. Temperature range from about 1740 to 1830°F, as judged visually.

Leopard Spot. Flaw on the surface of **Tin Plate** (see) in the form of a spot of dull appearance.

Leopoldi Furnace. Furnace, operating intermittently, for the roasting of mercury ores, the mercury vapor being condensed in brick condensing-chambers.

Lesco Steel. Series of low-alloy and high-alloy steels (Latrobe Electric Steel Co.).

Levelling. Flatting rolled metal sheet. See **Roller Levelling** and **Stretcher Levelling.**

Levigation. Grinding to a fine powder under water or other liquid. The term is often used also to denote separation of finer from coarser material by suspension in water.

Ley Pewter. Pewter (see) of very high lead content.

Li. Chemical symbol for **Lithium** (see).

Liberty Silver. See **Nickel Silver.**

Lichtenberg's Metal (or Alloy). See **Fusible Alloys.**

Liege Furnace. See **Belgian Furnace.**

Lift Hammer. See **Tilt Hammer.**

Lifting Table. Roll Table (see) which can be raised or lowered; often it can also be moved sideways.

Light Alloys. See **Light Metals.**

Light Burden. See **Burden.**

Light Case. Case (see) on metal with 0.01–0.02 inch penetration.

Light Cold-Rolled Sheet. Cold-rolled steel sheet of 0.0563 inch thickness or less.

Lightening. See **Blick.**

Light Goods. Standard steel products of relatively small weight or thickness.

Lighting. Starting a **Blast Furnace** (see), usually accomplished by turning a light air blast into the furnace, after filling the space in front of the tuyeres with oil-soaked wood; as the charge drops down in the furnace, the air blast pressure is raised.

Light Metals. Term loosely designating metals of low specific gravity. A specific gravity of 3 is considered normally the upper limit of the light metals. More arbitrarily, it has frequently been limited to the alkaline metals, the alkaline earth metals (despite the inclusion of barium, having a specific gravity of 3.75), beryllium, magnesium, and aluminum.

Light Products. See **Light Goods.**

Light Rail. Rail, for use on railroad service, weighing less than 60 pounds per yard.

Light Ruby Silver. Non-metallic sulfide mineral of silver and arsenic, proustite, corresponding approximately to $Ag_4As_2S_6$ in chemical composition.

Light Yellow Heat. Temperature approximating 2010°F, as judged visually.

Lime. Chemically, calcium oxide, CaO, resulting from the calcination of **Limestone** (see).

Lime Boil. Stage in the basic **Open-Hearth Process** (see) of steel manufacture, following the **Ore Boil** (see), in which limestone calcination becomes more rapid, with release of gaseous carbon dioxide, the lime being lifted to the top of the bath. At the same time, oxidation of carbon in the metal proceeds rapidly.

Lime Bright Annealed Wire. Steel wire annealed (see **Annealing**) by heating bright dry-drawn wire in pots with controlled atmospheres, to prevent surface oxidation.

Lime Coating. See **Lime Finish.**

Lime Finish. Coating of lime on hot-rolled and hot-rolled pickled steel sheets and similar steel products, for temporary protection against oxidation and to improve subsequent fabrication characteristics.

Lime Finished Sheet. Hot-rolled, annealed steel sheet with a final, lime-water treatment. See **Lime Finish.**

Lime Roasting Process. See **Huntington-Heberlein Roasting Process.**

Limestone. Naturally-occurring calcium carbonate, chemically essentially $CaCO_3$. Metallurgically, it is extensively used as a basic flux, the $CaCO_3$ being first calcined to the oxide, CaO, which then reacts as a typical basic oxide.

Liming. See **Lime Finish.**

Limited Austenitic Grain Size. Specification, in ordering or selling steel, limiting the metal to a stated range of grain sizes after a standard heat treatment. Steel is usually considered acceptable if 70% lies within the specified grain size limits. See **Austenitic Grain Size.**

Limiting Creep Stress. See **Creep Limit.**

Limiting Range of Stress. See **Endurance Range.**

Limonite. Iron ore representing, chemically, hydrated ferric oxide, or $xFe_2O_3.yH_2O$. Although quite common in the United States, limonite is not used extensively.

Linear Expansion. See **Thermal Coefficient of Linear Expansion.**

Lined Gold. Gold foil with base-metal backing.

Line Pipe. Seamless steel tubing for use in the pipe-line transportation of gas or oil products.

Line of Segregation. Portion of ingot, at the center of the section and near the bottom of the pipe (see **Pipe Cavity**), in which there is a segregation of low-melting constituents.

Linotype Metal. Type-casting alloy, with 11–12% antimony, 3–5% tin, and the remainder lead, generally used for casting on linotype machines. See **Type Metal.**

Lip. See **Lip Pouring.**

Lipowitz's Metal (or Alloy). See **Fusible Alloys.**

Lip Pouring. Pouring metal, usually for casting purposes, over the edge of the container. The latter normally has a "lip"—i.e., a lip-shaped zone—to facilitate pouring.

Liquation. Separating two or more constituents by heating to a temperature where one constituent melts away, while the others remain in solid form.

Liquation Furnace. Furnace especially designed for **Liquation** (see).

Liquefying. Melting.

Liquid Carburizing. See **Carburizing.**
Liquid (State). One of the three fundamental states of matter; in the liquid state, a material has a specific volume but its shape conforms to the confining vessel.
Liquidus. Line or surface in a temperature-composition diagram or model, indicating the equilibrium temperature of complete fusion when the metal is heated, or complete freezing when liquid metal is cooled.
Liquidus Curve. See **Liquidus.**
Liquidus Line. See **Liquidus.**
Liquidus Temperature. Temperature at which crystallization begins on cooling, and melting is complete on heating.
Liquor Finish. See **Liquor Finished Wire.**
Liquor Finished Wire. Steel wire, with a thin coating of tin-copper alloy, obtained by wet-drawing the wire while immersed in a copper-tin sulfate solution.
List Edge. Edge of hot-coated tin or galvanized plate or sheet leaving the pot last; the list edge inevitably has excess coating.
List Pot. Last pot of a series in tinning or galvanizing operations.
Lithium (chemical symbol Li). Element No. 3 of the periodic system; atomic weight 6.940. Relatively soft, white metal of the alkali group, malleable and ductile, reacts readily with water and corrodes quickly in air, tending to form the nitride; the lightest of all metals, with a specific gravity of only 0.534 (about half the density of water); melting point 367°F; boiling point 2437°F. Chemically, it is always monovalent. Since it reacts with water, it cannot, of course, be electrodeposited from aqueous solutions. Lithium may be isolated by electrolysis of the chloride. It is used, in fractional percentages, as a scavenger (generally as an alloy with calcium), particularly in copper-base metals and in some bearing metals, and is also utilized, in vapor form, as a protective atmosphere in heat-treating furnaces. The metal and its alloys can be obtained from Lithaloys Corp.
Lithographic Stock. **Tin Plate** (see) specially handled during tinning to assure suitability for lithographing on the metal.
Live Load. See **Moving Load.**
Lixiviation. See **Leaching.**
Load-Extension Curve. Curve resulting, in a tensile test, from plotting applied load versus resulting extension. See **Tensile Strength.**
Loading Range. Range of tensile or compressive values in which a testing machine will give accurate results.
Loadstone. See **Lodestone.**
Loam. Mixture of sand and clay, with natural organic matter added, for making metal-casting molds.
Loam Beater. Ramming tool used in forming a **Loam Mold** (see).
Loam Casting. Metal casting formed in a **Loam Mold** (see).
Loam Mold. Mold, for casting metal, in which a low grade of sand is used, with clay, backed by brick or similar material.
Local Action. Corrosion resulting from electric action due to chemical and physical differences within the same metal; certain areas will act electropositive to others, and will consequently tend to dissolve by electrolytic action.
Locomotive Firebox Quality (Plate Steel). Steel of low carbon content, suitable for the manufacture of locomotive fireboxes.
Lo-Cro. Series of low-alloy steels, with 4–6% chromium, and up to 1.25% tungsten or molybdenum (Crucible Steel Co.).
Lodestone. Magnetic iron ore.
Lode Tin. Tin obtained from lodes or veins in rocks. Cornish tin is of this type. The term is also used to describe the original ore.
Lo-Ex. **Low Expansion** aluminum **Alloy** (see), containing about 13% silicon, 2% nickel, 1% copper, and 1% magnesium (Aluminum Co. of America).
Loiseau Furnace. Gas-fired roasting and distillation furnace for processing zinc ores.
Longitudinal Bend Test. Transverse bend made on a longitudinal test speci-

men, cut in the direction of the rolling of the metal.

Longitudinal Specimen. Specimen, for test purposes, with its length parallel to the direction of rolling.

Longitudinal Test. Test, commonly used on steel specimens, to indicate that the longitudinal axis is parallel to the direction of the greatest extension of the material from which the test specimen has been taken. See also **Transverse Test.**

Longmaid Process. See **Henderson Process.**

Long Terne. Large steel sheet subjected to pickling and then coated with a lead-tin alloy by immersion in a fused bath of the alloy (see **Terne Plate**). Usually, long terne measures 60–144 in. in length, and 24–48 in. in width.

Long Terne Full Finish Sheet. Cold-rolled steel sheet with a terne coating. See **Long Terne.**

Long Terne Sheet. See **Long Terne.**

Long-Time Deformation Curve. See **High Temperature Test.**

Long-Time High Temperature Test. See **High Temperature Test.**

Long-Time Test. See **High Temperature Test.**

Looper. Device for taking up the slack between stands of a **Continuous Rolling** (see) mill.

Looping. (a) Fusion of ore during **Calcination** (see). (b) See **Looping Mill.**

Looping Continuous Mill. Continuous Mill (see) in which the rolls are divided into groups of two or four stands, with the metal making 180-degree loops between groups.

Looping Mill. Rolling mill in which the metal rod or sheet is looped from one two-roll stand to the next, usually at its side.

Looping Rod Mill. See **Looping Mill.**

Loop Mill. See **Looping Mill.**

Loosening Bar. Apparatus used for loosening a pattern from a casting **Mold** (see).

Loose Pack Rolling. Method of rolling wide steel sheets, in which steel bars are rolled down, at first singly and then after doubling, about half way, followed by pickling, cleaning, and then reheating before finishing of rolling.

Lorenz's Law. Ratio of the thermal and electrical conductivities, of all metals, is proportional to the absolute temperature.

Losing the Iron. Loss of iron from a **Blast Furnace** (see) due to leakage into the foundations.

Lost Wax Process. Process of casting metal, particularly for jewelry and artistic purposes, by surrounding a wax model with plaster, followed by melting out the wax model, and using the resulting plaster mold for the metal casting.

Lotus Metal. Lead-base bearing metal (Lumen Bearing Co.), with about 15% antimony and 10% tin.

Loup. See **Puddle Ball.**

Low-Alloy High-Strength Steel. See **Mild Alloy Steel.**

Low-Alloy High-Strength Structural Steel. Steel, of greater-than-normal tensile strength, used for structural or similar purposes, the increased strength being achieved by fractional percentages of alloying elements, such as manganese, silicon, nickel, etc.

Low-Alloy Steel. See **Alloy Steel.**

Low-Alloy Tool Steel. Tool Steel (see) of pearlitic structure with one or more alloying elements, commonly including chromium, in fractional percentages.

Low Blast. Air blast at low pressure.

Low Brass. Brass (see) of about 20% zinc content, highly ductile, and, therefore, extensively used for parts requiring drawing or a high degree of mechanical deformation. It lends itself to highly polished surfaces.

Low-Carbon Cast Steel. Cast steel [see **Cast Steel** (a)] of less than 0.22% carbon content.

Low-Carbon Ferro-Chromium. See **Ferro-Chromium.**

Low-Carbon Ferro-Manganese. See **Ferro-Manganese.**

Low-Carbon Ferro-Titanium. See **Ferro-Titanium.**

Low-Carbon Nickel-Molybdenum Vanadium Steel. See **Nickel-Molybdenum-Vanadium Steel.**

Low-Carbon Steel. Steel containing 0.03–0.30% carbon. These steels may be made commercially by any of the standard processes, except the crucible method.

Low-Carbon Steel Wire. Steel wire with carbon content up to the 0.03–0.15% range.

Low-Chromium Carbon Tool Steel. See **Low-Alloy Tool Steel.**

Lower Inwall. See **Blast Furnace.**

Lower Punch. In powder metallurgy, the lower unit in a die assembly, forming the bottom of the die cavity.

Low-Expansion Alloys. Aluminum with alloying constituents balanced to produce relatively low thermal expansion; usually, about 12% silicon is present, with low percentages of copper, magnesium, nickel, and iron. See also **Low-Expansion Steel.**

Low-Expansion Steel. Iron-base alloys, with 30–60% nickel, so called because of low coefficient of thermal expansion. See **Invar.**

Low-Iron Ferro-Manganese. See **Ferro-Manganese.**

Low-Level Bath. Pickling bath, during killing operation. See **Killing** (b).

Low-Manganese Steel. Steel, usually made by the open-hearth process or by electric furnace procedure, with 1–7% of manganese.

Low-Melting-Point Alloys. See **Fusible Alloys.**

Low-Metalloid (Quality) Steel. Steel ordered or sold on the basis of specifications calling for a total of "metalloids" (carbon, phosphorus, sulfur, manganese, and silicon) kept below special limits. Usually, these special limits are either 0.10% total, or 0.25% total. See **Ingot Iron.**

Low-Metalloid (Quality) Strip Steel. Strip made from open-hearth steel, in which total "metalloids" (carbon, manganese, phosphorus, sulfur, and silicon) are maintained within unusually low limits. Three grades are standard: #1, 0.10%; #2, 0.15%; and #3, 0.25% total metalloids.

Low-Metalloid Steel. See **Ingot Iron.**

Low-Phosphorus Pig Iron. Pig Iron (see) with phosphorus below 0.03%, used for steel castings.

Low Red Heat. Temperature between about 1000 and 1300°F; a material at this temperature is dull red in color.

Low Shoulder. See **Round.**

Low Steel. Steel of relatively low carbon content. See **Steel.**

Low-Temperature Impact Test. Charpy Impact Test (see) on steel at temperatures of –50°F and –75°F.

Low-Temperature Test. Test of physical properties of a material at temperatures significantly below room temperature.

Low-Tungsten Steel. See **High-Carbon Low-Tungsten Steel.**

L Series. Series of X-ray spectrum frequencies of an element, arising from change of electrons from different higher quantum states to the state of principal quantum number 2.

L.S.G. See **Metal Gage.**

Lu. Chemical symbol for **Lutecium** (see).

Lubricating. In powder metallurgy, the addition of a lubricating material to a metal powder, to facilitate pressing.

Lubrico. High-Lead Bronze (see) with about 21% lead, used for bearings, etc. (Buckeye Brass & Mfg. Co.).

Lucero. Corrosion and heat-resistant nickel-base alloy series (Driver-Harris Co.), with approximately 30% copper, and, optionally, about 5% manganese; used for electrical resistance wire and pumps, valves, etc.

Luce and Rozan Process. Modification of the **Pattinson Process** (see), in which the crude molten lead is stirred by means of live steam.

Luckhart Furnace. Shaft furnace, with a wood-burning fireplace at the bottom of the shaft, operating on a continuous basis, for the roasting of mercury ores.

Lüder's Lines. Long marks, appearing on the surface of certain metals, in the direction of the maximum shear stress, when the metal is subjected to deformation beyond the yield point.

Ludlite. Stainless steel sheet, integrally bonded to a fireproof, heat-insulating backing (Ludlum Steel Co.).

Ludlum Furnace. Tilting, direct-arc, electric furnace, with three electrodes in a straight line, and using three-phase alternating current. See **Arc Furnace.**

Ludwik Hardness Tester. Device for measuring **Hardness** (see) of a material, in which a 90° cone is indented into the surface, using constant load.

Lukens Clad Steel. Composite (clad) metals (see **Cladding**) with steel as the principal metal, coated with Inconel, Monel, or nickel, manufactured by Lukens Steel Co. The coating usually comprises from about 10% to 20% of the total.

Lukens (Steel). Series of low-alloy steels containing low percentages of nickel, chromium, or manganese, manufactured by Lukens Steel Co.

Lumen Alloy. Series of copper-base alloys, brasses and bronzes (see **Brass** and **Bronze**), manufactured by Lumen Bearing Co.

Lumen Bronze. Not a bronze, but a zinc-base, bearing alloy, with about 10% copper and 4% aluminum (Lumen Bearing Co.).

Lumen Metal. Tin-base bearing alloys (Lumen Bearing Co.), with 4–10% copper, and 5–8% aluminum.

Lumite. See **Stewart Lumite.**

Lurgi Metal. Lead-base, bearing metal of foreign manufacture, containing about 3% barium and fractional percentages of calcium and sodium.

Lurmann Front. In a **Blast Furnace** (see), an arrangement of castings through which iron and slag may be tapped; the system avoids the use of a forehearth.

Lusterite. Extremely hard stainless steel of about 18% chromium and 0.9% carbon, used for knives and surgical instruments (Latrobe Electric Steel Co.).

Lute. Pasty material, usually clay or a clay mixture, for sealing holes and joints in equipment, or for coating surfaces for protection from direct flame.

Lutecine. See **Paris Metal.**

Lutecium (chemical symbol Lu). Element No. 71 of the periodic system; atomic weight 174.99. It is one of the less common of the rare earth elements and has as yet not been isolated in sufficient purity to gain true indication of physical characteristics. Chemically, it is trivalent. Like all rare earth metals, it cannot be electrodeposited.

Lydian Stone. See **Touchstone.**

Lynite. Aluminum-base alloy, made by Aluminum Co. of America, with 9–13.5% copper, and optional additions of manganese, magnesium, and silicon; used principally as a piston alloy, and also for other automotive purposes.

M

Ma. Chemical symbol for **Masurium** (see).

MacArthur and Forest (Cyanide) Process. Treating gold ore with a solution of potassium cyanide, followed by cementation with aluminum or zinc, or by electrolytic deposition.

Macco (Steel). Series of simple carbon and low-alloy and high-alloy steels, manufactured by P. F. McDonald & Co.

MacDougall Furnace. Original form of vertical, cylindrical, **Roasting** (see) furnace, mechanically rabbled, with six hearths superimposed over one another; the ore travelled from the center to the circumference of the top hearth, then dropped to the hearth below, etc.

MacHempite. Series of iron-base alloys, with up to 4% nickel, up to 4% manganese and up to 1.25% chromium; carbon is in the 0.4–3.0% range (Mackintosh-Hemphill Co.).

Machinability. Property of metals and alloys, permitting them to be subjected to machining operations, such as cutting, drilling, etc. The degree of machinability is different for various metals and alloys. Generally, metals having only a single phase tend to tear rather than to machine cleanly; coarse-grained metals are, in general, more easily rough-machined than fine-grained ones; fine-grained metals are better for the finer finishes.

Machinability Test. Test to determine **Cutting Hardness** (see) and ease of machinability of a material.

Machine Cast Pig. Pig iron which has been cast into mechanically-actuated metallic molds.

Machine Cutting. Cutting or milling bars to exact size, after preliminary rough cutting by use of machine tools.

Machine Molding. Making **Molds** (see), for the casting of metal, by machine methods.

Machinery Brass. Red **Brass** (see) of about 16% zinc content, modified with about 1% tin.

Machinery Steel. See **Machine Steel.**

Machine Steel. Open-hearth steel, with carbon normally not over 0.30%. More broadly, any steel not to be subjected to heat treatment, used in the manufacture of machine parts.

Machine Straightening. Method of reducing the **Camber** (see) of bar, strip, or sheet metal.

Machining. Mechanical cutting, drilling, or similar treatment of any material.

Mach's Alloy. See **Magnalium.**

Macro Etching. See **Etching.**

Macro-Etch Test. See **Etch Test.**

Macrograph. Reproduction of an object or metallographic structure, at actual size, reduced, or not enlarged more than about 10 diameters.

Macroscopic. Visible to the naked eye (or practically so—requiring very little magnification).

Macroscopic Deep Etch Test. See **Deep Etch Test.**

Macroscopic Structure. See **Macrostructure.**

Macro-Segregation. Gross differences in composition, existing in a mass of metal such as ingot, casting, or forging.

Macrostructure. Gross structure of metal, as shown by examination of a polished surface, without a microscope.

Magazine Feeder. Device for holding metal bars or slabs, and discharging the units individually as required for further processing.

Magnaflux Testing. Testing steel by disclosing, magnetically, the presence of edges and surfaces, to determine mechanical and structural discontinuities.

Magnalite. Aluminum-base casting alloy, with about 2.5% copper, 1.5% nickel, 1.5% magnesium, and 0.5% zinc.

Magnalium. Series of aluminum-base alloys, with 5–30% magnesium, and, optionally, small quantities of copper, nickel, tin, and lead. The alloys, lighter than aluminum, are used principally for casting purposes.

Magne-Gage. Gage, measuring surface magnetic characteristics, used for the determination of the thickness of non-magnetic coatings on iron and steel, or of magnetic coatings on non-magnetic base metals (American Instrument Co.).

Magnesia. Chemically, magnesium oxide, MgO, resulting from the calcination of natural magnesite or any other carbonate of magnesium; extensively used as a basic refractory.

Magnesite. Ore of magnesium, $MgCO_3$, used metallurgically, after calcination to MgO, as a basic refractory, and also, in more recent processes, as a source of magnesium.

Magnesium (chemical symbol Mg). Element No. 12 of the periodic system; atomic weight 24.32; white, tough, easily machined metal which is quite ductile and malleable above 660°F, but only slightly malleable at room temperature. It tarnishes easily in air, and is quite susceptible to salt-water corrosion, and therefore requires surface protection when used as a structural metal. Specific gravity 1.75, lightest of all atmospherically stable metals; melting point 1204°F; boiling point 2025°F. Chemically, it is always divalent. It cannot be electrodeposited. Magnesium is isolated commercially by electrolysis of the dehydrated chloride, and, more recently, by carbon and by ferrosilicon reduction of the oxide. Extensively used in aeronautics and similar fields, in pyrotechnics (because of the extremely actinic quality of the light resulting from burning of the metal in air), and in incendiary bombs. In metallurgical practice, it is frequently used as a deoxidizer and scavenger in copper-base alloys and in nickel. It is a common constituent of many aluminum-base alloys. Its principal source has been the Dow Chemical Co., though recently new manufacturers are entering the field.

Magnesium-Nickel. Magnesium and nickel alloy readily, and master alloys, usually containing about 50% of each, are used for the scavenging of nickel-base alloys.

Magnesium Powder. Magnesium, in finely powdered form, used principally for photographic, pyrotechnic, and incendiary purposes.

Magnetic Aging. Magnetic changes occurring in certain types of steel, on standing at room or sub-normal temperatures. See **Aging.**

Magnetic Intensity. See **Magnetizing Force.**

Magnetic Iron Ore. See **Magnetite.**

Magnetic Method. See **Critical Point Test.**

Magnetic Permeability. Ratio between magnetic induction in a substance and the causative magnetic field.

Magnetic Scale. Listing of metals, in the sequence of magnetizability.

Magnetic Separator. Device, operating on magnetic principles, for separating magnetic materials (such as iron) from non-magnetic ones.

Magnetic Test. Test of the characteristics or soundness of a metal specimen, as determined by magnetic methods.

Magnetic Transformation Point. See **Curie Point.**

Magnetism. See **Ferromagnetism** and **Diamagnetism.**

Magnetite. Magnetic iron ore, Fe_3O_4, found only in low grades in the United States, but comparatively pure in Sweden.

Magnetizing Field Intensity. See **Magnetizing Force.**

Magnetizing Force. Number of magnetic lines of force intersecting a unit area of material.

Magnet Steel. Steel capable of intensive magnetization. At first, hardened carbon steels were used; then chrome and tungsten steels. Cobalt steels came into use during the first World War; more recently, the aluminum-nickel-cobalt (see **Alnico**) steels have proven of still higher magnetic value.

Magno. Nickel-base electric-resistance alloy, containing 4–5% manganese (Driver-Harris Co.).

Magnolia Metal. Lead-base bearing metal, with about 15% antimony, tin and other elements being optional (Magnolia Metal Co.).

Maillechort. Copper-base hardware alloy, with about 13.5% zinc, 17–19% nickel, and up to 3.5% iron.

Making Bottom. Building up a furnace bottom in an open-hearth or blast furnace, after preliminary heating. In the case of the open-hearth furnace, the acid or basic material is scattered over the area of the bottom, and heated to sintering temperature for a number of hours; this step is repeated until the desired bottom thickness has been attained.

Maldonite. Native mineral alloy of gold and bismuth, corresponding approximately to Au_2Bi in chemical composition.

Maletra Furnace. Reverberatory furnace, rabbled by hand, used for roasting finely divided ores, without the application of outside heat.

Malleability. Ease of flattening a material, when subjected to rolling or hammering. Primarily, the malleability of a metal depends on its toughness and tensile strength; coarsely crystalline structures are, in general, not malleable. For this reason, any alloy constituent or treatment which increases grain coarseness will adversely affect malleability. Gold is the most malleable metal (sheets as fine as 0.000004 inch have been made). In order of decreasing malleability, after gold, are: silver, copper, tin, platinum, lead, zinc, iron, nickel. Many metals originally believed to be non-malleable have proven to be excellent in this respect when prepared in highly purified form.

Malleabilized Casting. See **Malleabilizing.**

Malleabilizing. Process of annealing white cast iron to convert its chemically combined carbon, at least partially, into free graphitic carbon in rounded form, the latter being known as temper carbon.

Malleable Brass. See **Muntz Metal.**

Malleable Casting. See **Malleable (Cast) Iron.**

Malleable (Cast) Iron. Cast iron in which the free carbon has been physically changed as compared to its form in gray cast iron, so that the metal became malleable. Malleable cast iron is made from white iron castings by the annealing process called **Graphitizing** (see). The metal is packed in iron oxide and held at a temperature of about 1600°F for several days, followed by slow cooling in the furnace.

Malleable Iron Casting. See **Malleable (Cast) Iron.**

Malleable Pig (Iron). Pig Iron (see) with phosphorus below 0.20%, and chromium absent, used for making **Malleable Cast Iron** (see).

Malm Process. Treating complex sulfide ores, particularly those of zinc, with dry chlorine in a tube mill at a temperature of about 160°F, followed by heating on a multiple-hearth furnace at about 750°F. The resulting metal chlorides, after solution in water, or hot brine, are then converted to metal. In the case of zinc, this is accomplished by electrolysis of the fused chloride, the by-product chlorine returning to the process.

Mandrel Test. Test on galvanized steel wire (see **Galvanizing**) to determine adherence of the coating; the wire is wound around a mandrel of a diameter which is a pre-assigned multiple of the diameter of the wire itself.

Manganese (chemical symbol Mn). Element No. 25 of the periodic system; atomic weight 54.93. Lustrous, reddish-white metal of hard, brittle, and, therefore, non-malleable character. When pure, it is unaffected by normal atmospheric influences; melting point 2277°F; boiling point about 3640°F; specific gravity 7.2. Chemically, manganese shows a great variety of valences. Principal among them are: divalence (manganous) and trivalence (manganic); in the important permanganates, the metal has a valence of seven. In pure form, manganese can be isolated by metallic electrodeposition from a solution of a pure salt, such as the sulfate, followed by fusion under vacuum. The metal is used in large quantities in the form of **Spiegel** (see) and **Ferromanganese** (see) for steel manufacture as well as in manganese steels, and many copper-base alloys.

Manganese-Aluminum. Master alloys of manganese and aluminum, often containing 25% manganese, used for addition to aluminum alloys as hardening agents.

Manganese-Boron. Alloy containing 20–25% boron and 60–65% of manganese, used as a deoxidizing and degasifying agent.

Manganese Brass. See **Manganese Bronze.**

Manganese Bronze. Copper-base alloy series, with zinc normally approximating 40%, manganese up to 4%, and small amounts of tin and aluminum. High-tensile manganese bronze is lower in zinc, approximately 20%, and higher in aluminum. Many other modifications are known. Characterized by good strength and toughness, and excellent resistance to salt-water corrosion.

Manganese Casting Brass. Yellow brass with low percentages of manganese; other metals may also be present in small amounts.

Manganese-Copper. Alloys of manganese and copper, with manganese about 10–50%, and optional additions of iron, aluminum, silicon, zinc, etc.; generally, resistant to heat and corrosion.

Manganese-Molybdenum Cast Steel. Cast steel [see **Cast Steel** (a)], with good creep resistance, particularly at higher temperatures; carbon approximates 0.30%, manganese 0.90–1.75%, and molybdenum 0.15–0.45%.

Manganese-Molybdenum Steel. Low-alloy steel, usable where resistance to fatigue and impact is necessary, with 1.1–1.8% manganese, and 0.15–0.40% molybdenum.

Manganese-Nickel. See **Nickel-Manganese.**

Manganese-Silicon. Master alloy of manganese and silicon, low in iron, with 20–25% manganese; used for adding these elements to non-ferrous metals.

Manganese Steel. Steel, usually made in the acid open-hearth or electric furnace, containing more than 7% manganese. Because of extreme toughness and resistance to abrasion, these steels are extensively used in railroad crossings, switches, etc. Hadfield manganese steel, substantially austenitic in character, contains about 12% manganese and about 1.0–1.2% carbon. Medium manganese steel, also known as "Perman" steel or pearlitic manganese steel, to differentiate it from Hadfield steel, contains approximately 1.5–2.0% manganese and 0.3–0.8% carbon; 0.1–0.6% molybdenum is an optional constituent.

Manganese Sulfide. Compound, formed by interaction of manganese with the sulfur always present in iron-base alloys; chemically MnS. When in this form, sulfur in steel is practically devoid of its usual objectionable property of causing hot shortness.

Manganese-Titanium. Series of master alloys of manganese and titanium, with up to about 31% titanium and 53% manganese, for deoxidizing steel and non-ferrous alloys; the commercial alloys contain small percentages of aluminum, silicon, and iron, in addition to manganese and titanium.

Manganese-Vanadium Steel. Low-alloy steel, of good strength and ductility, with 1.45–1.75% manganese, and approximately 0.15% vanadium.

Manganiferous Cementite. **Cementite** (see) in which part or most of the iron has been replaced by manganese.

Manganin. Copper-base alloy, containing approximately 12% manganese and 4% nickel, manufactured by Driver-Harris Co. and others. Because of high electrical resistance and low temperature coefficient, extensively used in electrical instruments where constant resistivity is of importance.

Mangano (Steel). **Non-Deforming Steel** (see) with about 1.6% manganese and 0.20% chromium; carbon approximates 1.00% (Latrobe Electric Steel Co.).

Manhes Process. Treatment of copper matte in a blast furnace; the air blast, passing through the molten matte, oxidizes the sulfur of the matte to gaseous sulfur dioxide.

Manipulator. Device, in **Rolling Mill** (see) operations, for manipulating the ingot and rolled metal between various passes.

Mannesmann Machine. See **Mannesmann Process.**

Mannesmann Mill. See **Mannesmann Process.**

Mannesmann Piercer. See **Mannesmann Process.**

Mannesmann Process. Making seamless tubing by piercing a solid round billet. Two specially-shaped heavy rolls, mounted on an angle, rotate the billet and force it over a stationary mandrel, located between the rolls.

Mannheim Gold. Copper-base alloy, used for cheap jewelry, containing 7–20% zinc, and up to about 9% tin.

Man-Ten Steel. High-strength manganese steel with 1.25–1.75% manganese, about 0.20% copper and 0.20% silicon, and 0.30% carbon (United States Steel Corp.).

Mantle. See **Blast Furnace.**

Manufacturers' Wire. Steel wire for industrial conversion into manufactured products.

Manufacturing Terne. See **Short Terne.**

Marine Boiler Quality (Plate Steel). Steel, of low-carbon content, suitable for the manufacture of marine boilers. See **Boiler Steel.**

Marine Boiler Steel. See **Boiler Steel.**

Marine Bronze. Yellow **Brass** (see) of about 41% zinc, modified with fractional percentages of aluminum, nickel and tin.

Marine Metal. Metal used as anti-corrosive sheathing for ships, most commonly copper or a high-copper alloy.

Market Lead. Commercial desilverized lead, most commonly in pig form, ready for market.

Market Pot. The last in the series of silver-refining pots of the **Pattinson Process** (see).

Market Wire. Low-carbon annealed-in-process steel wire, used in bright-annealed and metal-coated finishes.

Marriner Process. Treating gold or silver ores by dead-roasting, grinding, and then agitating with a cyanide solution. See **Cyaniding** (b).

Marsh Ore. See **Limonite.**

Marsh Test. Delicate chemical test for the presence of arsenic.

Martempering. Heat treatment of steel, resulting in a martensitic structure (see **Martensite**) with a minimum of residual hardening strains. The steel is quenched into a medium capable of extracting heat at a rapid rate and then transferred to a constant temperature bath (or, with smaller pieces, quenched directly into the latter), at a temperature (called M_s point) just above that at which the austenite transforms into martensite. When the piece has equalized in temperature, it is removed and allowed to cool in air.

Marten (Hardness) Scale. Early procedure for hardness measurement, in which a diamond point, under constant load, was used to make a scratch in the surface being tested, and the width of the scratch used as a measure of the hardness.

Martens-Heyn (Hardness) Test. Test in which a ball indentor is pressed into the surface to a standard depth, the pressure required being a measure of the hardness.

Martensite. Structure existing in steel as a transition stage in the transformation of austenite. It is the hardest constituent of steel of eutectoid composition and represents a solution of carbon or Fe_3C in very fine alpha iron, with either carbon or Fe_3C in chemical dispersion. Obtained by quenching hypereutectoid steel in cold water, it is the chief constituent of hardened carbon tool steels. Martensite is magnetic.

Martensitic Nickel Steel. See **Nickel Steel.**

Martensitic Steel. See **Martensite.**

Martin Process. See **Siemens-Martin Process.**

Martin Steel. High-carbon (about 0.70%) steel, with about 0.40% silicon.

Mass Copper. Metallic **Copper** (see), occurring native as a mineral in large masses.

Mass Effect. Gradual decrease of hardness, in quenched metal, particularly steel, from the surface of the metal to the center, due to delay in heat transfer.

Master Alloy. Frequently, it is more convenient to add an alloying element in the form of a rich alloy than as a pure metal. Such an alloy, containing a large amount of the desired element, is known as a master alloy.

Masurium (chemical symbol Ma). Element No. 43 of the periodic system; atomic weight approximately 99. Only recently discovered, extremely rare, relatively little is known about the element and its compounds. Because of its position in the periodic system, it is assumed to resemble manganese.

Matching. In rolling metal in packs, the process of placing sheets together for proper rolling. See **Pack Rolling.**

Match Plate. See **Bottom Board.**

Mat (Finish). Dull surface finish, without luster, usually deliberately created on metal by mechanical or chemical methods.

Mathesius Metal. Lead-base bearing alloy, with about 3% calcium and up to 2% strontium or barium.

Matrix. Background material in which a constituent is present or is formed. See also **Matrix Metal.**

Matrix Metal. In powder metallurgy, any ingredient in a powder mixture which melts during sintering and thereby acts as a cementing medium. See also **Matrix.**

Matte. Intermediate, semi-metallic, sulfur-containing product, obtained in the smelting of sulfide ores, especially those of copper and nickel.

Matting. Process of forming a **Matte** (see).

Maturing. See **Aging.**

Max-El Steel. Machine Steel (see), containing about 1% manganese and 0.2% molybdenum (Crucible Steel Co.).

Maximum Load. See **Ultimate Strength.**

Maximum Stress. See **Tensile Strength.**

Maxite. High-speed alloy tool steel (Columbia Tool Steel Co.), with about 18% tungsten, 4% cobalt, 4% chromium, and 2% vanadium.

Mayari Pig (Iron). Pig iron, containing as alloying elements, found in the original natural Cuban ores from which the iron was reduced, about 1% nickel, 2% chromium, and fractional percentages of titanium and vanadium (Bethlehem Steel Co.).

Mayari Steel. Steel made from **Mayari Pig Iron** (see).

Mazlo. Series of magnesium-base alloys (American Magnesium Corp.).

McAdams Alloy. Series of aluminum-base alloys, with up to 55% copper, up to 43% chromium, up to 23% zinc, up to 8% cadmium, up to 8% tin, up to 5% antimony and up to 1% silver.

McFarland and Harder Alloy. Series of corrosion-resistant alloys of copper-nickel base (copper up to 55%, nickel up to 59%) with 10–43% chromium.

McGill Metal. Aluminum bronze (about 9% aluminum) containing about 2% iron (McGill Mfg. Co.).

McInnes (Steel). Series of low-alloy and high-alloy tool steels, manufactured by McInnes Co.

McQuaid-Ehn Test. Standard test for the determination of grain size and other properties of steel, particularly grain-growth rate of austenite at carburizing temperatures. The sample is carburized at 1750°F in a solid carburizer which will yield a hypereutectoid case; the sample is then cooled slowly enough to develop a pearlitic structure with a free cementite network around the pearlitic grains. Abnormal steel will yield free ferrite and lamellar pearlite instead of normal pearlite grains. For comparison, grain size charts have been set up, to allow classification of steels according to their grain size as determined by this test.

Meadow Ore. See **Limonite.**

Mechanical Mixture. Mixture of two or more substances, susceptible of mechanical or physical separation.

Mechanical Properties. See **Physical Properties.**

Mechanical Puddler. Stirring and rabbling device for agitating molten metal or slag, particularly in **Puddling** (see).

Mechanical Puddling. See **Mechanical Puddler.**

Mechanical Rabble. Stirring and rabbling device for agitating materials during processing, particularly in **Roasting** (see).

Mechanical Repeater. See **Repeater.**

Mechanical Testing. See **Physical Testing.**

Mechanical Tubing. Metal tubing of special, non-circular shape, such as rectangular, oval, etc.

Mechanical Twins. Twin Crystals (see) formed in certain metals, such as iron, as a result of deformation.

Mechanical Working. Plastic deformation or other physical change to which metal is subjected, by rolling, hammering, etc.

Medium Alloy Steel. See **Pearlitic Alloy Steel.**

Medium Annealing. Annealing on cold-rolled steel, most commonly tubing, in which a very light, loose scale is formed, with a black, oxidized surface underneath.

Medium-Carbon Ferro-Carbon-Titanium. See **Ferro-Carbon-Titanium.**

Medium-Carbon Ferro-Manganese. See **Ferro-Manganese.**

Medium-Carbon Steel. Hypoeutectoid steel with carbon from 0.30% to the eutectoid value, 0.85%, made by any of the standard processes.

Medium Case. Case (see) on metal, with penetration between 0.02 and 0.04 inches.

Medium Chill. See **Chilled Rolls.**

Medium Heavy Case. Case (see) on metal, with penetration between 0.04 and 0.06 inches.

Medium High-Carbon Steel Wire. Steel wire with carbon content from the 0.24–0.30% range to the 0.36–0.44% range.

Medium Low-Carbon Steel Wire. Steel wire with carbon content from the 0.11–0.16% range to the 0.24–0.30% range.

Medium Manganese Cast Steel. Cast steel of high strength and toughness [see **Cast Steel** (a)], with 0.20–0.50% carbon and 1.0–2.0% manganese.

Medium Manganese Steel. See **Manganese Steel.**

Medium Orange Heat. Temperature approximating 1200°F.

Medium Steel. See **Medium-Carbon Steel.**

Meehanite (Metal). Special cast iron of high resistance to heat and corrosion, manufactured by Meehanite Metal Corp.

Melonite. See **Tellurnickel.**

Melotte's Alloy (or Metal). See **Fusible Alloys.**

Melt. (a) To convert a solid substance into the liquid state. (b) Quantity of material melted during any specified period.

Melting Finery. See **Finery.**

Melting Furnace. See **Crucible Furnace.**

Melting Heat Steel. Steel, with about 1.50% carbon, made by the carburization of wrought iron.

Melting Hole. See **Crucible Furnace.**

Melting Point. Temperature at which a substance changes from the solid to the liquid state. Pure substances have a sharp and distinct melting point; most alloys have a melting range, with certain constituents melting before others. The freezing point of a substance is, theoretically, identical with the melting point, the freezing range with the melting range.

Melting Pot. Pot of metal or other material, used for the melting of metal. See also **Crucible.**

Melting Range. See **Melting Point.**

Mendelejeff Group. See **Periodic Group.**

Mendelejeff Law. See **Periodic System.**

Mender. Tin or terne plate, too imperfect to be sold, but usable if run through the metal pot again.

Merchant Bar (Iron). In the manufacture of wrought iron, **Muck Bar** (see) lengths of 2–3 foot are bound together and raised to white heat; this is followed by rolling, and the product is known as merchant bar, single rolled iron, single refined iron, or Number 2 iron.

Merchant Iron. See **Merchant Bar (Iron).**

Merchant Mill. Small rolling mill used for making a variety of products, of varying cross-section.

Merchant Roll. Finishing roll of a **Merchant Mill** (see).

Merchant Train. **Rolling Mill** (see) for the manufacture of **Merchant Bar Iron** (see).

Merchant Wire. Steel wire, usually in annealed or galvanized finish, used for general direct-consumer purposes such as baling, bracing, etc.

Mercoloy. Copper-base alloy, with about 25% nickel, 10% zinc, 2% iron, 2% lead, and 1% tin; used for valves (Merco Nordstrom Valve Co.).

Mercury (chemical symbol Hg). Element No. 80 of the periodic system; atomic weight 200.61. Heavy, bright silvery liquid, of 13.55 specific gravity; freezes at –38°F and boils at 674°F. It is substantially unaffected by air or moisture. Chemically, it is monovalent (mercurous) or divalent (mercuric). It can be electrodeposited from any soluble salt, but this process is not used commercially, except for co-depositing mercury with cadmium. Mercury is obtained from its ores, usually cinnabar, HgS, by heating in a current of air, followed by condensation of the vapor. Alloys of mercury, called amalgams, are used in dentistry, mirrors, etc.; the phenomenon of amalgamation is also utilized in the recovery of gold and silver from their ores. Unalloyed, the metal is extensively used in scientific instruments, thermometers, barometers, etc., and for fluorescent bulbs; mercury boilers, using the metal instead of water, represent an interesting development.

Mercury Furnace. Furnace for the roasting of mercury ores, with vaporization and condensation of the resulting metallic mercury.

Mercury Gatherer. Equipment for stirring mercury that has suffered flouring (see **Floured Mercury**) to enable it to coalesce again into fluid form.

Merica's Solution. Etchant for nickel-base metals, consisting of equal parts of 70% nitric acid and 50% acetic acid.

Merit Number. Tensile strength of a metal specimen multiplied by its percentage elongation, frequently indicative of the quality of hot-rolled metal.

Merrillite. High-purity zinc dust (Merrill Co.).

Merrit Plate. See **Bloomery.**

Mesh. In powder metallurgy, the finest screen size through which all of the sample under consideration will pass; some part of the sample, of course, will also pass through finer screens.

Mesh Fraction. In powder metallurgy, the part of a powder sample which passes through a specified screen but is retained on a finer screen.

Mesothorium. Isotope (see **Atomic Weight**) of radium, chemically identical with it, but having a much shorter half-life. Mesothorium decomposes into radiothorium, chemically identical with thorium.

Metal. Element (or, more loosely, also mixture of elements) having characteristic metallic lustre, and high electric and thermal conductivity, without decomposition.

Metal Alkyl. Compound of a metal and an organic radical, devoid of oxygen.

Metal Arc Welding. Arc Welding (see) utilizing metal electrodes, the metal melting and acting as filler or binder.

Metal Bath. Bath of fused lead or a low-melting metal mixture, for use in the heat treatment of steel or other chemical process requiring controlled heating.

Metal Blasting. See **Blast Cleaning.**

Metal Cementation. See **Metallic Cementation.**

Metal Dip Brazing. See **Dip Brazing.**

Metal Foil. Metal sheet, thinner than about 0.010 in., and thicker than 0.001 in. The most common foils are those of aluminum, lead, and tin.

Metal Gage. Standard thickness or diameter of metal sheet or wire. American Screw Gage is used on machine screws. American Wire Gage, also called A.W.G., B. & S. G., or Browne & Sharp Gage, is used on copper-base and other non-ferrous metals. Birmingham Wire Gage, also called B.W.G., and Stubs' Iron Wire Gage, is used on seamless tubing and spring steel. British Imperial Gage, also called B.I.G., L.S.G., and Legal Standard Gage, is commonly used in England for copper wire. Circular Mill Gage, also called Edison Gage, is based on the circular mill and used for wire. Galvanized Sheet Gage, abbreviated G.S.G., is based on pounds of zinc coating per square foot of steel sheet on galvanized steel sheet. Music Wire Gage, abbreviated M.W.G., is used on piano and music wire. Stubs' Steel Wire Gage is used on steel drill rod. Tin Plate Gage, abbreviated T.P.G., is based on pounds of tin coating per square foot of steel sheet on tin plate. Twist Drill Gage, also called Morse Gage, is used on drill rod. United States Standard Gage, abbreviated U.S.S.G., is used on steel sheets and light plates. Washburn & Moen Gage, also called Roebling Gage, National Wire Gage, United States Steel Wire Gage, or simply Steel Wire Gage, is used on steel wire generally. Zinc Gage is used on rolled zinc.

Metal Leaf. Metal in extremely thin form.

Metalled. Surfaced or ballasted with stone. The term is non-metallurgical, and is part of highway and railway terminology.

Metallic Cementation. Process by which one metal is absorbed into the surface of another. See **Sherardizing, Siliconizing,** and **Calorizing.**

Metallic Coating. Covering a base metal with another, such as by galvanizing, spraying, cementation, etc.

Metallic Conduction. Conduction of electricity in metals, by electronic action, without physical change in the material. See also **Electrolytic Conduction.**

Metallic Hardening Process. Patenting (see) process in which the metal, after heating in the standard manner, is drawn into a lead bath at a relatively low temperature.

Metallic Hardness. See **Hardness.**

Metallic Lustre. Surface sheen, characteristic of most metals when compact and free of surface oxidization. Finely divided and powdered metals are usually black, and lack metallic lustre.

Metallic Oxide. Binary compound of a metal and oxygen.

Metallic Shielded Arc Electrode. See **Shielded Arc Welding.**

Metallic Soap. Compound of a fatty acid with a metal.

Metallic Sulfide. Oxygen-free compound of one or more metals with sulfur.

Metallic Surface Absorption. See **Metallic Cementation.**

Metalliferous. Containing metal.

Metallized Wood. (a) Wood impregnated with molten metal; of course, only **Fusible Alloys** (see) or metals can be used for this purpose. (b) Wood coated with metal by **Metal Spraying** (see).

Metallizing. (a) See **Metal Spraying.** (b) See **Metallized Wood.**

Metallography. Science of the structure of metals and their constituents, principally as revealed by microscopy.

Metalloid. (a) Element intermediate in lustre and conductivity between the true metals and non-metals. Arsenic, antimony, boron, tellurium, and selenium, etc., are generally considered metalloids; frequently one allotropic modification of an element will be nonmetallic, another, metalloid in character. Obviously, no hard and fast line can be drawn. (b) In steel metallurgy, metalloid has a specialized, even if erroneous, meaning; it covers elements commonly present in simple steel: carbon, manganese, phosphorus, silicon, and sulfur.

Metallurgical Coke. Coke (see) of sufficient strength and chemical purity for use in metallurgical reduction operations.

Metallurgical Fume. See **Fume.**

Metallurgical Smoke. Mixture of gases, flue dust, and fume, emanating from a metallurgical process or furnace.

Metallurgy. Science of metals, their extraction from ores, processing, and utilization.

Metal Notch. Opening in a furnace, out of which molten metal is drawn.

Metal Protection. Method of preventing rusting or corrosion of the base metal, as accomplished by plating, galvanizing, spraying, etc.

Metal Spray Gun. See **Metallic Spraying.**

Metal Spraying. Process in which metal wire is fed into an oxy-hydrogen or oxy-acetylene blast, the metal being volatilized and projected, at high speed, against any object held in its path. The object may be of metal or any other material. Any metal which can be volatilized in the flame and which will not react unduly with oxygen or the products of combustion can be treated in this manner. A modification of the process uses an electric arc, by which practically all metals can be volatilized.

Metastability. State of false stability, such as supersaturation, continuing to exist as long as not disturbed, but returning to normal stability when disturbed.

Metcalf Test. Test on steel, to determine heat-treating characteristics; one end of a bar of the steel under test is heated to the point of scintillation, the other end being kept relatively cool. After cooling in still air, the steel is broken at every inch, and visual examination made of each fracture.

Meteoric Iron. Elementary iron, frequently alloyed with nickel, as found in meteors. See **Aerolite.**

Meteorite. See **Aerolite** (b).

Mg. Chemical symbol for **Magnesium** (see).

M. H. Alloy. Series of copper-base alloys, most of them bronzes of 5–16% tin content, with up to 25% lead and optional low percentages of zinc, nickel, and iron (McCallum-Hatch Bronze Co.).

Michiana Alloy. Series of heat-and-corrosion-resistant, cast, ferrous-base alloys, with chromium up to 28% and nickel up to 20% (Michiana Products Corp.).

Microanalysis. Ascertaining chemical composition from quantities not more than one-tenth as large as in customary laboratory practice.

Microcharacter. See **Microhardness.**

Microcrystalline. Possessing extremely small crystals, e.g., of microscopic size or smaller.

Micro-Etching. See **Etching.**

Microhardness. Hardness of microscopic constituents of a metal, as measured by any special apparatus, such as the micro-character, which carries a tiny diamond point, or the AB microhardness tester and the Knoop hardness tester, in which a tiny diamond indentor measures the hardness.

Micron. One-thousandth of a millimeter (0.00003937 inches).

Microscopic Structure. See **Micro-Structure.**

Microscopic Test. Examination of metal, with or without special treatment and etching, with the aid of the microscope.

Micro-Segregation. Composition differences occurring within crystal due to dendritic growth.

Micro-Structure. Detailed structure of metal, usually as determined by examination of an etched and polished surface under the microscope.

Middle Inwall. See **Blast Furnace.**

Middle Roll. Central roll in a **Three-High Mill** (see).

Midvale (Steel). See **Midvaloy.**

Midvaloy. Series of steels, of varying composition and for various purposes, manufactured by Midvale Co.

Mil. See **Circular Mil.**

Mild Alloy Steel. Steel of low-alloy content (see **Low-Alloy Steel**) and relatively high strength.

Mild Chill. See **Chilled Rolls.**

Mildew Bronze. Bronze, usually in cast form, chemically treated to simulate the effect of mildewing and long exposure to natural influences.

Mild Steel. (a) Hot-rolled simple steel, of up to about 0.40% carbon content. (b) See **Ingot Iron.**

Mil-Foot. Wire length of one foot, with a diameter of one mil (see **Circular Mil**).

Mill. See **Rolling Mill.**

Millard Metal. See **Avialite.**

Mill Bar. Bar rolled from a bloom or puddle bar, for further rolling into **Merchant Bar Iron** (see).

Mill Cinder. Slag resulting from **Puddling** (see) furnace operations.

Mill Edge. Normal, irregular edge, obtained on metal sheet or strip after rolling, without filing or edge rolling.

Miller Indices. Indices, h, k and l, which define a crystal face, h, k and l being small whole numbers which are the numerators of the reciprocals of the intercepts on the three coordinate axes of reference when the reciprocals have been reduced to a lowest common denominator.

Miller Process. Treatment of gold and silver bullion in crucibles, in a short cylindrical furnace, with chlorine gas. Base metals are removed as volatile chlorides, and then the silver is chlorinated to silver chloride, which floats on the residual gold.

Mill Furnace. Furnace for intermediate reheating of iron and steel blooms, bars, etc., between rolling or forging operations.

Milling. In powder metallurgy, mechanical mixing and/or grinding of a metal powder; a ball mill is frequently used for this purpose.

Mill Marking. Marks on worked metal to identify processing history.

Mill Roll. See **Rolling Mill.**

Mill Scale. Scale resulting from manufacture of metal at high temperatures. See **Scaling.**

Mill Spring. See **Roll Spring.**

Milwaloy. Series of corrosion-resistant, high-alloy steels (Milwaukee Steel Foundry Co.), with chromium up to 30% and nickel up to 15%.

Milwaukee (Steel). See **Milwaloy.**

Minargent. Copper-base alloy series, high in nickel (32–40%), with fractional percentages of aluminum and other elements. It is usable as a silver substitute.

Mine Pig (Iron). Pig Iron (see) made entirely from ore, without addition of cinder, as contrasted with crude pig iron.

Mineral. Solid inorganic material, occurring in nature, particularly if of economic value.

Mineral Cotton. See **Mineral Wool.**

Mineralogical Hardness. See **Moh's Hardness.**

Mineral Wool. Fibrous insulating material, resulting from the action of a jet of steam on molten slag.

Mine Tin. See **Lode Tin.**

Minium. See **Red Lead.**

Minofor. Tin-base alloy, with about 19% antimony, 10% zinc, 4% copper, and iron optional up to 1%; has been used for bearings.

Minus Mesh. That part of powdered material which passes through a specified screen.

Mira Metal. Corrosion-resistant, copper-base alloy, with about 16% lead, 7% antimony, and fractional percentages of zinc, tin, and nickel.

Mirror Finish. See **Best Bright Finish.**

Mirror Iron. See **Spiegel.**

Misch Metal. Alloy, containing approximately 50% cerium, with the remainder other rare earth metals, such as lanthanum, neodymium, etc. Alloyed with iron, misch metal is almost universally used as a pyrophoric alloy in cigarette lighters and similar devices.

Misco (Steel). Series of steels, of diversified composition and for various purposes, manufactured by Michigan Steel Casting Co. Important among them are the high-nickel, high-chromium steels of excellent chemical and heat resistance, with 8–65% nickel and 16–30% chromium.

Miscrome. High-chromium (from 16% to 30%) iron alloy series, with about 0.25% carbon, very resistant to chemical corrosion (Michigan Steel Casting Co.).

Miter Iron. Fagot (see **Fagotting**) of iron bars surrounding a central bar, as used for forging into a wrought bar.

Mitis Iron. Wrought Iron (see) treated with aluminum to achieve deoxidation.

Mixed Crystal. Crystal in which atoms of a second substance replace, in part, atoms of the principal substance.

Mixture. See **Mechanical Mixture.**

ML Steel. High-carbon, alloy tool steel (Ludlum Steel Co.), with about 18% tungsten, 4% chromium, 2% vanadium, 1% molybdenum, and about 0.80% carbon.

M. M. M. Nickel-base alloy, with about 26% copper, 10% tin, 2% iron, and fractional percentages of manganese and silicon, used for steam valves (Manning, Maxwell & Moore, Inc.).

Mn. Chemical symbol for **Manganese** (see).

Mo. Chemical symbol for **Molybdenum** (see).

Mock Gold. (a) Copper-platinum alloy, with 20–30% platinum, and optionally up to 4% zinc. (b) Nickel-base alloy, with about 12% each of platinum, silver, and copper.

Mock Lead. Non-metallic mineral of zinc, zinc blende, approximating ZnS in chemical composition.

Mock Platinum. Imitation platinum, most commonly a copper-zinc alloy of about 45% copper content.

Mock Silver. (a) See **Britannia Metal.** (b) Aluminum-base alloy, containing about 10% tin and 6% copper, with fractional percentages of phosphorus.

Mo-Cut (Tool) Steel. High-carbon, alloy tool steel (Braeburn Alloy Steel Co.), with about 8% molybdenum, 4% chromium, 1.5% tungsten, 1% vanadium, and 0.70% carbon.

Modified Aluminum Alloy. Aluminum containing 8–13% silicon, treated, in the molten condition, with sodium or sodium fluoride, to improve physical characteristics.

Modified Carbon Tool Steel. Carbon Tool Steel (see) to which small fractions of vanadium have been added in order to control austenitic grain size, or chromium in order to increase hardenability.

Modified Gun Metal. Gun Metal (see) modified by addition of lead.

Modulus of Elasticity. Force which would be required to stretch a substance to double its normal length, on the assumption that it would remain perfectly elastic, i.e., obey Hook's Law, throughout the test. Also definable as the ratio of stress to strain within the perfectly elastic range.

Modulus of Rigidity. Ratio of unit shearing stress to the displacement per unit length caused by it.

Modulus of Rupture. Breaking force per unit area of a test bar.

Moebius Process. Electrolytically separating gold from silver, by using the crude alloy as anode in an aqueous bath of silver nitrate, with thin, pure silver sheets as cathode. Silver deposits on the cathode, and is scraped off; gold goes into the anode mud.

Moffet-Jumbo Hearth. Hearth (see **Ore Hearth**), in which cooling of the hearth walls and sides is accomplished by a combination of air and water.

Moffet Ore Hearth. **Ore Hearth** (see) used in smelting ores.

Mogul (Tool) Steel. Alloy tool steel (Jessop Steel Co.), with about 8% molybdenum, 4% chromium, 2% tungsten, and 1% vanadium.

Mohawk Steel. High-alloy steel (Ludlum Steel Co.) used principally as a hot-die steel. Composition approximates 14% tungsten, 3.5% chromium and 0.60% vanadium and carbon in the 0.50% range.

Moh Hardness. See **Moh's Hardness.**

Mohr Platinum. See **Platinum Black.**

Moh's Hardness. Arbitrarily chosen list of ten minerals, the numerical values rising with increasing hardness. Graphite = 1. Gypsum = 2. Calcite = 3. Fluorspar = 4. Apatite = 5. Feldspar = 6. Quartz = 7. Topaz = 8. Sapphire = 9. Diamond = 10. Each can be scratched by all the minerals following it in the list.

Moh's Scale. See **Moh's Hardness.**

Moissan Process. Process for the electric-arc-furnace reduction of chromic oxide to chromium by means of carbon, using a hearth of calcium chromite.

Mol (or Mole). Molecular weight of a substance in grams.

Mold. Form or cavity into which molten metal can be poured.

Mold Board. See **Bottom Board.**

Mold Core. See **Core** (b).

Mold Drag. See **Drag.**

Molders' Rule. Ruler, with measurements sufficiently elongated to compensate for the heat expansion and contraction of metal; such rules are used in metal casting so as to correct dimensions to normal temperatures.

Mold Facing. See **Facing.**

Mold Gate. See **Gate.**

Molding. Mold-making operations, for use in the casting of metal. See **Mold.**

Molding Box. See **Flask.**

Molding Hole. Excavation in the sand floor of a foundry, for the pouring of large, crude castings.

Molding Loam. See **Loam Mold.**

Molding Machine. Machine for making loam molds, utilizing small patterns.

Molding Sand. Sand used in making molds for the casting of metal.

Mold Taper. Taper of cast metal, allowing removal from the mold.

Mold Wash. Liquid suspension, e.g., of clay, lime, or powdered metal, used to coat the interior surfaces of ingot or other casting molds.

Molecular Depression. Lowering of freezing point caused by a gram-molecule of a substance of 1000 g of solution.

Molecular Orientation. Preferred arrangement of certain axes or planes of molecules with respect to a given axis or plane.

Molecule. Smallest unit of a substance which can exist in itself and cannot be divided without changing the chemical properties of the substance.

Mol Fraction. Number of mols of a component divided by the sum of the mols of all components in a homogeneous mixture or solution.

Moly-. See **Molybdenum-.**

Molybdenum (chemical symbol Mo). Element No. 42 of the periodic system; atomic weight 95.95. Hard, tough metal of grayish-white color, becoming very ductile and malleable when properly treated at high temperatures; melting point 4748°F; boiling point about 6600°F; specific gravity 10.2. Chemically, it is principally trivalent (molybdous), tetravalent (molybdic), and hexavalent (in the molybdates). Electrodeposition is quite difficult, and has not been commercialized. Pure molybdenum can best be obtained as a black powder, by reduction of molybdenum trioxide or ammonium molybdate with hydrogen. From this powder, ductile sheet and wire are made by powder metallurgy techniques; these are used in radio and related work. In ferro-alloy form, molybdenum is used in many tool steels. The metal is obtainable from Fansteel Metallurgical Corp. and General Electric Co.

Molybdenum Carbide. Compound of molybdenum and carbon, analogous to **Tungsten Carbide** (see) and used, like the latter, for cemented carbide tools.

Molybdenum Cast Iron. Cast steel [see **Cast Steel** (a)] having high strength, particularly at elevated temperatures; molybdenum content ranges normally from about 0.3 to 0.8%; carbon usually exceeds 0.30%.

Molybdenum Steel. Steels containing up to about 10% molybdenum, usually in combination with other constituents, such as chromium, nickel, manganese, etc. In the higher range of molybdenum, the steels are usable as high-speed tool alloys, with the molybdenum replacing part, but usually not all, of the tungsten normally used in such steels, as they retain their hardness and strength at high temperatures. Molybdenum is also used in cast steel and cast iron, particularly when such castings require high-temperature strength. The low-molybdenum steels, usually made by the basic open-hearth process, contain 0.40–1.00% molybdenum.

Molybdenum-Vanadium Cast Steel. Cast steel of high strength [see **Cast Steel** (a)], containing approximately 0.3% molybdenum and 0.1% vanadium.

Mo-Lyb-Die (Steel). Series of abrasion-resistant die steels, manufactured by A. Finkl & Sons.

Molyte. Fused mixture of lime, silica, and molybdenum oxide, often used to introduce molybdenum into steel.

Mo-Max. High-speed steel of 0.60–0.85% carbon, used for cutting tools, dies, punches, etc.; alloying elements include 1.25–2.00% tungsten, 7.75–9.25% molybdenum, 3.50–4.00% chromium, and 0.90–1.50% vanadium (Cleveland Twist Drill Co.).

Monaca (Steel). Series of carbon tool steels, manufactured by Pittsburgh Tool Steel Wire Co.

Mond Metal. Nickel-base alloy with approximately 26% copper and 4% manganese (Mond Nickel Co.).

Mond Process. Process for extracting and purifying nickel; nickel oxide, obtained by treating matte, from a Bessemer converter, with dilute sulfuric acid, is reduced to metal by hydrogen and then treated with carbon monoxide at room temperature. The resulting nickel carbonyl is then decomposed by heating to about 270°F, depositing nickel on small nickel pellets.

Monel. See **Monel Metal.**

Monel-Clad Steel. Clad metal (see **Cladding**) in which a heavier layer of steel is coated with a lighter coating of **Monel Metal** (see).

Monell Process. Two-step modification of the basic **Open-Hearth Process** (see) of steel manufacture in which limestone and ore are heated together till pasty, followed by addition of molten pig iron; the slag which forms is brought to a foaming state by the oxidation of the carbon in the iron and runs out of the furnace through special slag notches.

Monel Metal. Nickel-copper alloy, with small amounts of other metals, made by International Nickel Co., by direct smelting of a natural mixture, the nickel being approximately twice the copper in weight percentage. Because of excellent corrosion resistance and relatively easy machining characteristics the alloy is extensively used, particularly in food and chemical work.

Monkey. See **Cinder Cooler.**

Monkey Cooler. See **Cinder Cooler.**

Monnier Process. Roasting of copper sulfide ores with sodium sulfate, followed by leaching and precipitation or cementation of the copper.

Monochromatic Ray. Light (or any other ether wave, such as X-ray) of a single wave-length.

Monoclinic (Crystal System). Crystal form in which all three axes are unequal, with one of them perpendicular to the other two, which intersect at an oblique angle.

Monometallic. Consisting of a single, elementary metal, as opposed to **Alloy** (see).

Monotron Diamond. See **Monotron (Hardness) Test.**

Monotron (Hardness) Test. Method of determining hardness of materials, in which the **Hardness** (see) is measured by the load required to indent a standard hemispherical diamond to a standard depth.

Monotype Metal. See **Linotype Metal.**

Monovalent. Having a chemical **Valence** (see) of one.

Montefiore Furnace. Liquation Furnace (see) for recovering zinc metal from **Blue Dust** (see).

Moore Furnace. See **'Lectromelt Furnace.**
Moorewood Machine. Apparatus for feeding steel plates into a tin or terne bath. Rolls move the metal into the bath, and another set of rolls remove it. See **Tin Plate** and **Terne Plate.**
Morass Ore. See **Limonite.**
More-Jones Alloy. Series of copper-base, tin-base, and lead-base alloys, manufactured by National Bearing Metals Corp.
Morgan Mill. See **Continuous Mill.**
Morse Gage. See **Metal Gage.**
Mosaic Gold. (a) Non-metallic mineral, stannic sulfide (tin disulfide), used in imitation gilding work. (b) High-zinc brass, with zinc approximately 37%.
Mosaic Silver. Alloy of mercury (see **Amalgam**) with bismuth and tin, used as an imitation silver.
Mosaic Structure. Structure presumed to exist, conforming to a general orientation, with units intermediate between atomic size and crystal or grain size.
Moseley's Law. Square root of the frequency of a given line of an element in the X-ray spectrum is directly proportional to the atomic number of the element.
Moseley Number. See **Atomic Number.**
Moss Gold. Metallic **Gold** (see) in the form of **Dendrite** (see).
Moss Silver. Metallic **Silver** (see) in the form of **Dendrite** (see).
Mossy Zinc. Aggregates of zinc obtained by pouring molten zinc into water.
MO Steel. Series of molybdenum steels (up to 1% of molybdenum) with fractional percentages of chromium and manganese (Climax Molybdenum Co.).
Mother Metal. Molten metal from which a metallic phase of unique composition has separated out, leaving a composition different from the original.
Motor Grade Sheet. Silicon steel sheet used in the manufacture of small motors. See **Electric Sheet.**
Mottle. In terne plate, pattern of fine lines, formed during freezing of the metal coating. The type of mottle can be controlled by the rate of cooling. See also **Mottled Iron.**
Mottled Chill. In cast iron, that part of the chill, next to the **Clear Chill** (see), which has a mottled appearance. It lies between the light clear chill and the gray center.
Mottled Iron. See **Pig Iron.**
Mottled Pig (Iron). See **Pig Iron.**
Mottled Plate. (a) Defect in **Tin Plate** (see), in which dull areas appear, due to excessive flux. (b) See **Mottle.**
Mo-Tung. High-carbon, alloy tool steel (Universal-Cyclops Steel Corp.), with about 8% molybdenum, 4% chromium, 1.5% tungsten, 1% vanadium, and 0.80% carbon.
Mould. See **Mold.**
Mountain Brown Ore. See **Limonite.**
Mouray (Aluminum) Solder. Zinc-base alloy, used for soldering aluminum; it contains 5–12% aluminum and 2–8% copper.
Mouth. Small opening at the top of a Bessemer converter (see **Bessemer Steel**) for charging and discharging.
Movement. Change in dimensions of metal as a result of heat treatment.
Movement Test. Test to determine dimensional changes suffered by steel as a result of heat treatment.
Moving Cathode Cell. See **Moving Cathode Furnace.**
Moving Cathode Furnace. Furnace for electrolytically isolating metals melting at relatively low temperatures, such as calcium, in which the cathode, usually centrally located, is gradually raised during the operation of the furnace, so as to permit solidification of the metal.
Moving Load. Dynamic load or stress sustained by a structure or structural element.
M_s Point. See **Martempering.**
Muck Bar. Initial product, obtained in the **Puddling** (see) process of wrought iron manufacture, after reduction of the ore and squeezing out of much of the slag by passing through rolls. Muck bar is usually approximately ¾ inch thick, and varies from 2½ to 8 inches in width.
Muck Iron. Puddled iron (see **Puddling**) prior to squeezing and rolling.
Muck Mill. **Rolling Mill** (see) for rolling puddled ball, after squeezing, in

the manufacture of wrought iron. See **Puddling.**

Muck Rolls. First set of rolls in a rolling mill operating on **Muck Bar** (see) iron.

Muck Train. Rolling mill train operating on **Muck Bar** (see).

Mucky Hole. Tapping hole in a furnace from which the metal does not flow freely.

Mud Gun. Mechanical device for plugging a tapping hole with clay. See **Tap Hole.**

Muffle. See **Muffle Furnace.**

Muffle Annealing. Annealing (see) of metal conducted in a **Muffle Furnace** (see).

Muffle Furnace. Furnace in which the charge is contained within a refractory container called the muffle, which is heated from the outside.

Muffle Roasting. Roasting (see) in a furnace in which heating is indirect, and products of combustion are kept away from the charge.

Multiple Drawing. See **Multiple Wire Drawing.**

Multiple Refining. See **Multiple System.**

Multiple System. Connecting anodes and cathodes in a cell of an electrolytic process, especially electrolytic refining, in parallel, i.e., all anodes of the cell are electrically connected together, and all cathodes together. The separate cells are, of course, usually in series. See also **Series System.**

Multiple Wire Drawing. Form of **Continuous** wire **Drawing** (see) without intermediate blocks to draw the wire through the dies; in general, not as satisfactory as continuous drawing.

Mu Metal. Nickel-base, high-permeability alloy of foreign manufacture. It contains 17–20% iron, about 5% copper, and low percentages of manganese or chromium.

Muntz Metal. Brass (see) containing 59–62% copper and the remainder zinc; shows excellent hot-working and fair cold-working properties. Although it can be welded by oxy-acetylene torch, arc and resistance welding methods are not readily usable.

Murex. Flux-coated welding rods, of many different metallic and flux compositions, marketed by Metal & Thermit Corp.

Mushet Special Steel. Early type of self-hardening steel, with 7–12% tungsten, 2–3% manganese, about 1% silicon, fractional percentages of chromium, and 1.5–2.0% carbon.

Mushet Steel. Alloy tool steel with 4–6% tungsten, 1.5–2.0% carbon, and small quantities of manganese and chromium.

Music Wire. High-quality, simple steel wire of approximately eutectoid composition, normally patented and cold-drawn; usually made by the acid open-hearth or electric furnace process, and used for spiral springs, piano and other musical instrument strings, etc.

Music Wire Gage. See **Metal Gage.**

M. W. G. See **Metal Gage.**

N

N. Chemical symbol for **Nitrogen** (see).

Na. Chemical symbol for **Sodium** (see).

Naco (Steel). Low-alloy casting steel containing about 1.5% manganese and about 0.35% carbon (National Malleable & Steel Castings Co.).

Nada NT Alloy. Casting bronze containing about 3.75% nickel, 3.75% tin, and 0.75% lead (Nassau Smelting & Refining Co.).

Nahnsen Process. Process used in Silesia for purifying impure zinc by electrolytic refining. See **Electrorefining.**

Nail Plate. Obsolete type of steel plate, rolled to a thickness suitable for cutting into steel nails. Nails are no longer made by this process.

Narrow Cold-Rolled Strip. See **Cold-Rolled Strip.**

Narrow Hot-Rolled Strip. See **Hot-Rolled Strip.**

Narrow Strip. See **Cold-Rolled Strip** and **Hot-Rolled Strip.**

Nascent (State). State of an element when first released, in atomic form, from a compound, frequently characterized by abnormally great chemical reactivity.

NA (Steel). Series of corrosion-resistant iron-nickel-chromium alloys, with chromium between 15 and 30%, and nickel up to 70% (National Alloy Steel Div.).

Nathusius Furnace. Electric **Arc Furnace** (see), usually three-phase, of the direct series arc type and with one electrode or series of electrodes imbedded in the hearth.

National Emergency Steels. Low-alloy steels, developed as substitutes because of shortage of nickel and other critical metals, with manganese increased to 1.60%, 0.15–0.50% carbon, molybdenum up to 0.40%, and optional amounts of nickel and chromium, but limited to 0.20–0.70%.

Nationalloy. Series of high-nickel, high-chromium steels (see **Stainless Steel**), used for chemically resistant products, manufactured by National Forge & Ordnance Co.

National Steel. Series of steels of diverse composition, manufactured by Blaw-Knox Co.

National Wire Gage. See **Metal Gage.**

Native. See **Native Metal.**

Native Element. See **Native Metal.**

Native Metal. Metal found chemically free in nature. Only the less reactive metals are ever found native; among the more usual are: Gold, silver, platinum, and other members of the platinum family.

Native Steel. Steel resulting from burning coal reacting with an iron ore deposit.

Natrium. Latin name for sodium, used in medicine and pharmacy.

Natural Rolled Width Edge. See **Mill Edge.**

Natural Round Edge. See **Mill Edge** and **Number 2 Edge.**

Natural Steel. Steel after any hot-working operation, with subsequent cooling in still air.

Naumanite. Black, lustrous mineral, chemically a silver-lead selenide.

Naval Brass. Copper-base alloy with about 39% zinc, and about 1% tin. It is excellent for all hot-working operations, and also lends itself to cold-working. Owing to resistance to salt water corrosion, it has found extensive application in naval work.

Naval Bronze. See **Naval Brass.**

Nb. Chemical symbol for niobium (see **Columbium**).

Nd. Chemical symbol for **Neodymium** (see).

Neck. (a) See **Fantail.** (b) Portion of roll (see **Rolling Mill**) resting in the bearing.

Necking. Characteristic of ductile metals, when under tension, of narrowing cross-section at some point. Fracture, when it finally occurs under further tension, normally takes place in the zone of necking.

Necking Down. See **Necking.**

Needle. Elongated particle of needle-like appearance.

Needle Antimony. See **Crude Antimony.**

Needle Wire. Tool steel, akin to **Drill Rod** (see), used for making needles, etc.

Negative Segregation. See **Inverse Segregation.**

Neloy (Steel). Low-alloy casting steel with about 1.2% manganese and about 0.35% carbon (National Erie Corp.).

Neodymium (chemical symbol Nd). Element No. 60 of the periodic system; atomic weight 144.27. Rare earth metal, isolated by electrolysis of the anhydrous trichloride in a bath of fused alkaline chlorides. Neodymium is yellowish-white, quite stable in dry air, but slightly reactive in moist atmosphere; melting point 1544°F; boiling point unknown; specific gravity 6.95. Chemically, it is trivalent. Like all other rare earth metals, it cannot be electrodeposited. Because of its rarity, it has no commercial use.

Neogen. Alloy of silvery color, with 58% copper, 27% zinc, 12% nickel, 2% tin, and fractional percentages of aluminum and bismuth.

Neor (Steel). Abrasion-resistant, high-alloy steel, with about 13% chromium and fractional percentages of nickel, silicon and manganese (Darwin & Milner, Inc.).

Neoytterbium. See **Ytterbium.**

N. E. Steels. See **National Emergency Steels.**
Net Calorific Value. Heat produced on complete combustion of a gram of fuel under specified conditions, water resulting from the combustion being in vapor form (in contradistinction to **Gross Calorific Value,** which see).
Net Heating Value. See **Net Calorific Value.**
Network Structure. Structure in which a thin continuous phase surrounds the grain boundaries of a dispersed phase.
Neumann Bands. Lines or narrow bands running across ferrite crystals in steel, deformed by impact or other cold-work. Although normally appearing as narrow twinned bands, they are frequently bent.
Neutral. Neither acid nor basic (alkaline).
Neutral Flame. In gas torch welding with a carbon-containing fuel, a flame with fuel proportioned to oxygen or air so as to yield perfect combustion.
Neutral Flux. Material used as a **Flux** (see), or to make slags more fusible, which has neither acidic nor basic properties. Fluorspar is the most common material of this type.
Neutral Point. (a) In chemical analysis, the point where acidity and alkalinity are equal or completely balanced. (b) In rolling of metal, the point within the metal being rolled where the speed of the metal flowing equals the peripheral speed of the rolls.
Neutral Refractory. See **Refractory.**
Nevastain Steel. Series of high-chromium, corrosion-resistant steels (Ludlum Steel Co.), with nickel optional up to 21% and chromium about 12–26%.
Nevyasmskite. Natural mineral alloy of iridium and osmium, frequently containing other members of the platinum family.
Newnam Hearth. Mechanically-rabbled, water-cooled furnace, used in the direct smelting of lead ores. See **Scotch Hearth.**
New Sand. Freshly mixed **Molding Sand** (see).
Newton's Alloy. See **Fusible Alloys.**
Newton's Metal. See **Fusible Alloys.**
Ney Oro. Series of gold-base, platinum-base, and silver-base alloys, manufactured by J. M. Ney Co.
Ni. Chemical symbol for **Nickel** (see).
Niag. Nickel Silver (see), approximating 47% copper, 41% zinc, 9% nickel, and 3% lead, optionally with fractional percentages of other metals.
Niagara (Brand). Series of master ferro-alloys, for addition to steel, manufactured by Pittsburg Metallurgical Co.
Nib. In powder metallurgy, a semi-finished compact.
Ni-Carb Process. See **Dry Cyaniding.**
Niccolite. See **Arsenical Nickel.**
Nichroloy. Nickel-chromium-iron alloy series, with optional low percentages of manganese, aluminum, and vanadium; used for electric resistance wire and heat-resistant products, as well as chemical equipment.
Nichrome. Nickel-base alloy series made by Driver-Harris Co., with 11–22% chromium, smaller amounts of silicon and manganese, and optionally some iron. Combining electrical resistivity with resistance against oxidation at high temperatures, Nichrome has been extensively used as an electric-resistance heating alloy.
Nick and Bend Test. Test on metal, involving nicking a specimen transversely and breaking it, followed by visual examination of the fractured surface, to determine macroscopic flaws.
Nickel (chemical symbol Ni). Element No. 28 of the periodic system; atomic weight 58.69. Silvery-white, slightly magnetic metal, of medium hardness and high degree of ductility and malleability and resistance to chemical and atmospheric corrosion; melting point 2651°F; boiling point about 5250°F; specific gravity 8.90. Chemically, it is normally divalent (nickelous), but in a few compounds also trivalent (nickelic). The metal can be isolated, in pure form, by hydrogen reduction of the oxide. It is extensively used for electroplating, the sulfate bath being most common, as a hydrogenation catalyst, and for coinage. As alloying agent, it is of great importance in iron-base al-

loys such as the nickel steel and stainless steel, and in copper-base alloys such as Monel metal. Nickel and its simpler alloys are obtainable from International Nickel Co.

Nickel Alloy Steel. Steel containing appreciable amounts of nickel, but less than 30%. See **Nickel Steel.**

Nickel-Aluminum. Master alloy, commonly with about 20% nickel, used for adding nickel to aluminum alloys.

Nickel Aluminum Bronze. Copper-base alloy series, generally available in rolled form, with 8–10% nickel and up to 2% total of aluminum and tin.

Nickel Brass. See **Nickel Silver.**

Nickel Bronze. Series of bronzes with nickel replacing part of the tin content; tin is commonly from 1% to 20%, nickel 3% to 40%, aluminum up to 30%, lead up to 3%, and zinc up to 12%.

Nickel Carbonyl. Volatile compound, $Ni(CO)_4$, used in the extraction and purification of nickel by the **Mond Process** (see).

Nickel Cast Iron. See **Alloy Cast Iron.**

Nickel Cast Steel. Cast steel of high strength and impact resistance [see **Cast Steel** (a)], with 0.5–5.0% nickel.

Nickel-Chromium Alloys. Series of alloys of nickel and chromium, with nickel predominating; iron is commonly a third constituent; used for heat-resistant products and electric resistance heating wire.

Nickel-Chromium Cast Steel. Cast steel of high strength, ductility, and fatigue resistance [see **Cast Steel** (a)], with about 0.30% carbon, 1.75–2.25% nickel, and 0.7–0.9% chromium.

Nickel-Chromium-Molybdenum Steel. (a) Low-alloy steel with from 0.12–0.50% carbon, 0.40–0.70% nickel, 0.40–0.60% chromium, and 0.15–0.30% molybdenum. (b) Low-alloy, air-hardening steel series of 0.20–0.50% carbon, 0.50–0.80% manganese, 1.50–4.25% nickel, 0.50–1.25% chromium, and up to 0.80% molybdenum content.

Nickel-Chromium-Molybdenum-Vanadium Steel. Low-alloy steel, with 1.75% nickel and fractional percentages of chromium, molybdenum, and vanadium, usually oil-quenched and tempered; used in forgings.

Nickel-Chromium Steel. Series of low-alloy and high-alloy steels, containing both nickel and chromium, the nickel most commonly about twice the content of chromium; characterized by hardness and toughness. Typical of the low-alloy ranges are nickel 1.10–1.40%, chromium 0.55–0.90%, and nickel 3.25–3.75%, chromium 1.40–1.75%. High-alloy members are most frequently of the stainless type. See **Stainless Steel.**

Nickel-Chromium-Tungsten Steel. Low-alloy steel, with about 3.5% nickel, 1% tungsten, and 0.8% chromium, used for large forgings and products subjected to elevated temperatures.

Nickel-Chromium-Vanadium Steel. Low-alloy steel, with about 3.5% nickel, and fractional percentages of chromium and vanadium, used for large forgings and products exposed to elevated temperatures.

Nickel-Clad Steel. See **Cladding.**

Nickel-Cobalt Alloy. Series of alloys of nickel and cobalt in which, normally, cobalt is in the 20–30% range and the alloy is silver-white in color.

Nickel-Copper Alloy. See **Copper-Nickel Alloy.**

Nickel-Copper-Chromium Steel. Low-carbon steel, with fractional percentages of nickel, copper, and chromium; largely used in the manufacture of pressure vessels exposed to subnormal temperatures.

Nickelene. Copper-base alloy with 5–30% nickel, 10–35% zinc, and optional additions of up to 10% lead and 2% tin.

Nickel, Grade A. See **"A" Nickel.**

Nickel, Grade D. See **"D" Nickel.**

Nickel, Grade E. See **"E" Nickel.**

Nickeliferous. Containing nickel.

Nickelin. Series of copper-base alloys of foreign manufacture, with 3–40% nickel, up to 10% aluminum, up to 20% iron, up to 20% zinc, and up to 3% manganese.

Nickel-Iron (Storage) Battery. See **Alkaline Storage Battery.**

Nickel-Manganese. Nickel with up to 15% manganese, used for **electric-**

resistance wire, spark-plug wire, and to reduce susceptibility to sulfur gases.

Nickel-Manganese Cast Steel. Cast steel of high strength and toughness [see **Cast Steel** (a)], with 0.30–0.40% carbon, 1.00–1.75% nickel, and 1.00–1.50% manganese.

Nickel-Molybdenum Cast Steel. Cast steel of high strength [see **Cast Steel** (a)], with 0.5 to 2.0% nickel and 0.2 to 0.6% molybdenum.

Nickel-Molybdenum-Iron. Acid-resistant alloy series, in which nickel is usually predominant (up to 60%), with the remainder approximately equal amounts of molybdenum and iron; up to 15% chromium and up to 3% manganese are optional.

Nickel-Molybdenum Steel. Low-alloy steel series, with about 1.5% nickel, 0.20% molybdenum and up to 0.50% carbon; used for forging and strong-cored case-hardened products.

Nickel-Molybdenum-Vanadium Steel. Low-carbon steel, used for low-temperature pressure service, with 2.5–4.0% nickel, about 0.50% molybdenum, about 0.20% vanadium, about 0.30% manganese, and less than 0.08% carbon.

Nickeloid. (a) Zinc sheet coated with nickel (American Nickeloid Co.). (b) The term has also been used for certain proprietary copper-nickel alloys.

Nickeloy. Aluminum-base alloy with about 4% copper and 1.5% nickel.

Nickel Phosphor Bronze. Phosphor Bronze (see), containing low percentages (usually about 1%) of nickel.

Nickel Plate. Product of **Nickel Plating** (see).

Nickel Plating. Process of coating metal objects with nickel, in which the object to be coated is made cathode (negative electrode) in an electrolytic bath containing a decomposable nickel salt, usually the chloride or sulfate; nickel metal is commonly used as anode (positive electrode). Cobalt plating is analogous, since nickel and cobalt resemble each other very closely.

Nickel Silver. Copper-base alloys with 17–32% zinc, and about 5–30% nickel. (Note the complete absence of silver despite the name.) Used primarily for ornamental work because of white color, high strength, lustre, good workability and chemical resistance.

Nickel Steel. High-strength steel, usually made in basic open hearth furnaces, containing more than 0.5% nickel and most commonly in the 3–4% range; used for structural elements, like bridge members in the low-nickel ranges, and automobile parts in the higher ones. For very high nickel-content steels see **Invar** (about 36% nickel) and **Platinite** (about 45% nickel). Up to 10% nickel, the steels tend to be pearlitic, from 10% to 25% martensitic, and above 25%, the steel is always austenitic.

Nickel-Tantalum (Alloy). Alloy series of nickel and tantalum, prepared by powder metallurgy technique, commonly containing about 70% nickel; quite hard, but malleable and ductile.

Nickel-Titanium. Series of **Master Alloys** (see) for addition to nickel-base metals, containing up to 50% titanium; commercially, the alloys contain small percentages of aluminum, silicon, and iron, in addition to nickel and titanium.

Nickel-Tungsten. Nickel-tungsten alloys containing 25–50% nickel.

Nickel-Vanadium Steel. Steel, used for high-strength castings, with about 1.3% nickel, 0.10% vanadium, about 0.30% carbon, 0.60% manganese and 0.40% silicon.

Nickel-Zirconium. Alloys, with 40–50% nickel and 25–30% zirconium, used as deoxidizers for steel and non-ferrous metals, particularly nickel.

Nicral. (a) Series of aluminum-base alloys, with fractional percentages of chromium, magnesium, and copper (Nicralumin Co.). (b) Series of high-chromium and high-chromium-nickel steels (Midvale Co.).

Nicrosilal. Heat-resistant, non-distorting alloy cast iron of foreign manufacture, with about 18% nickel, 5% silicon, and 3% chromium.

Niello (Silver). Bluish-black alloy of silver, with copper, lead, and bismuth, partially converted to sulfides, used for decorative purposes.

Nigger Head. Lumps of unfused raw material which tend to remain unmolten during open-hearth steel manufacture (see **Open-Hearth Process**).

Ni-Hard. Alloy cast iron, with approximately 4.5% nickel and 1.5% chromium added to the common constituents. It has relatively high toughness and resistance to corrosion.

Nilvar. High-nickel iron alloy (Driver-Harris Co.), with about 36% nickel, showing extremely low coefficient of temperature expansion.

Niobium. An older name for **Columbium** (see).

Ni-Resist. High-nickel cast iron, with 18–22% nickel, and 2–4% chromium, in addition to the common cast iron constituents; copper may be added up to 6%, and then the nickel is frequently reduced to about 13%.

Nirex. Nickel-base electrical-resistance alloy, with about 14% chromium and 6% iron (Driver-Harris Co.).

Nirosta. Stainless Steel (see), manufactured by a number of organizations, of the general 18% chromium-8% nickel type.

NI-Stainless Steel. High-alloy corrosion and heat-resistant steel, with about 14% chromium and 2% nickel (Latrobe Electric Steel Co.).

Nital. Solution of nitric acid in methyl or ethyl alcohol, 1.5% by volume, used as etching agent in ferrous metallography.

Ni-Tensiliron. Low-alloy cast iron, with 1–4% nickel, 1.2–2.8% silicon, and about 3% total carbon (International Nickel Co.).

Nitralloy. Series of ferrous alloys, controlled by Nitralloy Corp., and licensed to others as well, which lend themselves effectively to **Nitriding** (see). Nitralloy steels commonly contain chromium, molybdenum, and about 1% aluminum.

Nitrard. High-alloy **Nitriding** (see) steel with about 12% chromium, 1% molybdenum, 1% vanadium, and about 1.50% carbon.

Nitric-Hydrofluoric Acid Test. Test to determine susceptibility of **Stainless Steel** (see) to intergranular corrosion and weld decay by heating, at 160°F, using a solution of nitric and hydrofluoric acids.

Nitride. Oxygen-free compound of nitrogen with another element, usually a metal.

Nitrided Steel. See **Nitriding.**

Nitriding. Process of surface hardening certain types of steel by heating in ammonia gas at about 935–1000°F, the increase in hardness being the result of surface nitride formation. Certain alloying constituents, principal among them being aluminum, greatly facilitate the hardening reaction. In general, the depth of the case is less than with carburizing.

Nitriding Steel. Steel, usually of low-alloy type, and particularly those containing up to 1.50% aluminum, capable of being surface-hardened by the **Nitriding** (see) process.

Nitrogen (chemical symbol N). Element No. 7 of the periodic system; atomic weight 14.008, valence 3 or 5. Colorless gas, liquefying at –320°F and solidifying at –346°F, representing about 79% of the atmosphere of the earth, it is quite inert at ordinary temperatures to all metals except lithium and the alkaline earths. At elevated temperatures, many metals, iron-base alloys, particularly, tend to absorb nitrogen in small quantities; the crystal structure of high-chromium steels is significantly improved by the presence of nitrogen in fractional percentages.

Nitrogen Case Hardening. See **Nitriding.**

Noble Metals. Metals stable to chemical and corrosion attack, as opposed to **Base Metals** (see). The noble metals comprise gold, platinum, and the platinum family.

Nodular. Knotted-like in shape.

Nodulizing. Preliminarily converting finely divided ores into nodules, or small rounded masses; frequently used in roasting.

Nogroth Metal. Low-alloy, cast-iron series with about 1.9% carbon, 1.5% nickel, 0.3% chromium and 0.3% manganese (Q & C Co.).

Nomag. High-alloy cast iron of foreign manufacture, containing about 10% nickel and 6% manganese.

Non-Aging Steel. Steel which does not show aging characteristics. See also **Aging Steel.**

Non-Bessemer Pig Iron. Pig Iron (see) with phosphorus above 0.10%.

Non-Changeable Steel. See **Non-Deforming Steel.**

Non-Conducting Bottom Furnace. Electric **Arc Furnace** (see), with its bottom made of electrically non-conducting materials. Most electric furnaces are of this type.

Noncorrodite. High-chromium, corrosion–resistant steel (about 22% chromium) with low percentages of silicon and manganese, and fractional percentages of nickel and copper (Millbury Steel Foundry Co.).

Non-Deforming Steel. Oil-hardening, low-alloy steel series, usually with about 1.5% manganese, and, optionally, other alloying elements, characterized by stability of dimensions when heat-treated; extensively used for tools, dies, etc.

Non-Deforming Test. See **Deforming Test.**

Non-Destructive Test. Test, which does not affect the material being tested, or make it unusable, such as X-ray examination, electromagnetic test, etc.

Non-Electrolyte. Compound which, when dissolved in an appropriate solvent, like water, does not conduct electricity. Sugar and alcohol are typical non-electrolytes.

Non-Ferrous Alloys. See **Non-Ferrous Metals.**

Non-Ferrous Metallurgy. Metallurgy of all metals and alloys except iron and iron-base alloys.

Non-Ferrous Metals. Metals or alloys free of iron or comparatively so.

Non-Magnetic Steel. Steels, of varying composition, such as the 14%-manganese type or the austenitic stainless series, which are not magnetic.

Non-Metal. Any element having substantially no metallic characteristics, as opposed to **Metal** (see).

Non-Metallic Inclusion. See **Sonim.**

Non-Pressure Welding. Welding in which no pressure is used to effect joining of the metals. See also **Pressure Welding.**

Nonscalloping Quality Strip Steel. Strip steel ordered or sold on the basis of absence of uneveness, or ears, on the edges of the steel, when subjected to deep drawing. See **Hot-Rolled Strip Steel.**

Non-Selective Freezing. Solidification of more than one metal together from a molten mixture of metals, when gradually cooled. See **Selective Freezing.**

Non-Selective Process. Refining process which is not selective (see **Selective Process**).

Non-Shrinking Steel. See **Non-Deforming Steel.**

Non-Standard Steel. Steel which does not conform to chemical and/or physical specifications. See **Standard Steel.**

Norbide. Boron carbide (Norton Co.) approximating the formula B_4C, used for **Cemented Carbide** (see) tools.

Normalized Steel. See **Normalizing.**

Normalized Wire. Soft steel wire, of coarse pearlitic structure, drawn from hot-rolled rods, after annealing at a temperature above the critical range and with rather short, regulated cooling.

Normalizing. Annealing of iron-base alloys in which the metal is heated to about 100°F above the Ac_3 critical temperature range, with subsequent cooling in still air.

Normalizing Furnace. Furnace, batch or continuous, designed for **Normalizing** (see) steel.

Normalizing Temperature. See **Normalizing.**

Normal Pearlite. See **Pearlite.**

Normal Segregation. Segregation (see) of alloying or undesirable constituents in the last portion of ingot solidifying.

Normal Steel. Steel not showing ferrite divorcement. See **Abnormal Steel.**

Normal Structure. See **Normal Steel.**

Northrup Furnace. See **Ajax-Northrup Furnace.**

Norway Iron. See **Swedish Iron.**

Nose. (a) Constriction of the diameter of a Bessemer converter (see **Bessemer**

Steel) near the top. The opening at the top is known as the mouth. (b) Mass of frozen material around the inner end of a tuyere (see **Tuyere**), protecting the tuyere.
Notch. Hole in the side of a furnace, for the withdrawal of fluid slag or metal.
Notch Brittleness. Weakness of metal to stresses applied by impact. See **Impact Test.**
Notched Bar Test. Impact test, conducted on a notched bar specimen. See **Izod Test** and **Charpy Test.**
Notch Effect. See **Notch Brittleness.**
Notch Embrittlement. See **Notch Brittleness.**
Notch Plate. See **Bottom Board.**
Notch Sensitivity. See **Notch Brittleness.**
Notch Toughness. Resistance of metal to stresses applied by impact. See **Impact Test.**
Nouvelle Montagne Furnace. Double zinc distillation furnace, akin to the **Belgian Furnace** (see), having a single fireplace.
N. P. L. Alloy. Series of aluminum-base alloys, specified by the National Physical Laboratory of England.
Nucleus. See **Crystal Nucleus.**
Nugget. Lump or small mass of metal, such as gold or silver, as found in nature.
Number 1 Edge. Rolled edge, round or square, on strip metal.
Number 1 Finish. See **Dull Finish.**
Number 1 Temper. See **Hard Temper.**
Number 2 Edge. See **Mill Edge.**
Number 2 Finish. See **Regular Bright Finish.**
Number 2 Iron. See **Merchant Bar.**
Number 2 Temper. See **Half-Hard Temper.**
Number 3 Bar. See **Double Rolled Iron.**
Number 3 Edge. See **Slit Edge.**
Number 3 Finish. See **Best Bright Finish.**
Number 3 Iron. See **Double Rolled Iron.**
Number 3 Temper. See **Quarter-Hard Temper.**
Number 4 Edge. Type of edge, essentially round in shape, produced by edge rolling from either natural edge as hot-rolled, or from **Slit Edge** (see) material.
Number 4 Temper. See **Soft Skin-Rolled Temper.**
Number 5 Edge. Type of edge, approximately square, produced on strip metal by rolling or filing, after slitting to remove burr.
Number 5 Temper. See **Dead Soft Temper.**
Number 6 Edge. Square type of edge produced on strip metal by edge rolling square-edge, hot-rolled, strip metal.
Nuremberg Gold. Aluminum Bronze (see) of approximately 8% aluminum and 2.5% gold content; used for cheap jewelry.

O

O. Chemical symbol for **Oxygen** (see).
O. B. Alloy. Series of **Bronzes** (see), frequently with lead (up to 8%); tin and zinc are the most common alloying elements (Ohio Brass Co.).
O'Brien Furnace. Roasting furnace analogous to the **Herreshoff Furnace** (see), having a central shaft with rabbling arms horizontally attached to it.
Occlude. See **Occlusion.**
Occlusion. Absorption of relatively large quantities of gases by certain metals, the absorption of hydrogen by palladium being the most notable example. The physical state of the metal is an important factor in determining the magnitude and rapidity of occlusion.
Octavalent. Having a chemical **Valence** (see) of eight.
Odd-Side Board. See **Bottom Board.**
O.D.H. Tested Wire. Steel wire, used for upsetting into bolts and screws, made of selected stock and subjected to twisting tests to guard against seams and similar flaws which would militate against upsetting operations.
Offsize. Having measurements not meeting specifications.
Offtake. Large opening (usually there is more than one) in the top of a blast

furnace for the escape of gases. The offtakes lead to a single larger pipe, called the downcomer or downtake, or to a vertical uptake.

O. F. H. C. Copper. See **Oxygen-Free High-Conductivity Copper.**

O'Hara Furnace. Early form of mechanically-rabbled **Roasting** (see) furnace.

Ohio (Die) Steel. High-alloy die steel (Vanadium-Alloys Steel Co.) with about 12% chromium, 1.5% carbon, and vanadium, molybdenum, and cobalt less than 1% each. It resists wear well, and finds primary use in die fabrication.

Oildie Steel. Low-alloy steel (Columbia Tool Steel Co.), with about 0.8% carbon, 1.6% chromium and 0.5% tungsten, used principally for die manufacture because of its non-shrinking and wear-resistant properties.

Oiled Plate. Steel plate, temporarily protected from oxidation by a film of mineral or mixed oils.

Oiled Sheet. Steel sheet, temporarily protected from oxidation by a film of mineral or mixed oils.

Oiled Strip. Steel strip, temporarily protected from oxidation by a film of mineral or mixed oils.

Oil-Finish Plate. Terne or similar coated plate, leaving the coating bath with an appreciable coating of oil; alternatively, the oil may be added after metal coating.

Oil-Finish Terne Plate. See **Oil-Finish Plate.**

Oil-Hardening. Process of hardening high-carbon steels by **Quenching** (see) in oil, after heating.

Oil-Hardening Manganese Steel. See **Non-Deforming Steel.**

Oil-Hardening Steel. Steel adaptable to hardening by heat treatment and quenching in oil.

Oiling. Application of oil to steel to protect the surface against rusting.

Oilite. Graphited Bearing Metal (see) with a matrix metal approximating 10% tin, copper being the metallic base (Chrysler Corp.).

Oil Mark. See **Grease Mark.**

Oil-Quenched Steel. See **Oil-Quenching.**

Oil-Quenching. In the heat treatment of metals, the step of immersing the hot metal in oil in order to cool it rapidly (though not as rapidly as in water-quenching).

Oil Stain. Pattern formed on tin or terne plate by the chemical action of the hot palm oil.

Oil-Tempered Wire. High-carbon wire heated at the final size to above the critical range, quenched in oil, and finally brought into an oil or molten metal bath maintained at a temperature depending on the desired temper.

Oil-Tempering. See **Oil-Quenching.**

Oil-Well Tubing. Seamless steel tubing for use in the pumping of oil from oil wells.

Oker. Series of yellow **Brasses** (see) modified with fractional percentage of lead, tin, or iron.

Old Process Patenting. Patenting process in which the wire is heated in the open or in tubes, then passed into a lead bath, and finally cooled by drawing into the open air.

Old Sand. Molding Sand (see), unsatisfactory for further use because too friable and porous as a result of repeated heatings.

Old Silver. Silver articles chemically treated to give the appearance of age.

Oldsmoloy. Nickel Silver (see), used principally in cast form, approximating 39% zinc, 14% nickel, and 2% tin (Olds Alloys Co.).

Olsen Cup Test. See **Olsen (Ductility) Test.**

Olsen (Ductility) Test. Method of measuring the ductility of sheet metal which involves determination of the width and depth of impression, made by a standard steel ball, under pressure required to fracture the metal.

Olympic Bronze. High-strength silicon bronze, resistant to **Season Cracking** (see), with about 1% zinc and 2.75–4.25% silicon and optionally up to 0.5% lead (Chase Brass & Copper Co.).

Omega Brand Alloy. Series of copper-base alloys, of the **Phosphor Bronze** (see) and **Nickel Silver** (see) types (Riverside Metal Co.).

Omega Brand Phosphor Bronze. See **Omega Brand Alloy.**

One-Minute Wire. Galvanized wire (see **Galvanizing**) which will withstand for one minute, immersion in a standard, neutral copper sulfate solution.

One-Pass Cold-Rolled Stock. Hot-rolled light steel sheet after blue annealing and flattening. Note that, by present terminology, such metal is not called cold-rolled.

One-Piece Rolling. See **Single Rolling.**

Onion's Metal (or Alloy). See **Fusible Alloys.**

Ontario Steel. **Non-Deforming Steel** (see) of high-alloy character, with about 11% chromium, 0.80% molybdenum, 0.30% manganese, 0.30% silicon and 1.50% carbon; it is air-hardened and shows high resistance to abrasion (Ludlum Steel Co.).

Open Box Pass. Rolling mill in which the combination of grooves in the rolls forms a square or rectangular cross-section, the two rolls being at no point more than tangential to each other. See **Pass.**

Open Front. Blast furnace arrangement including a fore-hearth.

Open-Hearth Acid Pig (Iron). See **Bessemer Pig (Iron).**

Open-Hearth Basic Pig (Iron). See **Basic Pig (Iron).**

Open-Hearth Furnace. See **Open-Hearth Process.**

Open-Hearth Iron. See **Ingot Iron.**

Open-Hearth Process. Process of making steel by heating the metal in the hearth of a regenerative furnace. In the basic open-hearth steel process, the lining of the hearth is basic, usually magnesite; whereas in the acid open-hearth steel process, an acid material, silica, is used as the furnace lining and pig iron, extremely low in phosphorus (less than 0.04%), is the raw material charged in.

Open-Hearth Slag. See **Basic Slag.**

Open-Hearth Steel. See **Open-Hearth Process.**

Opening Floor. See **Cooling Floor.**

Open Mold. Mold that is substantially open at the top, such as molds in which ingots are cast, as distinguished from **Close Molds** (see).

Open Pass. Rolling mill in which the rolls are at no point more than tangential to each other. See **Pass.**

Open-Poured Steel. See **Rimmed Steel.**

Open Sand Casting. Casting of metal into cavities formed in sand, without the use of flasks.

Open Sand Mold. See **Open Mold.**

Open Steel. See **Semi-Killed Steel.**

Open Surface. Rough surface on **Black Plate** (see), resulting from imperfections in the original steel bars from which the plate was rolled.

Open-Topped Housing. Housing for rolling mill roll bearings, in which the top is a separate unit from the base and legs, and can be removed.

Operating Stress. Stress, averaged over the whole cross-section, to which a metal structural unit is subjected during service.

O. P. Patenting. See **Old Process Patenting.**

Optical Pyrometer. See **Pyrometer.**

Orange Heat. Temperature approximating 1740°F, as judged visually.

Orange Peel (Effect). Surface defect on rolled metal, caused by abnormal grain size and faulty cold-working.

Oranium Bronze. Series of aluminum bronzes, with 5–11% aluminum content, used for the manufacture of hardware, gears, bearings, etc.

Order-Disorder. Phenomenon that certain solid solution alloys exist in either the **Disordered State** (see) or the **Ordered State** (see), depending on heat treatment.

Ordered State. State of atomic distribution in certain solid solution alloy systems where the atoms of each metal are arranged so as to conform to a single lattice, though no separate phases appear. On reheating to above the critical temperature, the disordered state reappears.

Ordinary Carbon Steel. See **Carbon Steel.**

Ore. Naturally-occurring mineral from which metal can be profitably extracted.

Ore Beneficiation. See **Beneficiation.**

Ore Boil. In the basic **Open-Hearth Process** (see) for manufacturing steel, the first operating stage, during which carbon monoxide is evolved, causing foaming of the viscous slag.

Ore Down. Addition of iron ore, in the open-hearth process of steel manufacture, to reduce the carbon content of the steel.

Ore Hearth. Fireplace, made of cast iron, with at least three surrounding walls, and a tuyere at the back to convey air to the **Hearth** (see).

Ore Hearth Process. Direct smelting of high-grade lead sulfide ores, with air blown through a coke-ore mixture, while the mixture is stirred by hand or by mechanical means.

Oreide. Yellow or Red **Brass** (see) with up to 1.3% tin, used in hardware manufacture and for similar purposes.

Ore Process. See **Pig and Ore Process.**

Orientation. Arrangement of certain crystal axes or crystal planes in a polycrystalline aggregate with respect to a given direction or plane. If there is any tendency for one arrangement to predominate, it is known as the preferred orientation; in the absence of any such preference, random orientation exists.

Ormulu. Copper-base alloy, with 6–17% tin, and zinc optional up to 25% (Lumen Bearing Co.).

Orpiment. See **Yellow Arsenic** (b).

Orthorhombic (Crystal System). Form of crystal arrangement in which three axes, of unequal size, are perpendicular to one another.

Os. Chemical symbol for **Osmium** (see).

Oscillating Crystal (X-ray) Method. Method of X-ray examination of crystals which uses a single crystal oscillating at a constant speed, and a refracted beam of monochromatic X-radiation.

Osmiridium. Naturally-occurring noble metal alloy, in which osmium and iridium are predominant, with relatively small amounts of other noble metals of the platinum family. Because of its hardness, it is used for the tipping of pen points.

Osmium (chemical symbol Os). Element No. 76 of the periodic system; atomic weight 190.2. Very hard, brittle, crystalline metal of the platinum family, bluish-white in color, quite inert in massive form at room temperature, but oxidizes on heating or when in finely divided form; melting point about 4890°F; boiling point about 9900°F; specific gravity 22.48, the highest of all known elements. Chemically, it shows valences from two to eight. Osmium is found native, alloyed with iridium (osmiridium) and with other platinum-family metals; used as a catalyst and as an alloy with iridium for pen point tips.

Osmondite. Solid solution of carbon or iron carbide in alpha iron, present occasionally in hardened and tempered steel; formerly believed to represent a stage in the decomposition of **Austenite** (see) into pearlite.

Osmund Furnace. Medieval type of forge used for the manufacture of wrought iron.

Otiscoloy (Steel). Low-alloy, high-strength steel, with carbon not over 0.12%, 1.00–1.35% manganese, copper not over 0.5%, and nickel and chromium not over 0.1% each.

Otisel (Steel). Series of corrosion-resistant, high-chromium and chromium-nickel steels, with 12–30% chromium, and optionally 12–20% nickel; low percentages of tungsten or molybdenum are also optional (Otis Elevator Co.).

Ounce Metal. See **Composition Brass.**

Out-of-Round. Difference between maximum and minimum diameters of round metal, when measured at the same cross-section.

Out-of-Square. Greatest deviation of an end edge of a sheet from a straight line perpendicular to a side, and touching one corner.

Overblowing. Burning of steel in a Bessemer converter due to excessive air blast.

Overburdening. Charging a blast furnace with too great a proportion of ore and flux as compared with the amount of fuel (see **Burden**).

Overfill. See **Fin.**

Overheated Metal. See **Overheating.**

Overheating. Heating of metal to the point at which crystal growth adversely affects the characteristics of the metal, or oxidation occurs.

Overpoled Copper. See **Overpoling.**

Overpoling. Treatment of copper by **Poling** (see) for too long a period of time, and thus leaving too low a content of oxygen in the metal.

Overreduced Steel. Steel which has been subjected to too drastic a deoxidation, resulting in poor flow characteristics.

Oversize. See **Plus Mesh.**

Overstrain Aging. Age hardening (see **Aging**) of a metal, due to stressing beyond the **Elastic Limit** (see).

Overstressing. Subjecting metal to stresses exceeding the elastic limit of the metal, with resulting permanent deformation.

Overvoltage. Difference between voltage at which an elementary ion will discharge at a specific electrode, and the normal electrode potential of that element.

Oxford Impact Machine. Machine for determining impact strength (see **Impact Test**), operating on the principle of a pendulum, both anvil and pendulum being suspended by wires to avoid energy losses by friction.

Oxidation. Reaction of a substance with oxygen; more generally, loss of electrons by an element to another element. Oxidation and reduction are interrelated; for oxidizing one substance, another must be reduced.

Oxidation-Reduction Potential (Redox Potential). Potential difference set up at an electrode immersed into a reversible oxidation-reduction system; measure of the state of oxidation of the system.

Oxide. Compound of oxygen with another element.

Oxide Film. Thin film, on a metal, of the oxide of the metal. On steel and other metals which form colored oxides, such a film is often deliberately formed for appearance purposes.

Oxidizing. Theoretically, this term should be limited to processes of forming oxides on metals. Actually, oxidizing is the name for any process of forming a dull surface finish on a metal by heat or chemical means. In the case of silver, for example, oxidizing is achieved by forming a black sulfide coating. Oxidizing also means chemical union of a metal or any other element with oxygen, and the conversion of any compound to an oxide.

Oxidizing Agent. Substance which will release oxygen, particularly on heating, and which therefore can be used to yield oxygen to another substance.

Oxidizing Flame. In gas-torch welding with a carbon-containing fuel, a flame with oxygen (or air) in excess of that needed to produce a neutral flame, i.e., perfect combustion.

Oxidizing Roasting. Roasting (see) in which the metals are converted to the corresponding oxides.

Oxland-Hocking Furnace. Revolving furnace of cylindrical design, used in the roasting of sulfide ores.

Oxy-Acetylene Blowpipe. Blowpipe [see **Blowpipe** (b)], burning streams of acetylene and oxygen together.

Oxy-Acetylene Welding. See **Gas Welding.**

Oxygen (chemical symbol O). Element No. 8 of the periodic system; atomic weight arbitrarily assumed as 16.000. Colorless, odorless gas present free in the atmosphere to an extent of about 20%; in combined form, it is the predominant element of water and of rocks constituting the surface of the earth. It liquefies at −297°F and solidifies at −361°F; reacts readily with many metals, and less readily with others; at higher temperatures, with all except gold, silver, and the platinum family. Most metallic ores are oxygen-containing compounds or pass through an oxygen-containing stage during reduction to elementary form. Chemically, oxygen is always divalent. Elementary oxygen is used, under high compression, in steel cylinders, in oxy-hydrogen and oxy-acetylene welding and metal cutting.

Oxygen-Free (High-Conductivity) Copper. Electrolytic cathode copper, of about 99.98% purity, molten under conditions insuring substantial freedom from oxygen. Its conductivity is over 100% of standard.

Oxy-Hydrogen Blowpipe. Blowpipe [see **Blowpipe** (b)] burning streams of hydrogen and oxygen together.
Oxy-Hydrogen Welding. See **Gas Welding.**

P

P. Chemical symbol for **Phosphorus** (see).
Pa. Chemical symbol for **Protoactinium** (see).
Pack. Two or more partially-rolled sheets lying on each other, and subjected, together, to further rolling.
Pack Annealing. See **Annealing.**
Pack Carburizing. See **Carburizing.**
Packfong. See **Paktong.**
Pack Hardening. See **Pack Carburizing.**
Packing Material. In powder metallurgy, material in which compacts are imbedded during sintering.
Pack Mill. See **Pack Rolling.**
Pack Rolling. Passing more than one sheet or plate of metal simultaneously through a roll pass. Usually, this is limited to relatively thin metal.
Paint Adherence. Capacity of a metal surface, most commonly steel, to hold, without peeling, a film of paint; this varies with the metal, but depends mostly on the cleanliness and roughening of the surface.
Pair. In pack mill operation, a pack of two **Breakdowns** (see) for reheating and rolling.
Pair Furnace. See **Bar Furnace.**
Paktong. Modified **Nickel Silver** (see) tableware alloy, with 26–58% copper, 7–41% nickel, 16–45% zinc, and up to 2.6% iron.
Palau. Alloy of gold and palladium, commonly in the ratio of 80–20%, which was used as a substitute for platinum in laboratory ware and jewelry when platinum was significantly more expensive than gold.
Palladium (chemical symbol Pd). Element No. 46 of the periodic system; atomic weight 106.7. Steel-white metal, closely resembles platinum, although chemically and physically it is harder and of lower ductility and malleability; has the lowest melting point in the platinum family, 2829°F; the boiling point is believed to be about 7200°F, but may be much higher; specific gravity 11.40. Chemically, it is mostly divalent, but may also have four or other valences. It can be electrodeposited from a complex phosphate bath, but the process is not used commercially. Palladium is found native in combination with other platinum metals and is also obtained as a by-product of copper and nickel refining. In finely divided form, it is used as a catalyst; its remarkable occlusion of gases makes it suitable for use as a contact catalyst in automatic lighters. Palladium-gold alloys have been used as platinum substitutes in chemical equipment; they also find some use in dental and jewelry alloys.
Palladium Gold. (a) Gold containing up to 10% palladium used for jewelry. See **Porpezite.** (b) Jewelry alloy, approximating 40% copper, 31% gold, 10% silver, and 10% palladium in composition.
Pallas Iron. Non-metallic rock of highly basic nature, the iron oxide content exceeding the silica content.
Palm Oil Process. Process for manufacturing tin or terne plate in which molten palm oil is used as a flux on the surface of the fused metal. See **Tin Plate** and **Terne Plate.**
Pan. See **Amalgamation Pan.**
Pan-Amalgamation (Process). Method of recovering silver and gold from their ores, in which a cast-iron pan or barrel is used for contacting a slurry of the crushed ore with salt, copper sulfate, and mercury; the chemically-released silver and gold form an amalgam with the mercury.
Panelling. Surface defect, in the form of a crease, which appears on improperly cold-worked tin plate.
Pantal. Series of aluminum-base alloys, with fractional percentages of magnesium, manganese, silicon and titanium.
Paragon Steel. Non-Deforming Steel (see), with about 1.7% manganese, 0.60% chromium, 0.25% vanadium, and 1.00% carbon (Crucible Steel Co.).

Par Alloy. Series of iron-nickel-chromium alloys, used for heat- and corrosion-resistant products; nickel content is variable up to 65%, and chromium up to 30% (Crucible Steel Casting Co.).
Par-Exc Steel. Low-alloy steel (Vanadium Alloys Steel Co.) of tough structure, used as a hot-die steel and for small tools. Composition approximates 2% tungsten, 1.5% chromium, 0.3% vanadium, and 0.50% carbon.
Paris Metal. Copper-base alloy, with 6–16% nickel, up to 5% iron, up to 5% zinc, about 2% tin, and about 1% cobalt.
Parkerized Steel. See **Parkerizing.**
Parkerizing. Forming rust-resistant coating on iron or steel by immersion in a hot solution of acid manganese phosphate (Parker Rust-Proof Co.).
Parkes Process. Process of separating silver from lead, in which 1–2% zinc is stirred into the base lead, heated to above the melting point of zinc, and then the mass allowed to cool. The crust, forming on the surface, contains most of the silver. Repetition of the process removes practically all silver from the lead.
Parson's Alloy. Copper-base alloy, with about 41% zinc, 1% iron, and fractional percentages of tin, aluminum, and manganese.
Parson's White Brass. Tin-base bearing metals (Cramp Brass & Foundry Co.), with up to 35% zinc, up to 14% lead, up to 11% antimony, and up to 5% copper.
Partial Pyritic Smelting. Smelting of copper pyrite ores (usually when contaminated with considerable gangue) in a blast furnace; the ore is mixed with coke to facilitate oxidation. See **Pyritic Smelting.**
Partial Roasting. Roasting (see) in which only part of the sulfur is eliminated.
Particle Size. In powder metallurgy, the maximum dimension of individual particles, determined by screening or microscopic measurement.
Particle Size Distribution. In powder metallurgy, the percentages by weight of material in the various size ranges, measured by screens or the microscope.
Parting. (a) Separating silver from gold. (b) Zone of separation between parts of a **Flask** [see **Flask** (a)].
Parting Sand. Finely ground sand for dusting final surfaces in mold-making operations, for metal casting. See **Mold.**
Pass. Groove in a roll, or the combination of grooves in a pair of rolls, forming an opening of the shape of the section to be rolled; also used to indicate the step of running metal through such a pass.
Passive Element. See **Passivity.**
Passive Iron. See **Passivity.**
Passive Metals. See **Passivity.**
Passivity. Characteristic of certain metals, notably iron and chromium, of becoming resistant to many chemical reagents after treatment with strong oxidizing agents, such as concentrated nitric acid. A very thin film of oxide has been believed to be the cause of passivity, but this theory has not been found adequate, and the subject is still being investigated, especially from the viewpoint of atomic structure.
Patented Wire. Medium-carbon or high-carbon steel wire subjected to **Patenting** (see).
Patenting. Treatment of steel, usually in wire form, in which the metal is gradually heated to about 1830°F, with subsequent cooling, usually in air, in a bath of molten lead, or in a fused salt mixture held between 800 and 1050°F.
Patent Sheet. Hot-rolled, annealed steel sheet flattened by stretching.
Patera Process. Process of recovering silver from its ores, in which the ore is roasted with a small percentage of salt, and the mass then leached with sodium thiosulfate; the silver is later thrown out of the resulting solution by means of a soluble sulfide.
Patio Process. Process of recovering silver and gold from their ores, in which salt, copper sulfate, and mercury are contacted with a slurry of the ore, the chemically-released silver

and gold forming an amalgam with the mercury.

Pattern. (a) See **Diffraction Pattern.** (b) Form, usually made of wood, around which sand is tamped to form a mold for the casting of metal.

Patternmakers' Rule. See **Molders' Rule.**

Pattern Metal. Series of lead-tin-zinc alloys, with up to 40% of each; copper is optional up to 3%.

Pattinson Process. Process of removing silver from lead, in which the crude metal is cooled slowly to melt away relatively pure lead, leaving behind a silver-rich mass which can be handled separately by cupellation.

Pattinson's Pot. See **Pattinson Process.**

Pb. Chemical symbol for **Lead** (see).

P. C. E. See **Pyrometric Cone Equivalent.**

Pd. Chemical symbol for **Palladium** (see).

PDCP Copper. Copper (Phelps Dodge Copper Products Corp.) rolled into form directly from solid electrolytic copper, by high-temperature rolling under reducing conditions.

Peacock Copper. Non-metallic mineral, bornite, approximating Cu_3FeS_3 in chemical composition.

Pearce Turret Furnace. Roasting furnace, for use on sulfide ores, in which a narrow hearth is bent around a circle, with the ends not quite meeting.

Pearlite. Lamellar structure, resembling mother-of-pearl, barely resolvable in the finest microscopes; occurs in steel as a result of the transformation of austenite into aggregations of ferrite and iron carbide, Fe_3C (cementite). It contains approximately 0.85% carbon and is a constituent of annealed steels and both gray and white cast iron. Other structures containing the iron-iron carbide eutectoid, such as sorbitic pearlite and granular pearlite, do not show the mother-of-pearl luster.

Pearlitic Alloy Steel. Low **Alloy Steel** (see) of pearlitic structure or of **Pearlite** (see) with another constituent. Most common among the alloying elements are nickel, chromium, tungsten, molybdenum, and vanadium; the total alloying content seldom exceeds 7%.

Pearlitic Malleable Iron. White cast iron subjected to graphitizing, leaving much of the carbon in combined form (see **Combined Carbon**). Retention of the carbon is accomplished by the presence of certain alloying elements, manganese, chromium, or molybdenum, in fractional percentages, or by shortening the graphitizing cycle.

Pearlitic Manganese Steel. See **Manganese Steel.**

Pearlitic Nickel Steel. See **Nickel Steel.**

Pearlitic Steel. Steel, which, in the normalized or annealed state, shows a pearlitic structure or a structure of **Pearlite** (see) with ferrite or cementite.

Pearly Constituent. See **Pearlite.**

Pearlyte. See **Pearlite.**

Peeler. Apparatus for manipulating ingots, blooms, etc., in a heating furnace.

Peep Hole. Hole, in any apparatus, such as a cupola, etc., to enable observation of the progress of operation.

Peerless Steel. Series of high-alloy steels (Crucible Steel Co.) with about 9–15% tungsten, 2–3% chromium, and fractional percentages of vanadium.

Peirce and Smith Converter. Basic-lined **Converter** (see) for the treatment of copper ores. See also **Peirce-Smith Process.**

Peirce-Smith Process. Process of treating copper **Matte** (see) by blowing with air in a basically-lined converter; the iron of the matte is first fluxed away by use of silica.

Pelatan-Clerici Process. Continuous cyanide treatment of gold or silver ores, simultaneously electrolyzing the solution into a mercury cathode, thereby forming an amalgam.

Pelatan Furnace. Furnace for the roasting of finely divided, sulfur-bearing ores.

Pellin's (Hardness) Test. Method of measuring hardness, in which a hardened ball is allowed to drop on the specimen, and the diameter of the resulting indentation used as a measure of hardness.

Penalloy. Series of low-alloy casting steels, manufactured by Pennsylvania Electric Steel Casting Co.

Penang Tin. Pig **Tin** (see) of about 99.95% purity, obtained from the Penang mines in the Straits Settlements.

Penco (Steel). Series of carbon and low-alloy tool steels, manufactured by Peninsular Steel Co.

Penetration-Fracture Test. Semi-quantitative test, in which a steel specimen, heated to above the critical range and then quenched, is fractured and visually compared with fracture standards (Shepherd P-F standards). Generally, one side of the fracture is metallographically polished and etched, and the depth of hardness measured.

Penetration Hardness. Resistance of a material to penetration by a pointed instrument under pressure. See **Penetration Test.**

Penetration Test. Hardness test, in which a pointed or rounded testing element, under load, is forced into the material, inducing plastic flow.

Pen Stock. See **Tuyere Stock.**

Pentavalent. Having a chemical **Valence** (see) of five.

Pepper Blister. Tiny **Blister** (see) on galvanized or terne sheet.

Percentage Elongation. See **Elongation.**

Percentage Reduction. Percentage of reduction of the cross-section of a metal between rolling or drawing passes. Usually given as the **Draft** (see) divided by the cross-sectional area of the metal as it enters the mill or die.

Percussion Test. See **Drop Test.**

Percussion Welding. Resistance Welding (see) in which the weld area is pressed or hammered at the moment of welding.

Period of Constant Creep. See **Secondary Creep.**

Periodic Group. One of the nine groups into which all elements are classified in the **Periodic System** (see).

Periodic Law. See **Periodic System.**

Periodic System. Arrangement of all the chemical elements in the order of their increasing atomic numbers (essentially the same as their increasing atomic weights). Elements of similar properties are placed under each other, forming nine basic groups; the chemical valence varying from one group to another in regular sequence.

Periodic Table. See **Periodic System.**

Periscope Steel. High-alloy, corrosion-resistant steel, of approximately 20% nickel, 8% chromium, 1% silicon, and 1% manganese.

Peritectic Point. See **Peritectic Temperature.**

Peritectic Reaction. Reaction between a solid and a liquid phase, to yield a second solid phase.

Peritectic System. Constitution-temperature system, or diagram, in which there is a binary compound unstable in contact with a melt of its own composition; also known as Transition Type System, Unstable Compound System, and Hidden Maximum System.

Peritectic Temperature. Temperature in a **Peritectic System** (see) in which there is equilibrium between a solid and the remaining melt, the composition of which conforms to the peritectic point, at which this temperature line meets the liquidus curve.

Perlite. See **Pearlite.**

Permalloy. Iron-nickel series of alloys, with about 40–80% nickel, developed by Western Electric Co. for applications requiring high magnetic permeability and resistivity.

Permanent Hardness. Hardness in any alloy, conferred by the presence of the alloying elements, as opposed to **Temporary Hardness** (see).

Permanent Mold. Metal mold which can be used repeatedly for casting metals.

Permanent Set. Non-elastic deformation of metal under stress, after passing the elastic limit.

Perman Steel. See **Manganese Steel.**

Permeability. See **Magnetic Permeability.**

Permeability Alloy. Alloy of magnetic permeability significantly greater than that of pure iron. Commonly, iron-nickel alloys are meant, with or without additional constituents, such as cobalt and chromium.

Perminvar. Nickel-base alloy, with about 22.5% iron and 7.5% cobalt

(Western Electric Co.). See **Permeability Alloy.**

Permissible Working Stress. See **Safe Load.**

Permite. Series of aluminum-base alloys, with optional small percentages of silicon, magnesium, iron, nickel, etc. (Aluminum Industries, Inc.).

Pernot Furnace. Reverberatory furnace, with a circular revolving hearth, slightly inclined; used in the smelting or puddling of steel.

Perrin's Process. Method of dephosphorizing steel by lowering the carbon content to about 0.05%, removing the normal, first high-phosphorus slag on the metal, and then agitating the steel with a fresh highly-basic slag. See **Dephosphorization.**

Pershbecker Furnace. Shaft furnace, with fireplaces at opposite sides, for continuous roasting of mercury ores.

Peter Stubs' Gage. See **Birmingham Wire Gage.**

Petzite. Naturally-occurring alloy, a telluride of silver and gold of the general formula $(Ag,Au)_2Te$. Gold usually approximates 20% and silver 40%.

Pewter. Tin base alloy series, with lead at 10–15%, small quantities of antimony, and, optionally, copper; formerly in extensive use for household utensils and ornaments.

P-F Standards. See **Penetration-Fracture Test.**

P-F Test. See **Penetration-Fracture Test.**

pH. Measure of hydrogen ion concentration, expressed as the logarithm of the reciprocal of the concentration, i.e.,

$$pH = \log \frac{1}{[H]+}$$

The pH is a measure of acidity and alkalinity, neutrality being at pH 7; pH under seven indicates an acid solution and pH over seven an alkaline solution; the nearer the pH to seven the weaker is the acid or the alkali.

Phase. Homogeneous region of substantially constant chemical composition and physical state, separated from other phases by surfaces of discontinuity.

Phase Diagram. See **Constitution Diagram.**

Phase Equilibrium. See **Phase Rule.**

Phase Rule. Basic law defining heterogeneous equilibrium conditions; the number of phases present under the conditions plus the "degrees of freedom" of the system (independent variables, which must be arbitrarily set in order to study the system) equals the number of components, plus two (usually written simply as $P + F = C + 2$). In most metallurgical practice, where vapor pressures are relatively negligible, the vapor phase can be disregarded, cutting down one degree of freedom; the equation then takes the form of $P + F = C + 1$.

Philo (Brand). Series of ferro-master alloys (silicon, chromium, manganese), manufactured by Ohio Ferro Alloys Corp.

Phono Bronze. Copper-base alloy, with about 1.25% tin, used for electrical conductor wires (Bridgeport Brass Co.).

Phono Electric Alloy (Wire). See **Phono Bronze.**

Phosphate-Coating. See **Parkerizing.**

Phosphatizing. See **Parkerizing.**

Phosphide. Oxygen-free metallic compound of metal with phosphorus. When present in metals, phosphorus is usually in phosphide form.

Phosphor Bronze. Copper-base alloys, with 3.5–10% of tin, deoxidized with phosphorus. Because of excellent toughness and strength, fine grain, resistance to fatigue and wear, and chemical resistance, these alloys find general use as springs and in marine fittings.

Phosphor Copper. Copper-base alloys containing 10–14% phosphorus, used as master alloys for the addition of phosphorus to copper or its alloys.

Phosphorized Copper. Oxygen-free copper made by phosphorus addition to copper, with residual phosphorus at 0.02–.03%.

Phosphorized Steel. Steel containing phosphorus as an added constituent, or in excess of the normal fortuitous ranges.

Phosphorizing. Combining or treating with phosphorus.
Phosphor Tin. Master alloy of tin and up to 5% phosphorus, for addition to copper-base alloys in the manufacture of phosphorus bronzes.
Phosphorus (chemical symbol P). Element No. 15 of the periodic system; atomic weight 30.98. Non-metallic element occurring in at least three allotropic forms, obtained commercially by heating phosphate rock with coke and silica. The most common form, white phosphorus, is a waxy solid spontaneously inflammable in air and very poisonous; melting point 111°F; boiling point 536°F; specific gravity 1.82. Chemically, phosphorus shows trivalence (phosphorous) and pentavalence (phosphoric). In metallurgical practice, phosphorus often occurs as a fortuitous constituent. In steels, it is usually undesirable, with limits set in most specifications. However, ferrophosphorus (about 20% phosphorus) is occasionally utilized to raise phosphorus content in steel, primarily to prevent the sticking of light-gage sheets. Phosphor-copper (normally 10–14% phosphorus) is used as master alloy for phosphorus addition to copper-base alloys; for addition to white metals, phosphor-tin master alloy (3.5–5% phosphorus) is used. Both types of master alloys are intended for deoxidation and scavenging action.
Phosphorus Steel. See **Phosphorized Steel.**
Photomicrograph. Photograph of structure, as seen in the microscope, usually representing a magnification of ten diameters or greater.
Physical Metallurgy. Science of the physical and mechanical characteristics of metals.
Physical Properties. Factors such as tensile strength, fatigue limit, electrical conductivity, etc., which determine the uses of a material.
Physical Testing. Determination of the **Physical Properties** (see) of a material.
Physico-Chemical Test. (a) Test, by physical means, to indicate the chemical nature or composition of a material. (b) Test, involving chemical reaction or etching, to reveal metallic structure.
Piano Wire. See **Music Wire.**
Pickle. Solution used for **Pickling** (see).
Pickle Brittleness. See **Hydrogen Embrittlement.**
Pickled and Annealed Hot Mill Plate. **Black Plate** (see) subjected to pickling and then to annealing, or, first to annealing and then to pickling.
Pickled Finish. Surface **Finish** (see), achieved on metal, usually steel, by **Pickling** (see).
Pickled Strip (Steel). Hot-rolled strip steel, subjected to acid treatment to remove scale and oxide, followed by rinsing and drying.
Pickle Liquor. Exhausted **Pickling** (see) bath, ready for discarding.
Pickle Test. Immersing a metal sample, particularly steel, in dilute acid in order to render surface defects visible.
Pickling. Treating metal with acid, in order to remove scale and other surface imperfections. Usually, sulfuric or hydrochloric (muriatic) acid is used, frequently with an inhibitor added to minimize attack on the metal itself. Some of the hydrogen resulting from acid attacking the metal may be occluded, as in the case of steel, causing hydrogen embrittlement.
Pickling Acid. See **Pickling.**
Pickling Line. Sequence of operations in **Continuous Pickling** (see).
Pickling Machine. Apparatus facilitating **Pickling** (see), particularly **Continuous Pickling** (see).
Pickling Patch. Flaw in **Tin Plate** (see) due to faulty **Pickling** (see), leaving areas from which the oxide has not been completely removed.
Pickling Pit. Pit (see) caused by improper or excessive **Pickling** (see) action.
Pickling Rod. Long steel rod for cleaning out obstructions from a **Tap Hole** (see) during pouring of molten metal.
Pick Tongs. Tongs used in manipulating hot metal.
Picral. Solution of 4 g of picric acid in 100 ml methyl or ethyl alcohol, used

as an etching agent in ferrous metallography.

Piercer. (a) See **Piercing.** (b) See **Vent Wire.**

Piercing. Making tubes from solid metals by forcing a pointed plug or mandrel, called a piercer, through a heated billet.

Piercing Mandrel. See **Mannesmann Process.**

Piercing Process. See **Mannesmann Process.**

Pig. Casting, usually in block form, of any crude metal, such as crude cast iron or lead.

Pig Bed. Mass of sand in which cavities are made for the casting of crude pig metal.

Pig Boiling Process. Manufacturing wrought iron (see **Puddling**) using iron oxide as the bottom of the furnace, which reacts with carbon, and the gaseous reaction product causes boiling of the bath.

Pig Casting Machine. Machine for the mechanical casting of **Pig** (see) metal, usually pig iron.

Pigging Back. Adding **Pig Iron** (see) to excessively decarburized openhearth steel.

Pigging Up. Maintaining, or blocking, the carbon content of steel, in the **Open-Hearth Process** (see) of steel manufacture, by addition of pig iron.

Pig Hole. Hole in a furnace operating on steel, used for adding supplementary pig iron.

Pig Iron. Crude form of cast iron, product of blast-furnace ore reduction, cast into ingots called pigs. If much of the carbon in the metal is in chemically combined state, a light color results and the product is known as white pig iron; when most of the carbon is free, in simple dispersion, the resultant gray color accounts for the name gray pig iron. Intermediate types are known as mottled pig iron.

Pig Lead. Commercial, desilverized lead, in **Pig** (see) form.

Pig Metal. Metal in **Pig** (see) form.

Pig and Ore Process. Modification of the basic **Open-Hearth Process** (see) of steel manufacture, using pig iron and iron ore as the charge.

Pig and Scrap Process. Modification of the basic **Open-Hearth Process** (see) of steel manufacture, using pig iron and steel as the charge.

Pig Tin. Massive (block) tin in **Pig** (see) form. See **Tin.**

Pig Washing Process. Refining pig iron and removing most of the silicon and phosphorus, by oxidation in the liquid state, at low temperatures.

Pile. Wrought iron bars grouped together, prior to being heated and rerolled into a single bar.

Piler. Device for piling sheet metal, received from any processing.

Piling. See **Pile.**

Pillaring. Faulty operation of a **Blast Furnace** (see), causing the coarser portions of the raw material to segregate around the side walls of the furnace; the hot gases will, therefore, pass less readily through the center, leaving that region relatively cold.

Pillion. Tin metal, remaining in the slag after the first melting operation; usually recovered by crushing, screening, and washing.

Pill Press. In powder metallurgy, a press for making small compacts.

Pilz Furnace. Shaft furnace of circular or octagonal cross-section, usually increasing in size upwards; used for the smelting of lead ores.

Pimple Metal. Crude copper **Matte** (see) of about 78% copper content, obtained from the smelting of sulfide copper ores.

Pinch. Longitudinal **Lap** (see).

Pinchbeck Metal. Red **Brass** (see), with 6–12% zinc, used for cheap jewelry.

Pinch Effect. Tendency of a molten conductor, when subjected to high current densities, to contract its cross-section; used in certain types of induction furnaces.

Pine Tree Crystallite. See **Dendrite.**

Pine Tree (Crystal) Structure. See **Dendrite.**

Pinhead Blister. Small defect, at or near the surface of formed metal, resulting from a tiny gas bubble which failed to escape from the original molten metal.

Pinhole. (a) Tiny hole in a metal object, usually due to faulty casting technique. Pinholes are probably caused by the escape of occluded gas, chiefly hydrogen and carbon monoxide. (b) Tiny break in a metallic or non-metallic coating on a base metal.

Pin Metal. See **Pin Wire Brass.**

Pin Wire Brass. Yellow **Brass** (see), with about 62% copper and 38% zinc, used for the manufacture of pins.

Piobert Effect. See **Lüder's Lines.**

Pioneer Alloy. Series of high-temperature-resistant alloys, high in nickel (20–65%), with about 25% chromium, and up to 5% molybdenum, the remainder being iron (Pioneer Alloy Products Co.).

Pipe. Contraction cavity, essentially cone-like in shape, which occurs in the approximate center, at the top and reaching down into a casting; caused by the shrinkage of cast metal, particularly in ingot form. The depth of pipe, or shrinkage cavity, as it is more properly called, will vary with the nature of the metal, the size of the ingot, rate of cooling, etc. Pipe can be completely eliminated by the presence of sufficient blow holes. It can also be avoided by arranging the whole pipe in the riser.

Pipe Annealing. See **Annealing.**

Pipe Cavity. See **Pipe.**

Pipe Oven. Hot blast stove (see **Hot Blast Furnace)** containing pipes heated by gas in brick conduits.

Pipe Roll. Flannel-covered roll for the removal of oil film from **Terne Plate** (see).

Pipe-Segregation. Differences in composition, occurring around the pipe cavity of an ingot or casting.

Pipe Steel. Soft, low-carbon steel, usually made by the Bessemer or open-hearth process; suitable for welding and, therefore, for pipe manufacture.

Piping. See **Pipe.**

Pistol Pipe. See **Tuyere.**

Pit. (a) Small, deep hole in the surface of a material. (b) See **Soaking.**

Pitch. Distance between centers of a pair of rolls, when tight against each other, with barely a clearance between them.

Pitch Line. Center of gravity of a **Pass** (see) in a rolling mill.

Pitting. See **Pit.**

Pitting Corrosion Test. Test to determine susceptibility of a metal to pitting [see **Pit** (a)].

Pittsburgh Furnace. See **'Lectromelt Furnace.**

Placer Gold. See **Stream Gold.**

Plain Carbon Steel. See **Simple Steel.**

Plain Steel. See **Simple Steel.**

Planished Temper. See **Soft Skin-Rolled Temper.**

Planisher. See **Rolling Mill.**

Planisher Mill. See **Rolling Mill.**

Planisher Pass. See **Rolling Mill.**

Planisher Roll. See **Rolling Mill.**

Plastic Bronze. Bearing Bronze (see), sufficiently soft, because of lead content, to take the form of a running shaft. Composition approximates 65–84% copper, 5–10% tin, and 8–20% lead, with optional low percentages of nickel and zinc.

Plastic Deformation. Permanent deformation of material, under stress, due to plasticity.

Plastic Flow. See **Plasticity.**

Plasticity. Flowing, under stress, of certain materials, and probably of all materials under certain conditions. In general, this plastic state may be considered as semi-fluidity, intermediate between the completely solid state and the completely liquid; it appears to be due to slip along planes within the crystals of the metals.

Plastic Metal. Tin-base alloy, with about 10% copper, 9% antimony and 1.5% iron.

Plate. (a) Flat piece of metal more than about 6 inches wide, with thickness greater than $\frac{3}{16}$ inch. (b) See **Electrolytic Plate.**

Plate Amalgamation. Process of amalgamating gold or silver ores (see **Amalgamation Process**), in which the crushed ore, suspended in water, is contacted with surface-amalgamated copper plates.

Plate Mark. Mark on gold or silver articles indicating fineness, manufacturer, etc.

Plate Metal. Iron, cast into small pigs and then broken up into small pieces for alloying purposes.

Plate Mill. Mill stand or series of stands (see **Rolling Mill**) for the rolling of **Plates** (see).

Plate Roll. Rolling Mill (see) for the manufacture of sheets or plates, in contradistinction to mills with grooves, used in rolling rails, etc.

Platers' Brass. Brass **Anodes** (see) for use in brass electroplating.

Plates. In powder metallurgy, flat metal particles of appreciable thickness.

Plate Tongs. Tongs for manipulating metal plates during rolling and related processing.

Platina. See **Birmingham Platina.**

Platina Mohr. See **Platinum Black.**

Platine-Au-Titre. Silver-platinum alloy series, used for jewelry. Platinum ranges from 17 to 35%.

Plating. Coating metal objects with a thin film of another metal or alloy; the object is made cathode (negative electrode) in an electrolytic bath, containing a decomposable compound of the metal to be plated. Frequently, non-decomposable salts are present to increase conductivity, and addition agents to modify the crystalline nature of the deposit. The metal to be plated may be used as anode (positive electrode), for maintaining the electrolytic bath constant; or, alternatively, insoluble anodes may be used, in which case periodic adjustment of the bath is necessary.

Platiniferous. Containing platinum.

Platiniridium. Alloy of platinum and iridium, commonly found in nature together with other metals of the platinum family.

Platinite. Iron-nickel alloy, with 42–45% nickel; used as a substitute for platinum as an electrical lead-in through glass, because their coefficients of expansion are similar. Generally, platinite is copper-coated, to provide better contact with the glass.

Platinizing. Coating with platinum, most commonly by electrolytic methods.

Platinoid. Nickel silver alloy, used for jewelry, containing tungsten as an added ingredient. Typical compositions are: Copper 54–60%, zinc 20–24%, nickel 14–25%, and tungsten 2%.

Platinum (chemical symbol Pt). Element No. 78 of the periodic system; atomic weight 195.23. Lustrous, silvery-white metal, the principal member of the platinum family, very ductile and easily worked, both hot and cold; melting point 3223°F; boiling point about 8180°F; specific gravity 21.45. It is extremely inert chemically and remains untarnished in air; principally divalent (platinous) and tetravalent (platinic), though other valences are also known. Platinum occurs native, usually in association with other members of its family; it is frequently obtained as a by-product of copper and nickel metallurgy; extensively used as a catalyst, e.g., in the manufacture of sulfuric acid and in ammonia oxidation to nitric acid, for small chemical equipment, for accurate pyrometry, and for dental and jewelry alloys. It can be electrodeposited commercially; complex phosphate baths are used for this purpose.

Platinum Black. Fine black powder, consisting of metallic platinum, obtained by chemical reduction of a platinum salt. It is extremely reactive, occludes gases readily, and is commonly used as a catalyst in chemical processes.

Platinum Family (of Metals). Members of the last two transition series (Group VIII) in the periodic system, consisting of ruthenium, rhodium, palladium, osmium, iridium, and platinum. The first three are sometimes known as the light platinum metals, and the last three as the heavy platinum metals, because of the sharp difference in specific gravity.

Platinum-Iridium. Platinum hardened with up to 30% iridium, used for jewelry, electric contacts, etc. Most commonly, about 5% of iridium is used.

Platinum Lustre. See **Silvery Lustre.**

Platinum Metals. See **Platinum Family (of Metals).**

Platinum Plating. Coating metal objects with platinum; the object is made cathode (negative electrode) in an electrolytic bath containing a decomposable platinum salt, usually a complex phosphate. Platinum or carbon anodes are used as positive electrode; the platinum does not dissolve electrolytically, and the bath must, therefore, be periodically adjusted.

Platinum-Rhodium. Platinum alloy containing approximately 10% of rhodium; used frequently as one element in the standard platinum-versus-platinum-rhodium thermocouple.

Platinum Sponge. Black, highly-reactive, spongy form of metallic platinum, obtained by heating platinum compounds to the point of decomposition; occludes gases readily, and will cause the ignition of mixtures of air and combustible gases, a characteristic frequently used in automatic cigarette and gas lighters.

Platinum Thermometer. Resistance **Pyrometer** (see), using platinum wire as resistance.

Plattner Process. Process of recovering gold from its ores; the roasted ore is leached with chlorine-saturated water and the gold precipitated from the resulting gold chloride solution by ferrous sulfate, or hydrogen sulfide, and is collected in metallic form.

Plews Process. Obtaining antimony values from sulfide ores; the ore is heated in a rotary furnace, and antimony oxide volatized from the mass and then condensed in condensing chambers.

Plow Steel (Rope Wire). See **P. S. R. Wire.**

Plug Rolling. See **Plug Rolling Mill.**

Plug Rolling Mill. Two-High Mill (see) for manufacturing seamless tubing. The entry side has a pusher to force a small-diameter, thick-wall tube over a mandrel which extends through from the delivery side.

Plumbago. See **Graphite.**

Plumbers' Solder. See **Wiping Solder.**

Plumbiferous. Containing lead.

Plumbum. Latin name for lead, used in medicine and pharmacy.

Plumos Antimony. Non-metallic mineral, jamesonite or feather ore; a lead-antimony sulfide, chemically approximating $Pb_2Sb_2S_5$.

Plumrite. Series of brasses (Bridgeport Brass Co.), in both low-zinc and high-zinc ranges. See **Brass.**

Plunger Machine. Machine for batch **Pickling** (see), in which a group of equi-spaced horizontal arms is attached to a vertical, hydraulically-actuated piston. The material to be pickled is attached to an arm which, with rise and fall of the piston, dips the material into the pickling, washing, etc., baths.

Plunger Pickling. See **Plunger Machine.**

Pluramelt Process. Process for making clad steel in which the **Cladding** (see) metal (or constituents to form the cladding alloy), protected by a slag, is continuously fed onto a slab of the base metal, and both are subjected to electric arc action.

Plush Copper. Non-metallic mineral, chalcotrichite; a variety of cuprite, or native cuprous oxide, Cu_2O.

Plus Mesh. In powder metallurgy, the part of a powder sample which is retained on a specified screen.

Plyer. Device, in the cold-drawing of tubing, for gripping the tube and drawing it through the die and over the mandrel.

Ply Metal. See **Cladding.**

Ply Steel. Composite, consisting of two or more layers of steel of different compositions. See **Clad Steel.**

P. M. G. Copper-base alloy, with silicon up to 4%, and iron about 1.5% (Phelps Dodge Copper Products Corp.).

Pneumatic Process. See **Bessemer Steel.**

Po. Chemical symbol for **Polonium** (see).

Pocket. Cavity in a casting or metallic mass.

Pohlé-Croasdale Process. Recovering metallic values, as volatile chlorides, from ores containing gold, silver, copper, and lead, by roasting with salt, at a temperature of about 1900°F, usually with sulfur as an additional constituent. The smoke, containing the volatilized chlorides, is electrically precipitated and the mixed chlorides are then treated for metal recovery by fusion or electrolysis.

Pointing. Reducing the diameter of wire or tubing, on a short length, to facilitate entry into a die.

Point of Decalescence. See **Decalescence.**

Point of Recalescence. See **Recalescence.**

Poison Tower. Chamber for condensing the metal in the manufacture of arsenic.

Poisson's Ratio. Ratio of the transverse contraction per unit dimension of a bar to its elongation per unit length, within the elastic limit.

Polarization. In electrolysis, increase of solution resistance due to gas accumulation at the electrode, or chemical depletion in part of the solution. causes.

Poling. Insertion of wooden poles into molten copper or tin, to remove excess oxygen by the reducing action of the gases resulting from the burning of the pole. In copper, the metal, after poling, is in the tough pitch condition.

Poling Down. Preliminary **Poling** (see) of copper, under highly oxidizing conditions, to eliminate sulfur.

Polish Attack. Method of revealing structure of a metal for microscopic study; the etchant is combined with the polishing medium. No longer in common use.

Polished Finish. Surface **Finish** (see) produced by polishing the metal, most commonly steel.

Polishing. Final step in preparing a specimen for metallographic examination, prior to etching; a finely ground surface is subjected to a high degree of polish, most commonly by use of a **Polishing Wheel** (see).

Polishing Wheel. Rotating disc, usually cloth-covered, for use in polishing and buffing. In electroplating processes, the shaft, on which the disc rotates, is usually horizontal; in metallographic polishing, it is usually vertical.

Polonium (chemical symbol Po). Element No. 84 of the periodic system; atomic weight approximately 210. Because of its radioactivity and scarcity, little is known of the element, save that it vaporizes at about 1800°F and shows both trivalence and tetravalence, most of the chemical reactions being somewhat analagous to those of bismuth and tellurium. Polonium occurs in uranium ores, and is separated by chemical action. It can be electrodeposited, best from a solution acidified with hydrochloric acid. In minute quantities, it is used as an ionizing activator in spark plug wire.

Polycrystalline. Consisting of more than one crystal. Although large single metal crystals can be made by special techniques, metals are nearly always polycrystalline. In polycrystalline metals, the crystal axes may lie in random directions, yielding random orientation, or they may tend towards some preferred orientation.

Polymorphism. Characteristic of a material of existing in more than one crystallographic form. See also **Allotropy.**

Pompton Tool Steel. Series of **Carbon Tool Steels** (see), with carbon up to 1.4% (Ludlum Steel Co.).

Ponsard Furnace. Recuperative type of furnace, in which hot spent gases pass through tubular flues and pre-heat, through the flue walls, the incoming air.

Pony Rougher. See **Rolling Mill.**

Pope's Island Metal. Copper-base alloy series, of the **Nickel Silver** (see) type, with 14–20% nickel and about 14% zinc.

Porcelain Enameling Commercial Quality Steel Sheet. Steel sheet, ordered or sold on the basis of specifications governing use as a base for enameling or other vitreous coatings.

Porcelain Enameling Deep Drawing Quality Steel Sheet. Steel sheet, ordered or sold on the basis of specifications governing use as a base for enameling or other vitreous coatings, and also on the basis of deep drawing qualities.

Porcelain Gilding. Process of gilding porcelain, in which gold metal, suspended in a liquid, is applied to the porcelain, which is then heated till full adhesion is attained.

Porcupine (Cooling Rack). Rotating cooling rack, for use in the cooling and inspection of galvanized and similar sheets. The sheets lie on racks evenly spaced on a central horizontal hub.

Pore. In powder metallurgy, any cavity in a compact.

Pore Forming Material. In powder metallurgy, material added to metal powder for the purpose of volatilizing during sintering, thus producing desired porosity.

Porosity. (a) Unsoundness in metals, particularly when in cast form, due to the presence of blowholes and similar cavities. (b) In powder metallurgy, the characteristic, true in greater or lesser degree of all compacts, of pores being present throughout the mass.

Porosity Test. See **Deep Etch Test.**

Porpezite. Mineral alloy of gold and palladium, commonly containing also silver.

Porter. Bar attached to a forging billet, for manipulating during **Forging** (see).

Post-Pernot Furnace. See **Pernot Furnace.**

Pot. (a) Vessel for holding molten metal, particularly an iron-base vessel for low-melting metals and alloys. (b) See **Crucible.**

Pot Annealed Wire. Steel wire annealed (see **Annealing**), in coiled form, in pot furnaces, followed by slow cooling.

Pot Annealing. See **Annealing.**

Potassium (chemical symbol K). Element No. 19 of the periodic system; atomic weight 39.096. Very soft, malleable, silvery-white metal of the alkali family, reacting violently with water and, therefore, corroded rapidly by atmospheric conditions; melting point 146°F; boiling point 1400°F; specific gravity 0.86 (lighter than water). Chemically, it is always monovalent; since it reacts with water, it cannot be electrodeposited from aqueous solutions. Potassium may be isolated by electrolysis of its fused hydroxide. As a metal, it has little commercial utilization other than as a laboratory reagent. Its alloy with sodium is of interest as it is liquid at room temperature.

Potassium Amalgam. Alloy of potassium and mercury, analogous to **Sodium Amalgam** (see).

Potential. See **Voltage.**

Pot Lead. See **Graphite.**

Pot Liquation. Liquating the antimony values out of sulfide ores, by subjecting a pot containing the ore to heat, and permitting the easily-molten sulfide to run out through a hole in the bottom of the pot.

Pot Metal. Copper-base bearing metal, with 20–53% lead.

Potters' Lead. Non-metallic mineral; coarse **Galena** (see), used in manufacturing pottery glazes.

Pour. Quantity of metal which is cast at one time.

Pouring. See **Casting** (a).

Pouring Gate. See **Gate.**

Pouring Reel. Horizontal disc revolving at the delivery speed of the steel rod to be reeled; two rows of pins form an annular space into which the rod is drawn.

Pour Test. Simple test for determining proper pouring temperatures for molten metal. A small quantity of the metal is poured into a mold; no **Skull** (see) should remain in the spoon, and the poured metal should show perfect surface, free of wrinkles.

Powder. In powder metallurgy, particles of size, shape, and other characteristics such as to lend themselves to compacting, sintering, and any other operations necessary in powder metallurgy technique.

Powder Metallurgy. Art and science of manufacturing objects from pure, mixed, or alloyed metal powders, with or without added, non-metallic constituents. The object is shaped under pressure and then heated to form a compact mass; however, without complete fusion of the metal.

Powder (Method) Test. See **Magnaflux Test.**

Powder (X-ray) Method. See **Debye-Scheerer Method.**

Pr. Chemical symbol for **Praseodymium** (see).

Praseodymium (chemical symbol Pr). Element No. 59 of the periodic system; atomic weight 140.92. Silvery white rare earth metal, tarnishing in air, can be isolated by electrolysis of its anhydrous trichloride in a bath of fused alkali chlorides; melting point 1724°F; boiling point is not known; specific gravity 6.5. Chemically, it is trivalent. Like all other rare earth metals, it cannot be electrodeposited. There are no uses for the metal.

Precious Metals. Expensive metals in common use. Specifically, gold, silver, and platinum.

Precipitate. Material which separates out of a solution as a solid phase.

Precipitation. Separation, out of solution, of a constituent as a solid phase.

Precipitation Hardening. Hardening, in metal, resulting from the precipitation of a constituent from a supersaturated solid solution. See also **Age Hardening.**

Precipitation Hardening Magnet Steel. Series of substantially carbon-free magnet steels, in which hardness is achieved by precipitation hardening.

Precipitation Hardness. See **Precipitation Hardening.**

Precipitation Process. Recovering lead from its ores by heating with iron, or with iron-rich slag or minerals, most commonly in a shaft furnace.

Preece Test. Test for uniformity and thickness of zinc coating on galvanized iron and steel; a neutral copper sulfate solution is used, and the results are given in terms of the number of minutes of immersion which the coating will successfully withstand.

Preferred Orientation. See **Orientation.**

Preheating. Heating of metal preliminary to a second heating or special processing. Frequently, preheating to a moderate temperature over a relatively long period, is followed by another, more drastic heating for a shorter period.

Preheating Zone. Zone, in a **Continuous** heat-treating **Furnace** (see), in which the metal is given preliminary heating.

Premier Nickel Chrome. Nickel-base alloy, with about 23% iron and 12% chromium, used for heat-and-corrosion-resistant products (Alloy Metal Wire Co.).

Presintering. In powder metallurgy, preliminary heating of a compact at a temperature below that of the final sintering. Presintering is usually intended to give sufficient strength to allow handling, or to remove a binder or similar substance, by volatilization.

Pressed Bar. In powder metallurgy, a compact shaped in bar form.

Pressed Density. In powder metallurgy, the weight of a unit volume of powder, pressed into a standard finished compact at a specified pressure.

Press Forging. Forging in hydraulic presses, with deformation accomplished by pressure. The relatively longer period of stress, as compared with hammer forging, allows for more plastic flow.

Pressing. In powder metallurgy, forming the compact by means of pressure.

Pressing Crack. See **Slip Crack.**

Pressing-In Test. See **Indentation Test.**

Pressurdie (Steel). Series of complex alloy tool and die steels, with about 5% chromium, optionally about 5% tungsten, and fractional percentages of vanadium, molybdenum, cobalt and silicon (Braeburn Alloy Steel Co.).

Pressure Die Casting. See **Die Casting.**

Pressure Forging. Forming metal, normally at high temperature, by means of pressure.

Pressure Modulus. See **Bulk Modulus.**

Pressure Welding. Welding (see) in which the heated metals are joined by pressure.

Presto Steel. Low-alloy steel (Carpenter Steel Co.) with about 1.5% chromium and 1% carbon; used for roller and other anti-friction bearings.

Pricking Bar. Long bar for cleaning out **Tap Holes** (see) or **Tuyeres** (see).

Prillion. Tin metal extracted from the slag of a previous smelting operation.

Prillon. See **Prillion.**

Primary Austenite. Solid solution of carbon in **Gamma Iron** (see), separating out from the liquid phase during the freezing of an iron-carbon alloy,

with carbon in the 0.15 to 4.3% range.
Primary Cell. See **Battery.**
Primary Crystallization. Crystallization, occurring when metal solidifies from the liquid state.
Primary Metal. See **Virgin Metal.**
Primary Solid Solution. Solution of one metal in another, in the solid state, in which atoms of the solute are included in the crystal lattice of the solvent.
Prime. See **Primes.**
Primer Brass. See **Cartridge Brass.**
Primes. Metal products (such as sheet and plate steel), free from defects observable by macroscopic methods.
Prime Western Spelter. See **Prime Western Zinc.**
Prime Western Zinc. Low-grade of virgin zinc, containing about 98% zinc; lead is limited to 1.60% and iron to 0.08%, with no limitations set on cadmium and aluminum.
Prince's Metal. (a) **Brass** (see), with 17–39% zinc content. (b) Copper-base alloy, with about 15% antimony.
Print Test. Test of steel to determine segregation, of elements such as sulfur and phosphorus; a polished surface of the metal is treated with dilute acid and then covered with paper chemically sensitized to react with the dissolved component, yielding a colored spot.
Prism Test. Hardness test, in which two pieces of the tested material are pressed together under constant load.
Process Alloy. Alloy added to molten metal, in the furnace, ladle, or ingot mold, to interact with the metal and yield desired reactions, such as degasification, etc.
Process Annealing. See **Annealing.**
Processed Wire. See **Bright Soft Wire.**
Process Metallurgy. Science of isolating metals from their ores.
Process Wire. Wire, after preliminary drawing, awaiting further drawing or processing.
Producer. See **Gas Producer.**
Producer Gas. See **Gas Producer.**
Production Test. Test on tool steel, in which a tool made from the steel is judged on the basis of operating service life.
Pro-Eutectoid. Phase separating out of a solid solution prior to the final decomposition with formation of a eutectoid; the pro-eutectoid constituent tends to separate out at the grain boundaries of the original phase.
Progressive Fracture. Fracture (see) spreading from a small starting point, frequently at the surface, gradually through the material, until complete breakage results. Fatigue failures are of this type.
Progressive Hardening. See **Differential Hardening.**
Progressive Impact Test. Test of impact resistance of a material, in which the impact is progressively increased, usually by automatic means, until breakage of the specimen.
Projection Welding. Resistance **Welding** (see) in which heat is localized to limited areas of contact, in the form of projections on one or both materials being welded.
Prolong. Condenser of vertical or horizontal construction, usually made of sheet steel, for condensing zinc vapors to zinc metal.
Promal. Malleable Cast Iron (see), with about 0.4% combined carbon and 1.4% tempered carbon (Link-Belt Co.).
Promat Process. Process of electrodepositing zinc or cadmium, using solutions of organo-metallic compounds as electrolytes, and alternating current superimposed over the normal direct current (Poor & Co.).
Proof Load. See **Proof Stress.**
Proof Strength. See **Proof Stress.**
Proof Stress. Stress which will cause permanent deformation to any preassigned small, but measurable value; this is of value in designs, such as aircraft, where a low factor of safety is required.
Proof Test. Determining if a material or completed structure is satisfactory for prescribed utilizations.
Proof Weight. See **Proof Stress.**
Proplatinum. Nickel-base alloy series, used as a substitute for platinum and for jewelry, containing about 24% silver and about 4% bismuth. See also **Platine-Au-Titre.**

Proportional (Elastic) Limit. Point on a stress-strain diagram at which marked deviation from Hook's Law begins, i.e., where strain is no longer proportional to stress and begins to exceed the normal ratio.

Protoactinium (chemical symbol Pa). Element No. 91 of the periodic system; atomic weight 231. It is a rare element, of slow radioactivity, and relatively little is known about its characteristics. Theoretically, it should resemble tantalum.

Proton. Positive nucleus of the hydrogen atom.

Protoxide. Metal oxide whose oxygen content is the lowest of all oxides of that metal.

p.s.i. Pounds per square inch. Commonly used in conjunction with tensile and compressive testing to indicate stress magnitude.

P. S. R. Wire. High-quality, usually tinned and liquor-finished, steel wire; ordered and sold on the basis of quality and resistance to torsion, for use in rubber tire cables.

Pt. Chemical symbol for **Platinum** (see).

Puddle. (a) Bath of fluid cast iron in the **Puddling** (see) process. (b) Molten metal under the arc or torch in **Welding** (see) operations.

Puddle Ball. See **Puddling.**

Puddle Bar. See **Puddled Bar.**

Puddled Ball. See **Puddling.**

Puddled Bar. Wrought iron bar, rolled after squeezing out part of the slag from the puddled ball. See **Puddling.**

Puddle Iron. See **Puddling.**

Puddle Mill. See **Muck Mill.**

Puddler. Rabble used in **Puddling** (see).

Puddle Roll. See **Muck Mill.**

Puddler's Candles. In the **Puddling** (see) process, the appearance of bursts of flame (burning carbon monoxide) through the slag, during the period of rapid carbon elimination.

Puddle Train. See **Muck Mill.**

Puddling. Process of making wrought iron; pig iron is melted on a hearth, with iron oxide added; refining takes place through action of the iron oxide and the silica of the hearth. There is a preliminary boiling period, called the low boil, during which the greater part of the manganese and silicon is removed. The reaction raises the temperature of the metal, and carbon is eliminated rapidly, with long shafts of blue flame, bringing about the high boil. The elimination of the carbon raises the melting point of the metal, and, after the reaction has been completed, the metal freezes in the form of small iron particles. Despite the freezing of the metal, the slag remains liquid, and the metal can be collected, either by hand-actuated or mechanically-actuated means, into spongy metal balls. When composited, the puddled metal ball is subjected to mechanical squeezing, by means of a squeezer, or to hammering (shingling) which removes most, but not all, of the excess slag. See also **Wrought Iron.**

Puddling Furnace. See **Puddling.**

Puddling Iron. Pig iron of low silicon content for conversion into puddled or wrought iron in a puddling furnace.

Puddling Process. See **Puddling.**

Puddling Staff. See **Staff.**

Puffed Bar. In powder metallurgy, a **Cored Bar** (see) blown out of shape by gas pressure within the bar.

Puller-Out. Worker, in the crucible process of making steel, with the duty of removing crucibles, when ready for casting, from the crucible furnace.

Pulling (Slag). Raking off slag from contact with molten metal.

Pulling Test. See **Tensile Strength.**

Pull-Over Mill. Rolling mill of two-high construction (see **Two-High Mill**) in which the metal is returned to the operator by placing it on top of the upper roll.

Pull Test. See **Tensile Strength.**

Pulverization. In powder metallurgy, reduction in size of metal and other powders by mechanical means.

Pulverulent. Easily powdered or converted to dust.

Pump Tubing. Seamless steel tubing for use in the pumping of oil products.

Punch Steel. See **Chisel Steel.**

Pure Metal. No metal is absolutely pure in the sense of having only one element present; the term pure metal should be understood to be a relative one.

Purging. Replacing one atmosphere with another in a furnace.

Purging Chamber. See **Eutectrol Process.**

Pur-O-Tin. Anodes (see) of highly-purified tin, for use in the electroplating of tin; manufactured by E. I. du Pont de Nemours & Co. See **Tin.**

Purozinc. Anodes (see) of relatively pure zinc (total impurities about 0.20%), for use in the electroplating of zinc; manufactured by E. I. du Pont de Nemours & Co. See **Zinc.**

Purple of Cassius. Purple colloidal gold, formed by reducing gold chloride with stannous chloride.

Purple Gold. Gold containing approximately 21% aluminum, used in jewelry manufacture.

Pusher Furnace. Continuous Furnace (see) in which the material is pushed ahead through the furnace by the material moving behind it.

Pyrasteel. Nickel-chromium-iron alloys (Chicago Steel Foundry Co.), of excellent corrosion and heat resistance, with 4–35% nickel, 8–25% chromium, and low percentages of silicon.

Pyrite. Naturally-occurring iron sulfide, corresponding to the formula FeS_2. Because of its pale yellow color, it is also called fools' gold. See also **Pyrites.**

Pyrites. Naturally-occurring metal sulfides of high lustre, such as iron disulfide (pyrite), copper-iron disulfide (copper pyrite or chalcopyrite), and tin sulfide (tin pyrite or stannite).

Pyritic Process. See **Pyritic Smelting.**

Pyritic Smelting. Smelting sulfide ores, such as those of copper; free silica is added to the ore prior to smelting; a small proportion of coke may also be added to initiate the reaction; thereafter, the reaction maintains itself by the heat of oxidation.

Pyrocast. High-alloy cast iron, containing high percentages of nickel and chromium (Pacific Foundry Co.).

Pyrometallurgy. Science and technique of metallurgical operations conducted by means of heat action, as in roasting, smelting, etc.

Pyrometer. Instrument for measuring high temperatures. The electric-resistance pyrometer works on the principle of increased resistance of metals with increasing temperature. Platinum is commonly used in such pyrometers. Water pyrometers (or specific-heat pyrometers) involve the measurement of the temperature increase of a known amount of water when a known weight of metal, heated to the temperature to be measured, is plunged into it. The thermo-electric pyrometer is the most common type. Whenever two dissimilar metals are in contact, a change of temperature will give rise to an electric potential, which can be measured, in millivolts. Such a combination is known as a thermocouple; the point of contact of the metals, brought to the temperature to be measured, is known as the hot junction. The fundamental thermocouple materials have been platinum-versus-platinum-rhodium; cheaper types are represented by iron-versus-constantan, chromel-versus-alumel, and others. Radiation pyrometers involve the radiation heating of a delicate thermocouple, through a telescope sighted on the source of heat. Since the radiation varies as the fourth power of the absolute temperature, radiation pyrometers are most effective at high temperature ranges; their disadvantage is that correction must be made for the fact that ordinary materials are not perfect black-bodies, that is, they do not conform to perfect radiation requirements. In the optical pyrometer, temperature is measured by means of matching the intensity of the light against a lighted electric filament, which has previously been calibrated in terms of the current necessary to yield known temperatures.

Pyrometer Cone. See **Pyrometric Cone Equivalent.**

Pyrometric Cone Equivalent. Temperature-equivalence scale, used in connection with refractories, based on the softening of standard composition cones of known softening temperatures.

Pyrometry. Science and technique of temperature measurement. See **Pyrometer.**

Pyrophoric Alloys. Alloys which throw off sparks when struck or abraided; used in cigarette lighters, etc. The most common pyrophoric alloys are those of cerium-metal base (see also **Misch Metal**); commercially, iron is the most common alloying element.

Python Steel. Low-alloy **Shock-Resistant Steel** (see), of about 0.20% vanadium, 0.20% silicon, and 0.30% manganese; carbon is high, about 0.90%.

Q

Qualitative Analysis. Chemical analysis in which only the nature of the constituents is determined, and not their proportions.

Quality Factor. See **Merit Number.**

Q-Alloy. High-chromium steels, of high heat and corrosion resistance (General Alloys Co.) with 12–30% chromium and optional additions of up to 23% nickel.

Quantitative Analysis. Chemical analysis in which the relative proportions of the constituents are determined.

Quartation. (a) Separating gold and silver by treatment with nitric acid, which dissolves out the silver; the gold should not exceed 25%. (b) Adding silver to a gold-silver alloy, for bringing the gold content down to 25%, to make possible treatment with nitric acid, as discussed above.

Quarter-Hard Temper. Cold-rolled strip or sheet metal of medium-soft temper, suitable for forming and drawing.

Quaternary (Alloy). Alloy containing four elements, exclusive of impurities.

Quaternary Diagram. Constitution Diagram (see) for a quaternary alloy system.

Quaternary Steel. Steel containing two major alloying elements in addition to carbon.

Quaternary System. See **Quaternary Diagram.**

Queen's Metal. Tin-base type metal, with 7–17% antimony, 1–17% zinc, and, optionally, up to 17% lead and up to 3.5% copper.

Quench Aging. Gradual hardening, in steels of very low carbon (0.03–0.06%), after rapid cooling from about 2375°F.

Quenching. In the heat treating of metals, the step of cooling metals rapidly in order to obtain desired properties; most commonly accomplished by immersing the metal in oil or water.

Quenching Crack. Crack on the surface of quenched steel due to faulty processing. **Quenching** (see) cracks can usually be eliminated by removing the steel from the quenching bath when it has reached about 480°F and air-cooling thereafter.

Quenching Medium. Fluid medium, such as water, air, oil, etc., used for **Quenching** (see).

Quenching Nozzle. See **Brunorizing.**

Quenching Oil. Oil for quenching metals in heat-treating operations; it may be compounded, i.e., a mixture of various mineral, vegetable, fish, and animal oils. See **Quenching.**

Quenching Solution. Solution used as a **Quenching** (see) medium.

Quenching Tub. Vessel, containing water, for **Quenching** (see) steel.

Quick. See **Quickening.**

Quick Cutting Steel. See **High Speed (Tool) Steel.**

Quickening. Treatment of a base metal, such as copper, with mercury or a solution of a mercury compound, to facilitate adhesion of silver in silver plating.

Quickening Liquid. Solution of a compound of mercury, such as the nitrate or double alkali cyanide, for use in **Quickening** (see) operations.

Quicksilver. Common name for **Mercury** (see).

Quicksilver Cradle. Rocking apparatus, tilted, for contacting gold-bearing ore with mercury in amalgamation operations.

Quicksilvering. Coating or treating with mercury.

Quicksilver Water. See **Quickening Liquid.**

Quinary (Alloy). Alloy, containing five elements, exclusive of impurities.

Quinary Steel. Steel containing three major alloying elements, in addition to carbon.

R

Ra. Chemical symbol for **Radium** (see).
Rabble. Device, usually in the form of a rake at the end of a long arm, to facilitate **Rabbling** (see).
Rabbler. See **Rabble.**
Rabbling. Stirring a charge of ore or semi-molten metal, by hand or by mechanical means.
Rabbling Rake. Device, shaped like a rake, for **Rabbling** (see) by hand.
Rabbling Tool. See **Rabbling Rake.**
Radianite. High-chromium (about 17%) **Stainless Steel** (see), with about 0.75% carbon; used for cutlery (Latrobe Electric Steel Co.).
Radiant Tube Annealing Box. Annealing (see) box which is heated, inside, by means of tubes in which gas is burned; the hot tubes radiate their heat to the covered pile of metal, standing on the base of the box.
Radiant Tube Furnace. See **Radiant Tube Annealing Box.**
Radiation. See **Thermal Radiation.**
Radiation Pyrometer. See **Pyrometer.**
Radioactive Elements. Group of elements of high atomic weight, which give off radiations continually, while being decomposed. Radium and mesothorium are the elements used commercially for radioactive purposes. Others, having less strong radioactive properties, or being less available, include actinium, ionium, polonium, thorium, and uranium.
Radioactive Metals. See **Radioactive Elements.**
Radio Alloy. Series of copper-nickel alloys, used in the wiring of electrical instruments (Wilbur B. Driver Co.).
Radiographic Test. See **Radiography.**
Radiography. Examination of substances by means of X-rays; the X-rays are permitted to activate a fluorescent screen for temporary work, or photographic film may be used for permanent record.
Radiometallography. See **Radiography.**
Radio Steel. See **Armco Radio (Steel).**
Radium (chemical symbol Ra). Element No. 88 of the periodic system; atomic weight 226.05; has a high degree of radioactivity, decomposing ultimately into radio-lead. Silvery white metal, resembling barium chemically and physically; reacts with water and tarnishes readily in air; melting point 1760° F; boiling point 2084° F; specific gravity approximately 5. Chemically, it is divalent. Despite its extreme rarity, radium has been isolated; an amalgam, made by electrolysis of the chlorides into a mercury cathode, is heated in an atmosphere of hydrogen, leaving radium behind. The metal has no commercial utilizations.
Ragged Fracture. See **Fracture.**
Ragged Roll. Roll (see) with roughened surface.
Ragging. Series of horizontal grooves, cut into the surface of a **Roll** (see), to render possible increase in the **Angle of Bite** (see).
Ragging Mark. Flaw on the surface of steel when improperly rolled on rolls that have been subjected to **Ragging** (see).
Railroad Bronze. Series of copper-base alloys, used in railroad and other heavy equipment bearings, with 73–89% copper, and lead, tin, and zinc each up to 20%.
Rail Steel. Steel, with 0.50–0.80% carbon, and 0.60–1.00% manganese, used for the manufacture of railway rails.
Rail Test. Ductility test for rails in which a heavy weight is dropped from specific heights onto the rail supported on two points.
Rail Train. Rolling Mill (see) train for converting steel ingots and blooms into rail form.
Rain Chamber. Zone in which fumes are washed or condensed by a stream or spray of water.
Ramet. See **Vascaloy-Ramet.**
Rammelsbergite. Naturally-occurring, mineral alloy, chemically approximating nickel diarsenide, $NiAs_2$.
Rammer. Device, used in foundry practice, for compacting the material of a **Mold** (see).

Ramming. Packing of sand around a pattern [see **Pattern** (b)] in the casting of metals.
Randolph Process. Process of refining copper by a series electrolytic operation (see **Series System**), with horizontal electrodes, the top surface being anodic, the bottom surface cathodic, with the crude copper, to be refined, in between.
Random Orientation. Lacking **Preferred Orientation** (see).
Range. Lowest and highest values between which chemical or physical properties, temperature, etc., may vary.
Range of Stress. Total between limits of cyclic stresses to which a sample is subjected, particularly in a **Fatigue Test** (see).
Rankin Process. Recovering copper from its ores by treatment with nitric acid at about 250° F, yielding soluble copper nitrate.
Raoult's Law. Under ideal conditions, equi-molecular quantities of different materials, when dissolved in a given substance, have the same effect in lowering the freezing point. The law is valid only under rather limited conditions.
Rapid Steel. See **High Speed (Tool) Steel.**
Rappage. Oversize of a casting due to excessive **Rapping** (see) of the mold when removing the original pattern.
Rapping. See **Rapping Out.**
Rapping Out. Removing pattern from mold after completion of the mold. See **Pattern** (b).
Rare Earth Metals. Family of metals derived from a group of relatively scarce ores known as the rare earths ores. The metals resemble one another very closely chemically. They include: Yttrium, lanthanum, cerium, praseodymium, neodymium, illinium, samarium, europium, gadolinium, terbium, dysprosium, holmium, erbium, thulium, ytterbium, and lutecium.
Rare Elements. See **Rare Metals.**
Rare Metals. Elements which have not been closely studied because of scarcity or lack of interest. In addition to the **Rare Earth Metals** (see), the platinum family of metals, and the radioactive metals, the list would include such elements as beryllium, scandium, titanium, vanadium, gallium, germanium, selenium, zirconium, columbium, molybdenum, indium, tellurium, tantalum, tungsten, thallium, thorium, uranium, masurium, and rhenium.
Raschette Furnace. Shaft furnace for smelting iron, copper, or lead ores.
Rate of Oil Flow. In powder metallurgy, the rate at which oil passes through a sintered compact under standard test conditions.
Rattle Barrel. See **Tumbling Barrel.**
Rattler. See **Tumbling Barrel.**
Raw Ore. Ore that has been subjected to no chemical or heat treatment, such as calcining or roasting.
Razor Steel. High-carbon or alloy steel, capable of maintaining a keen edge, used for razors. Safety razor blade steel commonly contains 1.25% carbon and 0.25% chromium.
Razor Temper Steel. Tool Steel (see) with about 1.50% carbon.
Rb. Chemical symbol for **Rubidium** (see).
R.D.S. Steel. Low-alloy steel (Carpenter Steel Co.), used for small tools. Composition approximates 2% nickel, 1% chromium, and fractional percentages of manganese and silicon, and 0.75% carbon.
Re. Chemical symbol for **Rhenium** (see).
Reaction Alloy. See **Process Alloy.**
Reaction Process. See **Roasting-Reaction Process.**
Realgar. See **Red Arsenic.**
Reaumur Process. See **Malleable Cast Iron.**
Rebound Hardness. Hardness (see), measured by the rebound of a falling body under standard test conditions. See **Scleroscope Hardness.**
Recalescence. Phenomenon of sudden liberation of heat by iron and iron alloys during cooling through the so-called recalescence point, metallographically known as the Ar_1 point, approximately 1333°F. If the metal is carefully observed in the dark, a glow can be seen at the recalescence point.
Recalescence Point. See **Recalescence.**
Recarburization. (a) See **Bessemer**

Steel. (b) Restoration of carbon, after previous decarburization.

Recarburizer. See **Bessemer Steel.**

Reckoning. Counting of metal plates, prior to final packing for customer delivery.

Recoiler. Device for again coiling continuous metal after being uncoiled, most commonly to enable processing or working.

Recovery. (a) Return to an original state or condition. In metallurgy the term usually refers to the removal of work-hardening effects, without any visible change in the microstructure. (b) Proportion of the valuable components of a mineral obtained in any processing.

Recovery Oven. See **By-Product Oven.**

Recovery Temperature. Lowest temperature at which work-hardening can be removed at a commercial rate.

Recrystallization. Change in crystal size and structure which occurs in cold-worked metals on heating to a characteristic recrystallization temperature, or for a sufficiently long time at a somewhat lower temperature. New grains appear, and, as heating continues or temperature rises, these grains grow, forming a new crystal pattern. The approximate recrystallization temperature of some of the more common metals is as follows: Aluminum 300°F, cadmium 120°F, copper 390°F, gold 390°F, iron 840°F, lead 30°F, magnesium 300°F, nickel 1150°F, platinum 880°F, silver 390°F, tin 30°F, tungsten 2210°F, zinc 40°F.

Recrystallization Temperature. See **Recrystallization.**

Recuperative Furnace. Furnace operating on recuperative heating principles. See **Recuperator.**

Recuperative Heating. See **Recuperator.**

Recuperator. Apparatus for the recovery of heat from hot-spent gases which are used for the indirect pre-heating of incoming fuel or air, the stream of spent gases being separated from the other stream by being confined inside pipes or tubes, whereas the other stream surrounds the pipes or tubes, or vice versa.

Red Antimony. Cherry-red non-metallic mineral, kermesite, also known as kermes mineral, chemically approximating Sb_2OS_2.

Red Arsenic. Non-metallic mineral, realgar, chemically approximating arsenic monosulfide, AsS.

Red Brass. See **Brass.**

Red Casting Brass. High-copper alloy of good casting characteristics. The alloying elements are usually lead and tin, up to about 5% each, in addition to zinc, which commonly ranges from 4–15%.

Red Cobalt. Non-metallic mineral, erythrite, or cobalt bloom, a hydrated cobalt arsenate, chemically approximating $Co_3(AsO_4)_2.8H_2O$.

Red Cut Steel. Series of high-speed alloy tool steels (Vanadium Alloys Steel Co.) with about 18% tungsten, 4% chromium, up to 2% vanadium, and, optionally, up to 1% molybdenum and up to 5% cobalt.

Red Gold. Gold-base alloys, with copper as the alloying element; a typical example is the 25% copper alloy.

Red Hardness. Hardness of a metal, most commonly high-speed tool steel, which is retained even at red heat. See **Temper Hardening.**

Red Heat. Temperature between 1000°F and 1800°F.

Red Iron Ore. See **Hematite.**

Redistilled Zinc. See **Zinc.**

Red Lead. Not a metal. Really red oxide of lead, minium, chemically Pb_3O_4.

Red Manganese (Ore). Non-metallic mineral of manganese, either rhodanite, the silicate, $MnSiO_3$, or rhodochrosite, the carbonate, $MnCO_3$.

Red Metal. (a) See **Plastic Bronze.** Frequently, copper-lead alloys of very high lead content (up to 50%) are meant. (b) Crude copper matte, of about 48% copper content, obtained from the smelting of sulfide copper ores. (c) Colloquial name for **Copper** (see).

Redox. See **Oxidation-Reduction Potential.**

Red Phosphorus. See **Amorphous Phosphorus.**

Red-Searing. Breaking or cracking at **Red Heat** (see).

Red Short. See **Hot Short.**

Red Star Tungsten Steel. Low-alloy steels (Vanadium Alloys Steel Co.), used for tools, dies, etc. In addition to tungsten approximating 1.6% and carbon 0.90%, they contain fractional percentages of chromium and vanadium.

Reduced Iron. Iron powder, made by heating powdered iron oxide in a reducing gas, such as hydrogen.

Reducing. (a) See **Reduction.** (b) See **Reduction in Area.**

Reducing Agent. Substance that will react with oxygen, particularly on heating, and which, therefore, can be used to extract oxygen from another substance.

Reduction in Area. Permanent percentage change between the original area of the cross-section of a piece of metal, compared with the area of the cross-section after being subjected to stress.

Reducing Flame. In gas torch welding with a carbon-containing fuel, a flame with fuel in excess of the oxygen necessary for full combustion.

Reducing Furnace. Furnace, usually of shaft design, for reducing ores to metal.

Reducing Roasting. Heating oxygen-containing ore with a reducing agent, below fusion, to obtain a compound of lower or no oxygen content, or to obtain metal.

Reduction. (a) Removal of oxygen from, or addition of hydrogen to, a compound; more generally, increase in the number of electrons. See also **Oxidation.** (b) In metal working, reduction refers to decrease of one or more dimensions of shaped metal by drawing, hammering, etc.

Reduction Cell. Electrolytic unit in which metal is formed by electrolysis of a corresponding compound.

Reduction Furnace. See **Reducing Furnace.**

Red X Alloy. Series of aluminum-base alloys, with 8–13% silicon, about 1.5% copper, and fractional percentages of magnesium and manganese (National Smelting Co.).

Reel. Coiling machine for wire or strip metal.

Reese River Process. Process of pan-amalgamation [see **Pan-Amalgamation (Process)**], after preliminary roasting of the ores.

Refined Bar. See **Merchant Bar.**

Refined Iron. See **Merchant Bar.**

Refinement. See **Grain Refinement.**

Refinery. See **Finery.**

Refinery Fire. See **Finery.**

Refining. (a) Science and technique of treating crude metals to attain a higher state of purity. (b) See **Grain Refinement.** (c) See **Finery.** (d) See **Restoring.**

Refining Heat. See **Refining Temperature.**

Refining Temperature. (a) Predetermined temperature to which a metal is heated in order to achieve grain refinement. (b) See **Medium Orange Heat.**

Reflective Insulator. Polished metal foil, usually in crumpled form, used for insulation purposes.

Refractory. Heat-resistant material which withstands furnace conditions without melting, crumbling, or cracking. Other characteristics desired in a refractory are low expansion and contraction on heating and cooling, low heat conductivity, low permeability, high mechanical strength and chemical resistance. No refractory is completely resistant under all operating conditions. Acid refractories are those of silica or clay; basic refractories are the magnesia, lime, dolomite, and bauxite types; graphite and chromite are the only common "neutral refractories."

Regeneration. (a) Permitting a **Carburizing Compound** (see) to stand, exposed to the atmosphere, in order to absorb moisture and carbon dioxide and thus regenerate its powers. (b) See **Regenerator.**

Regenerative Furnace. Furnace operating on regenerative heating principles. See **Regenerator.**

Regenerative Heating. See **Regenerator.**

Regenerative Quenching. Process of

double-quenching carburized iron-base alloys; the original quench is from a high temperature, the second is from a lower temperature.

Regenerator. Apparatus for the recovery of heat from hot, spent gases in which the gases are conducted through open brick-work, or checkers, permitting the brick-work to heat up to nearly the temperature of the gases, then turning off the flow of spent gases, and passing either the raw fuel or air through the hot brick-work, to raise their preignition, initial temperature.

Regular Bright Finish. Semi-glossy finish on cold-rolled strip or sheet, produced by finishing rolls having a moderately polished finish.

Regular Carbon Cast Steel. Cast steel [see **Cast Steel** (a)], with carbon ranging from 0.22 to 0.35%.

Regular (Crystal) System. See **Cubic (Crystal) System.**

Regular Sketch Sheet. Sheet metal sheaied into four-sided shapes other than squares or rectangles.

Regular Tolerance. Tolerance (see) on a commercially acceptable product.

Regulus. (a) Metal, separating as a molten button or larger mass from the slag, in a furnace or crucible. (b) See **Matte.** (c) See **Regulus of Antimony.**

Regulus of Antimony. Metallic **Antimony** (see).

Regulus Metal. See **Antimonial Lead.**

Reheater. See **Reheating Furnace.**

Reheating Furnace. Furnace for heating metal ingots prior to hot-working.

Reinforcing Bar (Iron). Steel bar or rod, most commonly of square or round cross-section, for strengthening concrete against tensile stresses.

Reinforcing Rolls. See **Four-High Mill.**

Reith Alloy. Copper-base alloy, used for bearings, etc., containing about 11% tin, 10% lead, and 6% antimony.

Reladling. Transferring metal, usually steel, to a second ladle. Frequently, this is done in order to expose the metal to the action of a separate slag.

Relative Density. See **Specific Gravity.**

Relaxation. In creep, the application of stress under conditions allowing for the relief of such stress; it permits the **Initial Creep** (see) to be completed more rapidly.

Relief of Stress. See **Relaxation.**

Remanence. Magnetic induction remaining in a circuit after removal of the magnetizing force.

Rennerfelt Furnace. Indirect-arc electric furnace, using two-phase current by means of two electrodes through the sides of the furnace and one through the top, the latter acting as a common return. See **Arc Furnace.**

Renyx. Die-casting alloys (Allied Die Casting Corp.) of aluminum-base and of zinc-base type.

Repeated Impact Test. Test of resistance of a metal specimen to recurring impact stresses. See **Impact Test.**

Repeater. Semi-circular trough, open at the top, used as a guide in the rolling of metal rods, to guide the end of the rod from one pass to the next; the top opening permits looping action. See **Looping Mill.**

Replacement. Supplanting some of the atoms of one element in a compound or alloy by atoms of another element without alteration of basic structure or crystalline form. In the case of alloys, replacement occurs by substitution of atoms of a second element in the lattice of the solvent metal.

Republic Double Strength (Steel). Low-alloy steel, with fractional percentages of manganese, copper, nickel, and molybdenum (Republic Steel Corp.).

Republic (Steel). Series of simple and alloy steels as manufactured by Republic Steel Corp.

Rerolling Quality Billet. See **Rerolling Quality Steel.**

Rerolling Quality Bloom. See **Rerolling Quality Steel.**

Rerolling Quality Slab. See **Rerolling Quality Steel.**

Rerolling Quality Steel. Hot-rolled steel, in the form of billets, blooms, or slabs, suitable for further rolling to final shape, such as strip, plate, etc.

Residual Stress. Stress in a metal resulting from unequal heating or cooling. See **Stress Relieving.**

Resilience. Tendency of a material, when strained within its elastic limit, to return to its original shape, after removal of the stress.
Resist. Chemically inert material, used as a coating over part of a metallic surface to protect that part from chemical action, as in etching, etc.
Resistac. Copper-base alloy (American Manganese Bronze Co.), with about 88% copper, 10% aluminum, and 2% iron.
Resistal. Modified aluminum **Bronze** (see), with about 9% aluminum and 1% iron.
Resistance. See **Electrical Resistivity.**
Resistance Alloy. See **Resistance Wire.**
Resistance Brazing. Brazing (see) in which the material is heated by electrical resistance.
Resistance Butt Welding. Resistance **Welding** (see) in which fusion occurs over the whole cross-section of the materials being welded.
Resistance to Deformation. See **Deformation Resistance.**
Resistance Furnace. Electric furnace in which heat is generated by electric resistance.
Resistance Heating. Heating, accomplished by passage of electricity through material offering substantial resistance. See **Resistance Furnace** and **Resistance Welding.**
Resistance Percussion Welding. See **Percussion Welding.**
Resistance Pyrometer. See **Pyrometer.**
Resistance Thermometer. See **Pyrometer.**
Resistance-to-Wear Test. Test designed to determine resistance of a surface to abrasive or wearing influences.
Resistance Welding. See **Welding.**
Resistance Wire. Wire of a metal or alloy which, owing to its electrical resistance and resistance to oxidation at high temperatures, can be used for electrical heating wire. Nickel-chromium alloys are the best known materials of this type.
Resistivity. See **Electrical Resistivity.**
Resquared Plate. See **Resquaring.**
Resquared Sheet. Hot-rolled annealed steel sheet, finally specially sheared in length and width.
Resquaring. Shearing metal sheet, after completion of all processing, so as to avoid distortion.
Restoring. Converting coarse-grained steel into fine-grained metal by reheating from below 1275°F to slightly above that temperature; sometimes also called refining.
Restricted Chemical Composition. Chemical limitation, in ordering or selling metal, more restrictive than standard specifications.
Restricted Tolerance. Specification, in ordering or selling metal, directing tolerances on chemical or physical properties to be more limiting than the commercial restriction standard for such metal.
Restrictive Testing Requirements. Specifications, in the ordering or selling of metal, directing physical properties, such as tensile strength, elasticity, etc., to be more restrictive than the standard commercial limitations for such metal.
Restrictive Thickness Tolerance. Specification, in the ordering or selling of sheet or strip metal, directing thickness tolerances to be more restrictive than the commercial limitation standard for such metal.
Resulfurized Stock. Steel to which sulfur has been added, to facilitate machining.
Resurfacing. Adding metal to worn-down surfaces by means of welding techniques, using **Filler Metal** (see) to build up the surface.
Retardation Point. See **Critical Point.**
Retort. Vessel used for distilling volatile materials. In metal processing, metals such as zinc and mercury are frequently extracted from their ores by retort heating.
Retort Annealing. See **Box Annealing.**
Retort Carburizing. Modification of pack **Carburizing** (see) in which the articles are revolved in a retort with solid carburizing agents.
Retort Coke. See **Coke.**
Retort Gas. Gas resulting from the heating of coal in retorts, as in the

by-product process of **Coke** (see) manufacture.

Retorting. Volatilizing away mercury from gold and silver amalgams. See **Amalgam Retort.**

Retort Oven. See **By-Product Oven.**

Revalon. Copper-base alloy, most commonly used for tubing, with about 22% zinc and 2% aluminum (Revere Copper & Brass, Inc.).

Reverberatory Furnace. Furnace in which flame passes across the surface of the charge, lying in a shallow hearth, with a relatively close roof. Heating is both by direct contact with the gas and radiation from the roof.

Reverberatory Roasting. Roasting conducted in a **Reverberatory Furnace** (see).

Reverse Bronze. Copper-tin alloys in which tin is the predominant constituent; frequently used as bearing metals.

Reverse Nickel Steel. See **Ferro-Nickel (Alloys).**

Reversible Nickel Steel. See **Ferro-Nickel (Alloys).**

Reversing Mill. Rolling mill in which the direction of rolling can be reversed.

Reversing Roll. See **Reversing Mill.**

Rex-T-Mo (Tool) Steel. High-carbon, high-speed, alloy tool steel (Crucible Steel Co.), with approximately 8% molybdenum, 4% chromium, 2% tungsten and 1% vanadium.

Rex (Tool) Steel. Series of high-speed alloy tool steels (Crucible Steel Co.) with about 18% tungsten, 4% chromium or cobalt, and, optionally, 1% vanadium, as well as the usual fractional percentages of manganese and silicon.

Rextox. Electric rectifying material, consisting of copper with a layer of cuprous oxide formed, by heat, on the surface.

Rezistal. Stainless Steel (see) of the general 18% chromium, 8% nickel type, manufactured by a number of firms under license.

Rh. Chemical symbol for **Rhodium** (see).

Rhenania Furnace. Mechanically rabbled roasting furnace, having four hearths and a combination flue.

Rhenish Furnace. Furnace for heating retorts in the distillation of zinc. The retorts are suspended, usually in three double rows, in the combustion chamber. See also **Silesian Furnace.**

Rhenium (chemical symbol Re). Element No. 75 of the periodic system; atomic weight 186.31. Hard, but ductile and malleable, gray metal, stable in air, it has an extremely high melting point, 5733°F; its boiling point is unknown, but is reported higher than that of tungsten; specific gravity 20.53; mechanically, it lends itself to hot forging and rolling. Chemically, it shows a great variety of valences, resembling manganese, its congener. The metal may be made by reduction of its oxide; has only experimental use, although its high melting point would indicate uses competitive with tungsten. The metal can be electrodeposited from an acidified solution of a perrhenate, yielding a bright plate.

Rhodanizing. See **Rhodium Plating.**

Rhodite. See **Rhodium Gold.**

Rhodium (chemical symbol Rh). Element No. 45 of the periodic system; atomic weight 102.91. Highly reflective, silvery white metal of the platinum group, with a slightly reddish cast; normally, not workable cold, but can be forged above 1470°F and then given a moderate degree of cold work; has extremely high corrosion resistance, being immune to all acids; melting point 3571°F; boiling point above 8100°F; specific gravity 12.5. Chemically, its principal valence is three, but shows also other valences. Rhodium is found native, normally associated with other members of the platinum family or with gold. It is used principally as a hardener for platinum and palladium; its alloy with platinum is used for furnace windings, catalysts, thermocouples and similar small-scale chemical utilizations; greater outlet is for electrodeposition, by use of sulfate and phosphate baths; rhodium-plated jewelry has the advantage of high brilliance and corrosion-resistance.

Rhodium Gold. Mineral alloy of gold and rhodium, found in nature, with 34–43% rhodium.
Rhodium Plating. Coating metal objects, jewelry in particular, with rhodium, in which the object to be coated is made cathode (negative electrode) in an electrolytic bath, containing a decomposable rhodium salt, usually a phosphate or sulfate; rhodium or carbon anodes are used.
Rhombic (Crystal System). See **Orthorhombic (Crystal System).**
Rhotanium. Gold-base alloy series, with 10–40% rhodium, used principally for jewelry.
Richard's (Aluminum) Solder. Tin-base solder, for use on aluminum, with about 25% zinc and 4% aluminum.
Richard's Bronze. Corrosion-resistant bronze, with about 42% zinc, up to 2% aluminum, and about 1% iron.
Rich Gold Brass. See **Rich Gold Metal.**
Rich Gold Metal. High-copper **Brass** (see), with about 10% zinc.
Rich Low Brass. Red **Brass** (see), with approximately 15% zinc, generally used for drawing and stamping.
Rider Sheet. See **Conveyor Sheet.**
Ridge Fillet. Runner for molten metal. See **Runner** (a).
Rimmed Ingot. See **Rimmed Steel.**
Rimmed Steel. Low-carbon steel in which incomplete deoxidization permits the metal to remain liquid at the top of the ingot, resulting in the formation of a bottom and side rim of considerable thickness. This rim is of somewhat purer composition than the original metal poured. The remaining core is somewhat higher in impurities than the original metal, and is characterized by blowholes, scattered throughout, but relatively little pipe. If the rimming action is stopped shortly after pouring of the ingot is completed, the metal is known as capped steel.
Rimming (Steel). See **Rimmed Steel.**
Rim Toughening. Process of hardening and toughening the rim of a steel wheel, in which the wheel, after preliminary normalizing, is water-quenched, on the rim only, in a special quenching machine, and then the whole wheel is tempered and cooled.
Riser. (a) Reservoir for molten, excess metal at the top of a casting mold to supply metal in compensation for the shrinkage of the casting when solidifying. (b) Metal, after solidification, filling such a reservoir.
Rising. In the last stages of solidification of **Killed Steel** (see), the lifting of the mushy, central metal at the top (see also **Bleeding**) by a moderate evolution of gas, usually hydrogen or, occasionally, nitrogen.
Rita Steel. Series of tool steels, manufactured by Cannon-Stein Steel Co.
Riverside—. Series of copper-base alloys of various compositions, manufactured by Riverside Metal Co.
Rivet Steel. Mild Steel (see), suitable for the manufacture of rivets.
Road Metal. Non-metallurgical term, designating non-metallic rock material, used for macademizing roads.
Roak. Longitudinal defect in steel bars, sometimes accompanied by segregation and decarburization, caused by elongation of surface blowholes.
Roast. See **Roasting.**
Roaster. See **Roasting.**
Roaster Slag. Slag resulting from the smelting of sulfide copper ores by the **English Copper Process** (see).
Roasting. Process of heating ores under oxidizing conditions, usually to remove sulfur from sulfide ores, or to convert the ore to sulfate.
Roasting Cylinder. Rotating roasting furnace of circular cross-section.
Roasting Furnace. See **Roasting.**
Roasting Oven. See **Roasting Furnace.**
Roast-Reaction Process. Smelting lead sulfide ores in a reverberatory furnace; sulfur dioxide is formed and volatilized, leaving molten lead behind; usually accomplished in two steps: (1) oxidation of part of the sulfide to oxide and sulfate at about 950°F; (2) raising the temperature to about 1475°F to effect reaction between the unreacted sulfide and the oxygen-containing material.
Roast-Reduction Process. Process of smelting lead sulfide ores, in which the ore is first roasted to lead oxide, followed by reduction to metal in a shaft furnace.

Roast Sintering. See **Blast Roasting.**
Roast Stall. Roasting (see) furnace of compartment or stall construction, having flues at the back of the furnace.
Rocker Shear. See **Rocking Shear.**
Rocking Furnace. See **Detroit Rocking Furnace.**
Rocking Shear. Improved type of **Guillotine Shear** (see), with a specially curved blade, moving across the metal with a rocker-like movement.
Rockwell B. See **Rockwell Hardness (Test).**
Rockwell C. See **Rockwell Hardness (Test).**
Rockwell Hardness (Test). Standard method for measuring the hardness of metals. The hardness is expressed as a number related to the depth of residual penetration of a steel ball or diamond cone ("brale") after a minor load of 10 kilograms has been applied to hold the penetrator in position. This residual penetration is automatically registered on a dial when the major load is removed from the penetrator. Various dial readings, combined with different major loads, give "scales" designated by letters varying from A to H; the B and C scales are most commonly in use.
Rockwell Superficial Hardness (Test). Form of the **Rockwell Hardness Test** (see), in which the hardness of a surface is determined by use of relatively light loads.
Rod Iron. Iron or steel in the form of round bars or rods.
Rod Mill. Rolling Mill (see) for rolling metal into long rod form.
Rod Reel. See **Pouring Reel** and **Laying Reel.**
Roebling Gage. See **Metal Gage.**
Roe Furnace. Rocking furnace, of rectangular shape, for the manufacture of **Wrought Iron** (see). The rocking feature avoids the necessity for hand rabbling.
Roesing Lead Pump. Automatically-operating pump for pumping lead from the desilverizing kettle of the **Parkes Process** (see).
Roesing Wire. Series of wires suspended in a condensing chamber, for the collection of furnace fumes and dust.
Roesler Process. Treating gold to remove copper and, in part, silver, by fusion with sulfur or antimony sulfide, forming copper and silver sulfides.
Rohn Mill. Cluster rolling mill (see **Cluster Mill**) in which the working rolls are of small diameter.
Roke. See **Roak.**
Roll. See **Rolling Mill.**
Roll Body. Portion of roll (see **Rolling Mill**) which contacts the metal being rolled.
Roll Compacting. In powder metallurgy, compacting metal powder by means of a rolling mill.
Rolled Gold. See **Rolled Plate.**
Rolled Plate. Thin layer of gold **Cladding** (see) on base metal, usually made by soldering bars of metal together and then rolling the composite. Rolled plate has a thinner coating than **Gold-Filled** metal.
Roller. Operator feeding metal into **Rolling Mill** (see).
Roller Furnace. Continuous Furnace (see) in which the material moves through the furnace on rollers.
Roller Leveller. See **Roller Levelling.**
Roller Levelling. Passing cut lengths of metal strip or sheet through a series of staggered small rolls, so as to flatten the metal.
Roller Table. See **Runout Table.**
Rolling Mill. Equipment for rolling down metal to a smaller size or to a given shape. The first rolls are called the roughing rolls, and the whole stand called the rougher; these are used merely to reduce the size of the billets. Further reduction is carried out by the pony rougher. When the metal is to be shaped, this is done by stands known as strands, and there may be a series of these. Then comes the planisher pass, which does the final forming, and may also give the metal its surface finish. Last is the finishing pass, used for removing any irregularities and surface rolling imperfections. These rolls are generally of hard, chilled iron, to achieve optimum finish.

Rolling Pass. See **Pass.**
Rolling Plant. See **Rolling Mill.**
Roll-Knobbling. See **Spellerizing.**
Roll Latten. Polished Sheet **Brass** (see).
Roll Line. See **Roll Train.**
Roll Marks. Periodic marking or scoring received by a metal passing between the rolls, in a rolling mill whose surface is damaged.
Roll Pass. See **Pass.**
Roll Scale. Scale formed on iron or steel during hot rolling.
Roll Spring. Deflection of the axis of rolls from a straight line, as a result of pressure exerted by the metal being rolled. See **Rolling Mill.**
Roll Straightener. Machine, comprising vertical and horizontal rolls, for straightening metal wire.
Roll Table. Table, with surface consisting of a series of small rolls, onto which metal is run after rolling.
Roll Train. Grouping of roll **Stands** (see), connected end to end, or in series. See **Rolling Mill.**
Rol-Man (Manganese) Steel. High-manganese (11–14%) steel, with about 1% carbon (Manganese Steel Forge Co.).
Roman Brass. Copper-base alloy, with about 39% zinc and about 1% tin (Revere Brass & Copper Co.).
Roman Bronze. See **Roman Brass.**
Roofing Copper. Annealed sheet copper, used in construction work, and usually formed into shape at the point where needed.
Roofing Temper. See **Roofing Copper.**
Roofing Terne. See **Terne Plate.**
Roofing Tin. See **Terne Plate.**
Rope Wire. Patented steel wire (see **Patenting**), with carbon varying from 0.20 to 0.90%, used for manufacturing steel rope.
Ropp Furnace. Reverberatory furnace, for ore heating and roasting, with a long hearth on which the ore is moved forward by means of rabbles or rakes drawn by a continuous cable.
Rose Copper. Red-colored discs of copper, the color being due to the presence of cuprous oxide, formed by chilling the surface of molten copper by water sprays.
Rosein. Jewelry alloy with approximately 40% nickel, 30% aluminum, 20% tin, and 10% silver.
Rose's Alloy. See **Fusible Alloys.**
Rose's Metal. See **Rose's Alloy.**
Rosette Copper. See **Rose Copper.**
Rosette Fracture. See **Fracture.**
Rosin Jack. See **Mock Lead.**
Rosin Tin. Non-metallic mineral, a reddish variety of cassiterite, tin dioxide, SnO_2.
Rossie Furnace. See **Rossie Hearth.**
Rossie Hearth. Hearth (see **Ore Hearth**) in which cooling of the walls and sides is accomplished by a current of air; used in the roasting of galena.
Ross and Welter Furnace. Roasting furnace, of multiple deck construction, with a central hole for the ore to drop from one hearth to the next.
Rotary Puddler. Mechanical **Puddling** (see) arrangement, in which the furnace is rotated to facilitate puddling.
Rotary Rolling. Method of manufacturing large-diameter tubing and pipe, in which two cone-shaped rolls, with axes at an angle, act to force a small-diameter, thick-walled tube over a mandrel.
Rotary Shear. Apparatus with rotating blades for shearing the edges of metal strip, sheet, etc.
Rotary Squeezer. Squeezing machine (see **Puddling**) in which a rotating drum is mounted out of center in a cylindrical case.
Rotating Body Furnace. Batch furnace in which the heating body of the furnace is rotated, to enable more rapid transfer of heat to the work.
Rotating Crystal Method. See **Bragg (X-ray) Method.**
Rotator. See **Rotating Body Furnace.**
Rouge. Finely-divided hydrated iron oxide, extensively used in metal buffing and polishing.
Rougher. See **Rolling Mill.**
Roughing Down Rolls. See **Roughing Stand.**
Roughing Hole. Hole for the casting of slag, or unsatisfactory metal, from a furnace.
Roughing Mill. See **Roughing Stand.**

Roughing Pass. See **Roughing Stand.**
Roughing Rolls. See **Roughing Stand.**
Roughing Stand. Stand (see) of rolls for preliminary rolling; usually comprises horizontal and vertical edging rolls and horizontal reduction rolls.
Round. Metal, finished in round cross-section; the top and bottom of a round are the surfaces subjected to the roll pressure in the finishing pass; the sides are at 90° to the top and bottom, opposite the roll clearances. The shoulders lie between these two axes. The latter are generally not exactly identical; the longer diameter is called high shoulder, and the shorter one low shoulder.
Round Bloom. Bloom (see) of round or essentially-round cross-section.
Round-Corner Square. Cross-section shape of metal bars, with rounded corners.
Round-Edge Flat. Cross-section shape of metal bars, with edges rounded on a radius exceeding half the thickness.
Rozan Process. Modification of the **Pattinson Process** (see) of lead refining.
R R Alloy. See **Hiduminium.**
Ru. Chemical symbol for **Ruthenium** (see).
Rubber. Apparatus for amalgamating the gold content of a slime, in which the slime is rubbed against an amalgamated copper surface. See **Amalgamation.**
Rubidium (chemical symbol Rb). Element No. 37 of the periodic system; atomic weight 85.48. Very soft, silvery-white metal of the alkali family, closely resembling potassium both physically and chemically; it reacts violently with water, and corrodes in air; melting point 101°F; boiling point 1292°F; specific gravity 1.53. Chemically, it is mostly monovalent. The metal may be isolated by reaction of the hydroxide with aluminum in vacuo; has no significant commercial utilization. Since it reacts with water, electrodeposition from aqueous solutions is impossible.
Ruby Arsenic. See **Red Arsenic.**
Ruby Copper. Non-metallic mineral, cuprite, native cuprous oxide, chemically approximating Cu_2O.
Ruby Silver. Non-metallic mineral, proustite, silver-arsenic sulfide, chemically approximating $Ag_4As_2S_5$.
Ruby Zinc. (a) Non-metallic mineral, sphalerite, zinc sulfide, ZnS, in a transparent form of deep red color. (b) Non-metallic mineral, zincite, red oxide of zinc, ZnO, of deep red color.
Ruffles. Imperfection in rolled metal, such as strip or sheet, in the form of ruffled edges, usually due to the edges being thicker than the center of the metal.
Rumbler. See **Tumbling Barrel.**
Run. (a) Quantity of material handled in a single operation. (b) Complete cycle of operations, e.g., in a furnace.
Runner. (a) Metal casting, formed in the shape of a deep trough, or similar shape of sand or clay, for carrying away molten slag or metal from a blast furnace or other metal-melting furnace. (b) Channel, running horizontally in a mold from the bottom of the gate to the mold cavity. (c) See **Gate.**
Running Off. See **Casting** (a).
Running-Out Fire. See **Melting Finery.**
Running-Over. Rolling operation on a finishing mill for the purpose of reducing thickness of a pack (see **Pack Rolling**) to the point requiring one rolling step.
Running Stick. Round stick or rod, slightly tapered, used as a pattern for a **Gate** (see) in a mold.
Running Stopper. Leaking stopper in a ladle or other vessel containing molten metal.
Runoff (Slag). Tapping off excess slag after the **Ore Boil** (see), in the **Basic Open-Hearth Process** (see) of steel making.
Runout. Escape of metal from a mold during **Casting** [see **Casting** (a)] because of faulty making or joining of the mold.
Run-Out Fire. See **Melting Finery.**
Runout Table. Table, at the receiving end of rolls, for holding rolled metal.
Run Steel. See **Malleable (Cast) Iron.**
Ruselite. Aluminum-base, die-casting alloy series (Ruselite Corp.), with low percentages of copper, chromium, molybdenum, and titanium.

Rush Gold. Native gold covered with a coating of iron or manganese oxide.

Russell Process. Modification of **Patera Process** (see), in which cuprous sodium hyposulfite is used to leach the roasted ore.

Russia Iron. Highly-polished sheet iron, formerly made in Russia by hammering of plates in piles, with powdered charcoal between the plates.

Russian Tula. See **Niello (Silver).**

Rust. Iron oxide, commonly hydrated, formed by corrosion of iron (see **Rusting**). More loosely, any oxide on a metal resulting from corrosion by atmospheric conditions.

Rusting. Corrosion of iron, or iron-base alloys, forming iron oxide on the surface of the metal. See **Rust.**

Rust Inhibitor. Material or process, used on iron-base materials, which prevents or significantly retards **Rusting** (see).

Rustless Iron. See **Stainless Steel.** Rustless irons are always of low-carbon content, not over 0.15%.

Rustless (Stainless). Series of chrome and nickel-chrome steels, manufactured by Rustless Iron & Steel Corp. Chromium may be present up to 26% and nickel up to 13%; exhibit resistance to high temperatures and corrosive conditions.

Rusty Gold. Native gold, resisting amalgamation because of being coated with a film of iron oxide, silica, or similar non-metallic contaminants.

Ruthenium (chemical symbol Ru). Element No. 44 of the periodic system; atomic weight 101.7. Fairly hard, gray, brittle member of the platinum group, which can be forged with some difficulty above 2750°F; melting point above 4440°F; boiling point, above 8850°F; specific gravity 12.2. Has extremely high corrosion resistance, akin to that of iridium, and is immune to all acids. Chemically, it shows a variety of valences. Occurs with other members of the platinum group, but is extremely rare. It has no commercial utilizations.

Rycar (Steel). Simple steel (Joseph T. Ryerson & Son, Inc.) of about 1% manganese and about 0.15% carbon content, commonly used in cold-drawn form for screw stock and carburizing.

Ryerson Steel. Series of steels of varying formulae, manufactured for Joseph T. Ryerson & Son.

Ryolite Steel. Series of alloy tool steels, manufactured for Joseph T. Ryerson & Son.

S

S. Chemical symbol for **Sulfur** (see).

Sabeco Metal. High-Lead Bronze (see), with about 24% lead, used for bearing purposes (Fredericksen Co.).

Sable Iron. See **Russia Iron.**

Sack's Mill. Special type of **Universal** rolling **Mill** (see), used in rolling steel **Shapes** (see).

S.A.E. Abbreviation for Society of Automotive Engineers. This organization has specified common and alloy steels in accordance with a numerical index system allowing approximation of the composition of the metal. The last two digits always indicate the carbon content (usually within 0.05%). Most of the steels are specified by a four digit number; common steels are in the 10xx series, free cutting steels in the 11xx and 13xx series, manganese steels 13xx and T13xx, nickel steels 2xxx (the second digit representing the approximate percentage of nickel), nickel-chromium steels 3xxx, molybdenum steels 4xxx, low-chromium steels 5xxx, high-chromium steels 51xxx and 52xxx, chrome-vanadium steels 6xxx, tungsten steels 7xxx and 7xxxx, and silico-manganese steels 9xxx. Specifications have also been drawn for aluminum-base and copper-base alloys.

S.A.E. Steel Number—. See **S.A.E.**

Safalloy. See **Cerrosafe.**

Safe Load. Stress to which a given material may be subjected in common working practice. See also **Factor of Safety.**

Safe Stress. See **Safe Load.**

Safety Factor. See **Factor of Safety.**

Salamander. Metal, lying below the tapping hole in a **Blast Furnace** (see);

solidifies whenever the furnace is blown out.

Salge Metal. Zinc-base, bearing and die-casting alloy, with about 10% tin, 4% copper, 1% lead, and 1% antimony.

Salge's Bronze. See **Salge Metal.**

Salt. Product resulting from the reaction of an acid and a base.

Salt Bath. Bath of fused salt or mixture of salts, used for quenching or tempering. The salt mixtures in use vary in operating temperatures from 400° to 2375°F. Salt baths have the advantages of temperature uniformity, because of internal convection, and of complete exclusion of air.

Salt Bath Furnace. Electric furnace in which a heat-treating salt mixture is maintained in the molten state by its resistance.

Salt Spray Test. Test to determine susceptibility of a metal to corrosion by exposure of a specimen to a spray of salt solution, either continuously or with alternating periods of exposure and rest.

Samarium (chemical symbol Sm). Element No. 62 of the periodic system; atomic weight 150.43. Light-gray metal of the rare earth group, isolated by electrolysis of its trichloride in a bath of fused alkali chlorides; melting point about 2460°F; its boiling point is unknown; specific gravity 7.7. Chemically, it is principally trivalent like the other rare earths, but in some compounds it is divalent. Has no commercial utilizations. In common with the other rare earth metals, it cannot be electrodeposited.

Sampling. Taking a sample of a material for physical or chemical examination.

Samson (Steel). Series of low-alloy structural steels, of about 1.25% nickel and 0.60% chromium content (Carpenter Steel Co.).

Sand. See **Molding Sand.**

Sand Bed. Foundry floor, in which large castings are cast.

Sandblasted Finish. Surface **Finish** (see) obtained by subjecting a metal, most commonly steel, to **Sandblasting** (see).

Sandblasting. Cleaning surface of metal by air blast, using sand as abrasive.

Sand Bottom Furnace. Furnace, usually of the reheating type (see **Reheating Furnace**), lined with silica sand as a refractory.

Sand-Burned Casting. See **Sand Burning.**

Sand-Burning. Formation, on a sand-cast casting, of a hard surface, by reaction between the sand of the mold and the hot cast metal.

Sand-Cast. Casting, particularly iron casting, cast in a mold made of sand.

Sand Cast (Iron) Rolls. See **Sand Rolls.**

Sanderson (Steel). Series of carbon and low-alloy tool steels, manufactured by Crucible Steel Co.

Sand Hole. Hole in cast metal, formed as a result of loose sand in the mold or by loosening of sand during the casting.

Sand Iron Roll. See **Sand Roll.**

Sand Roll. Roll made of iron, cast in a sand mold, used as roughing and intermediate roll, and for wrought iron. See **Roll.**

Sands. Ore crushed or ground to a particle size which settles easily in water, as opposed to **Slimes** (see).

Sandusky—. Series of iron-base and copper-base alloys, manufactured by Sandusky Foundry & Machine Co.

Saniter Process. Selectively removing sulfur from pig iron, in the fused state, by contacting with a mixture of lime and fluorspar or calcium chloride.

Sap. Core of soft steel in **Blister Steel** (see) bars.

Sappy Fracture. See **Silky Fracture.**

Saratoga (Steel). Non-Deforming Steel (see), with about 1.5% manganese, 0.50% chromium, and 0.40% tungsten or molybdenum; carbon approximates 0.90% (Ludlum Steel Co.).

Satco Metal. Lead-base alloy, manufactured by National Lead Co., used for bearings, involving high speeds and loads. The lead content is approximately 97%; the remainder is primarily tin and alkali and alkaline earth metals.

Saturated Martensite. See **Hardenite.**

Saturated Solution. Any solution in stable equilibrium with excess of the solute.
Saturnine. Pertaining to lead.
Savelsberg Process. See **Blast Roasting.**
Say Ladle. Small ladle, at the end of a long handle, for taking samples of fused metal.
Sb. Chemical symbol for **Antimony** (see).
Sc. Chemical symbol for **Scandium** (see).
Scab. Imperfection on ingot or cast metal in the form of a shell or veneer, usually at the bottom of the ingot, resulting from improper pouring.
Scabby. See **Scab.**
Scabs. See **Dross.**
Scad. See **Nugget.**
Scaffold. See **Hanging.**
Scaffolding. See **Hanging.**
Scale. See **Scaling** (a).
Scale Breaker (Stand). Rolling Mill (see) stand, operating principally to loosen surface scale, which is later removed by a **Descaler** (see).
Scale Breaking. Subjecting metal sheet or strip to the action of a series of staggered rolls, so as to flex the metal sufficiently to loosen and break off adherent scale and oxide.
Scale Copper. Copper metal in the form of very thin flakes.
Scale-Removing Rolls. Set of rolls, frequently several stands in series, to slightly compress pipe and then re-straighten it, for loosening scale formed on the inside of the pipe.
Scaling. (a) Oxidation of metal due to heat, resulting in relatively heavy surface layers of oxide. (b) Removal of scale from metal.
Scaling Dip. Acid solution, usually containing sulfuric, nitric, and hydrochloric acids, for dipping brass articles, after pickling and prior to electroplating.
Scandium (chemical symbol Sc). Element No. 21 of the periodic system; atomic weight 45.10. Rare element, isolated, only in impure form, by electrolysis of its trichloride in a bath of fused alkali chlorides; chemically related to the rare earths; melting point stated to be 2190° F and boiling point 4350° F; specific gravity 2.5. Chemically, it is trivalent. Like the rare earth metals, it cannot be electrodeposited.
Scar. Imperfect spot on cast metal.
Scarf. Tapered edge of metal preparatory to joining by welding.
Scarfing. (a) Removing surface defects from ingots, etc. by cutting out the defect, most commonly by gas torch technique. (b) Bevelling or tapering the edges of strip or plate metal, usually to make lap-welded pipe.
Scarfing Torch. See **Scarfing** (a).
Scarf Weld. See **Scarf.**
Scavenge. Removing undesirable oxides, gases, etc. from metals, usually by adding a chemically reactive material.
Scavenger. See **Scavenge.**
S-C Curve. See **S-N Curve.**
Schneider Furnace. Furnace for treating zinc-lead ores, in which the zinc is distilled over and the lead recovered in molten form.
Schoen Mill. Special type of **Rolling Mill** (see) for rolling steel wheels.
Schoop Metallizing. See **Metal Spraying.**
Scleron. German aluminum-base alloy, containing fractional percentages of lithium, copper and other alloying constituents.
Scleroscope Hardness (Test). Standard method for measuring the hardness of metals; a diamond-pointed hammer drops from a fixed distance through a tube onto the smoothed metal surface, and the rebound measured. The scleroscope hardness value is empirically taken from the rebound distance, with a specified high-carbon steel as 100.
Scorching. Rising and falling of electrodes, actuated by automatic mechanism, during start of operations of an electric **Arc Furnace** (see), until the scrap or other raw material in the furnace is melted.
Scoria. See **Dross** and **Slag.**
Scorification. Heating gold or silver ores with lead and borax, to separate the metal values. The impurities unite, as a slag, with the lead oxide formed by oxidation of part of the lead; the gold and silver dissolve in the re-

mainder of the lead, and sink to the bottom.

Scorifier. (a) Crucible for use in the **Scorification** (see) process. (b) Furnace for treatment of precious metal sweepings.

Scotch Hearth. Hearth (see **Ore Hearth**) in which cooling of the walls and sides is accomplished by radiation from large iron castings; chiefly used for lead smelting.

Scotch Ore Hearth. See **Scotch Hearth.**

Scotch Pig. High grade of **Pig Iron** (see).

Scott Furnace. Shaft furnace for the recovery of mercury from its ores, containing a series of shelves at an angle; the ore drops down from one shelf to another while being heated, finally dropping down to a hearth; the volatilized mercury is condensed in a series of condensers behind the shelf tower.

Scouring. Chemical and physical attack on a furnace lining or hearth, particularly by slag action.

Scouring Cinder. Slag, usually of basic nature, which has a **Scouring** (see) action on a furnace lining or hearth.

Scovill—. Series of copper-base alloys of varying composition, manufactured by Scovill Mfg. Co.

Scrap. Material unsuitable for direct use or repair, and usable only as a raw material for reprocessing.

Scratch Brush Wire. Steel wire, suitable for the manufacture of brushes, in the .0025–.0045 inch range.

Scratched Plate. Tin Plate (see) and **Terne Plate** (see), whose coating has been scratched due to faulty manipulation.

Scratch Hardness. See **Moh's Hardness.**

Scratch Hardness Test. See **Moh's Hardness** and **File Test.**

Screen Analysis. Determination of relative percentages of material passing through or retained on a sequence of screens of increasing mesh size.

Screening. Separation of powdered or crushed material into fractions of approximately the same size by passing through a series of screens or sieves, of decreasing mesh size.

Screw Brass. Yellow **Brass** (see), with about 40% zinc and 2% lead added for easy machining.

Screw Bronze. Modified red **Brass** (see), with about 5% zinc, up to 1% tin, and up to 1% lead.

Screw Metal. See **Screw Brass.**

Screw Stock (Brass). Free-cutting **Brass** (see), with the addition of small quantities of lead.

Screw Stock (Steel). Free-cutting soft steel. The free-cutting quality is usually obtained by the presence of sulfur, in amounts up to 0.40%.

Scrubbing. Passing metal plate, sheet, or strip, after pickling, between revolving brushes in a flow of water, to remove chemicals. Scrubbing is, of course, followed by drying.

Scruff. Dross, in a tin-coating bath, consisting of a mechanical mixture of tin oxide and iron-tin alloy.

Scruff Guard. Metal plate apron, in tin plate manufacture, to prevent **Scruff** (see) from being carried over with the coated metal.

Scruffy Surface. Faulty surface on **Tin Plate** (see) resulting from the presence of **Scruff** (see) in the coating bath.

Scum. Dross (see) rising to the surface of molten metal. See also **Scumming.**

Scumming. Formation of a scum on a cyanide bath [see **Cyaniding** (a)] due to decomposition; the scum contains carbon and other products of decomposition.

S Curve. Curve showing the relationship between the time of decomposition of austenite and the temperatures at which decomposition takes place.

Se. Chemical symbol for **Selenium** (see).

Sea Coal. Powdered coal for **Facing** (see) of molds.

Seam. Crevice in rolled metal which has been closed by rolling or other working, but has not been welded into sound metal. Seams are caused by blowholes in the ingot, if subsequent rolling is inadequate to weld the metal together. They are also caused by folds, laps, and scratching during working.

Seamless Products. Class of products, including tubing, pressure cylinders, etc., manufactured by processes which avoid the presence of seams and joining edges.

Seamless Tubing. Tubing not made by welding the edges of rolled strip. Seamless tubing can be made by billet piercing or, in the case of the softer metals, by extrusion.

Seam Welding. Welding (see) in which the metals are joined along a line by passing them between two rotating electrode wheels pressing against the metals from opposite sides.

Season Cracking. Spontaneous, intercrystalline cracking, occurring in metals on standing, caused by internal stresses after cold working, coupled with external corrosion. Cold-worked brass is particularly subject to season cracking.

Seasoning. Relieving strains in cast iron by aging at room temperature.

Second. See **Seconds.**

Secondary Cell. See **Storage Battery.**

Secondary Creep. See **Creep.**

Secondary Crystallization. Crystallization change, accompanying a phase change, on cooling.

Secondary Flow Period. See **Secondary Creep.**

Secondary Hardness. See **Red Hardness.**

Secondary Inclusion. Inclusion in metal, resulting from compounds formed on the surface of the metal when heated to relatively high temperatures.

Secondary Metal. Metal, recovered from scrap or from metal previously formed into shape and later discarded.

Secondary Pipe. In long ingots, the phenomenon of two-section pipe occasionally resulting from the ingot bridging across the pipe, before solidification has been completed.

Second Forging. The second of any sequence of **Forging** (see) operations.

Second Law of Thermodynamics. Heat cannot pass, without expenditure of energy, from a colder to a warmer body.

Seconds. Metal products, such as plate and sheet steel, with moderate amounts of visible imperfections.

Seconds in Addition. See **Seconds Arising.**

Seconds Arising. Specification, in the ordering or selling of metal sheet or plate, directing the separate delivery of **Seconds** (see), additional to the full amount ordered.

Seconds Included. Specification, in the ordering or selling of metal sheet or plate, directing the separate delivery of **Seconds** (see), but their inclusion in the total delivery.

Section Strip. High-carbon, hot-rolled strip steel (see **Forging Quality Hot-Rolled Strip Steel**), used for the manufacture of agricultural cutter blades.

Segment Die. In powder metallurgy, a die which can be readily taken apart, to facilitate the removing of the compact from the die.

Segregate. Material which shows **Segregation** (see).

Segregation. Separation of various constituents of an alloy into localized areas; usually, the result of certain constituents freezing and crystallizing, whereas other constituents remain liquid; rapid cooling will, of course, minimize segregation.

Seizing. See **Galling.**

Selected Zinc. Metallic **Zinc** (see) with total impurities not over 1.25%. Lead is limited to 0.80%, iron to 0.04%, and cadmium to 0.75%, with no aluminum allowed.

Selective Carburizing. Carburizing (see) in which only certain areas are to be reacted and other areas are protected, or stopped off.

Selective Freezing. Solidification of a molten complex substance, in which one constituent freezes out separately leaving other constituents in the molten state.

Selective Process. Removing a limited number or amount of impurities from metal, which is usually in molten form.

Selective Surface Hardening. See **Tocco Process.**

Selector. Converter, for operation on copper ores, forming copper **Bottoms** (see).

Selenide. Oxygen-free compound of selenium with one or more **metals.**

Seleniferous. Containing selenium.

Selenium (chemical symbol Se). Element No. 34 of the periodic system; atomic weight 78.96, metalloid of the sulfur-family, existing in many allotropic forms. The metallic modification is a gray, crystalline, and brittle solid, unique in that its electrical conductivity increases with light; melting point 423°F; boiling point 1270°F; specific gravity, 4.81. Chemically, it resembles sulfur in showing divalence, tetravalence, and hexavalence; in all, selenium acts as a non-metal. Selenium is obtained as a by-product of copper refining; the slimes from the electrolytic refinery, when heated, yield selenium vapors, which condense in the flues, predominantly as the dioxide; this is dissolved in acid and then reduced to the element by sulfur dioxide. It is used as an alloying agent with copper and steel, to improve machinability and, in elementary form, for applications where its unique electrical characteristics are of value, as in photoelectric cells, for ruby-red coloring of glass, and as a special vulcanizing agent in rubber. It can be electrodeposited by electrolysis of selenious acid, but this process is not used commercially. Selenium is available from Anaconda Sales Co. and from American Smelting and Refining Co.

Selenium Copper. Copper, with approximately 0.5% selenium, which facilitates machining.

Self-Annealing Metal. Metal which crystallizes at room temperature or at temperatures spontaneously generated during working. Commercially important in this group are tin, lead, and zinc.

Self-Baking Electrode. See **Continuous Electrode.**

Self-Fluxing Ore. Ore containing natural impurities which, on heating, act as adequate flux during smelting.

Self-Hardening Steel. See **Air-Hardening Steel.**

Self-Lub. Compacted powdered iron, impregnated with oil, for bearing purposes (Keystone Carbon Co., Inc.).

Semi-Continuous Mill. See **Combination Mill.**

Semi-Finished Products. Metal in mechanically intermediate form, between large ingot and final rolled or worked form. Slab metal, blooms, billets, forging blocks, etc., are in this category.

Semi-Finished Steel. Steel in the form of billets, blooms, etc., requiring further working before completion into finished steel, ready for marketing.

Semi-Killed Steel. Steel which has been deoxidized, but not as completely as **Killed Steel** (see), resulting in the formation of blowholes in the ingot.

Semi-Metal. Obsolete term for non-malleable metals, such as bismuth, arsenic, etc.

Seminole Steel. Low-alloy, shock-resistant tool steel, of approximately 0.5% carbon content, with approximately 2% tungsten, 1.4% chromium, and 0.2% vanadium (Ludlum Steel Co.).

Semi-Plastic Bronze. Bearing Bronze (see) approximately 75–79% copper, 13–17% lead, and 7–9% tin.

Semi-Steel. Cast iron (not steel) of high quality, obtained by using a large percentage of steel scrap with the pig iron.

Separable Tin. See **Sparable Tin.**

Series Process. See **Series System.**

Series System. Connecting of anodes and cathodes in a cell of an electrolytic process, especially electrolytic refining, in series, i.e., anode to cathode, thus making a series of cells within the unit cell. See also **Multiple System.**

Set Copper. Molten copper saturated with copper oxide.

Set Temper Steel. Tool Steel (see) with a carbon content of about 0.90%.

Settling Head. See **Riser.**

Seward Process. See **Harvey Process** (a).

Seymourite. Copper-base alloy, of high corrosion resistance, with about 18% nickel and 18% zinc (Seymour Mfg. Co.).

Shackle. See **Grip.**

Shadowgraph. See **Radiographic Test.**

Shadrach. See **Salamander.**

Shaft. See **Blast Furnace.**

Shaft Furnace. Furnace of vertical design, usually charged at the top and tapped from the bottom.
Shakdo. See **Shakudo.**
Shaking Down. See **Poling** (b).
Shaking Out. Removing casting from **Mold** (see).
Shakudo. Copper-base alloy, with about 4% gold and 1% silver, for ornamental purposes.
Shallow Hardening Steel. Steel chosen on the basis of capacity for surface or case hardening.
Shank. Ladle with long handles, for hand manipulation.
Shape. Rolled or drawn metal of any cross-section other than square, rectangular, hexagonal, round, or oval.
Shape Bloom. Bloom (see) of any shape other than square, rectangular, or round.
Shaped Wire. Wire in which the cross-section is any shape other than round.
Shape Ingot. Metal **Ingot** (see), partially shaped as cast, intended for rolling directly to final shape.
Sharon (Steel). Series of high-chromium and **Stainless Steels** (see), manufactured by Sharon Steel Co.
Shatter Crack. See **Flake** (a).
Shaven Latten. Brass in very thin sheet form.
Sheaf. Pile of metal ingots.
Shear. (a) Deformation in which adjacent parallel planes slide across each other, so as to be displaced in a parallel direction. (b) Equipment for cutting metal by shearing.
Shear Bow. Defect in plate or sheet, in the form of bowing, resulting from faulty shearing; it can be avoided by the use of rocker type shears (see **Rocking Shear**).
Sheared Edge. See **Number 3 Edge.**
Sheared Mill Plate. See **Sheared Plate.**
Sheared Plate. Steel plate, rolled between straight horizontal rolls only, followed by trimming of all edges.
Shear Fracture. See **Fracture.**
Shearing. (a) Cutting. (b) Converting into **Shear Steel** (see).
Shearing Stress. Stress that tends to cut the specimen across its cross-section.
Shear Steel. Steel made from **Cement Steel** (see) by heating and hammering or rolling, followed by cutting into short lengths, reheating in a bundle, and hammering into a bar, called single-shear steel. For obtaining a higher grade steel, these bars may be broken in half, and the two parts reheated and again hammered, resulting in double-shear steel.
Shear Strength. See **Shear Stress.**
Shear Stress. Force of **Shear** [see **Shear** (a)] per unit area of cross-section. See also **Shearing Stress.**
Sheet. Flat piece of metal, more than 20 inches wide, with thickness between 0.003 and 0.188 inches.
Sheet Bar. Metal reduced from **Blooms** (see) to a thickness of 0.25 to 3 inches, having a width of 8 to 16 inches and a length approximating 30 feet, the edges being usually rounded. Prior to rolling into sheet, the sheet bars are cut into lengths, corresponding to the required sheet width, with allowance for trimming.
Sheet Furnace. Furnace for reheating sheet metal between passes of rolling mill operations.
Sheet Iron. Iron in **Sheet** (see) form.
Sheet Lead. Lead rolled into sheet form. See **Sheet.**
Sheet Pack. See **Pack Rolling.**
Sheet Steel. See **Sheet.**
Sheet Tin. Pure **Tin** (see) rolled into sheet form. (Note difference from **Tin Plate.**)
Sheet Zinc Gage. See **Zinc Gage.**
Sheffield Plate. Composite or clad metal, consisting of a thin coating of silver on a base metal, usually copper or a copper-base alloy.
Shepherd P-F Standards. See **Fracture Standards.**
Sherardized Coating. See **Sherardizing.**
Sherardizing. Method of coating iron with a thin, firmly adherent film of zinc. The iron is heated to around 675°F, in contact with extremely finely powdered zinc containing small amounts of zinc oxide.
Shielded Arc Welding. Arc Welding (see) in which the electric arc and the metal are protected from the atmosphere. The most common method in-

volves the use of metallic, shielded arc electrodes, which have a thick coating of fluxing materials. This coating shields the arc, and, as the metal is consumed, the melting flux also shields the molten metal.

Shielded Carbon Arc Welding. **Shielded Arc Welding** (see), utilizing a carbon electrode.

Shielded Metal Arc Welding. **Shielded Arc Welding** (see), utilizing a metal electrode.

Shimer Process. **Activated Bath** (see) cyaniding, in which calcium cyanamide is added to the cyanide bath.

Shingler. Apparatus for **Shingling** (see).

Shingling. Old process, essentially hammering, used on puddled balls (see **Puddling**) to form a bloom of iron.

Shingling Hammer. See **Shingling.**

Shingling Tongs. Tongs for manipulating puddled balls and blooms in **Shingling** (see) operations.

Ship Plate. Steel plate for use in marine construction.

Shirt. **Blast Furnace** (see) lining.

Shock-Resistance Test. Determination of resistance of a specimen to shock. See **Impact Test.**

Shock-Resistant Steel. Steel, usually of the low-alloy type, used for tools subject to shock. Fractional percentages of chromium, vanadium, and molybdenum are typical, with carbon approximating 0.50%.

Shock Test. See **Drop Test.**

Shoder. Pack of gold and gold-beaters' skins, in the second stage of gold-leaf manufacture.

Shoe. (a) Girder, firmly bolted to a foundation in a **Rolling Mill** (see), lying parallel to the rolls, to which the roll housing is clamped. (b) See **Sycee Silver.**

Shoe Die Steel. Low-alloy, shock-resistant steel, of about 0.55% carbon, 0.70% chromium and 0.40% molybdenum content (Ludlum Steel Co.).

Shore Monotron (Hardness) Test. See **Monotron (Hardness) Test.**

Shore Hardness (Test). See **Scleroscope Hardness (Test).**

Short. Brittle. See also **Cold Short** and **Hot Short.**

Short Coil. Coil of sheet or strip metal from 40 to 87½% of coil weight as specified when ordered.

Shorterizing. See **Flame Hardening.**

Shorter Process. See **Flame Hardening.**

Short Hole. Tap hole, in a furnace, which is inadequately stopped up and is in danger of breaking through prematurely.

Shorts. Product remaining on a screen or sieve, as opposed to fines [see **Fines** (b)].

Short Terne. **Terne Plate** (see) of approximately 20 by 28 inches size.

Short-Time Tensile Test. Test of strength of a material, which is conducted over a short period, therefore, not involving the factor of **Creep** (see).

Short-Time Test. See **High-Temperature Test.**

Shot. Metal formed into small, essentially-spherical units by allowing drops of the molten metal to fall from a height into water. In the case of lead, from which most shot is made, approximately 1% of arsenic is added to facilitate perfect formation of the spherical form. In other metals, such as nickel, where exact shape is not important, shot is obtained as small granules of irregular size and shape.

Shot Blasting. Cleaning surface of metal by air blast, using metal, usually steel shot, as abrasive.

Shot Copper. Copper in **Shot** (see) form.

Shot Lead. See **Lead Shot Metal.**

Shot Metal. See **Lead Shot Metal.**

Shot Sample. Granulated sample of metal, made by pouring the molten metal into water.

Shotting. Pouring refined iron into a ladle containing molten slag, in the **Byers Process** (see).

Shotty Gold. Granular gold, somewhat resembling **Shot** (see) metal.

Shoulder. See **Round.**

Shoving Up. Reduction in the operating size of a die by the accumulation, in the die hole, of particles from the metal being drawn. See **Drawing** (b).

Shrinkage. Decrease in volume as a result of any operation; in powder metallurgy, the term usually refers to the

decrease occurring during sintering.
Shrinkage Cavity. See **Pipe.**
Shrinkage Check. See **Check.**
Shrinkage Gage. See **Molders' Rule.**
Shrinkage Hole. See **Pipe.**
Shrinkage Rule. See **Molders' Rule.**
Shrink Hole. See **Pipe.**
Shutting Up. Compressing metal or closing pores in metal by hammering or pressing.
Si. Chemical symbol for **Silicon** (see).
Sickened Mercury. See **Floured Mercury.**
Sickening. See **Floured Mercury.**
Sick Tin. Tin which has been affected by the **Tin Pest** (see).
Sicromal. Series of low-sulfur, low-phosphorus iron-base alloys, resistant to furnace gases and temperatures, with chromium and aluminum as the alloying elements (American Rolling Mill Co.).
Sicromo Steel. Low-alloy steel (Timken Steel & Tube Co.) with excellent resistance to oxidation at high temperatures and, therefore, used for still and condenser tubes; contains about 2.5% chromium, 0.5% molybdenum and up to 0.15% carbon.
Side Blowing Converter. Bessemer converter (see **Bessemer Steel**) in which the air blast is admitted through the side wall, near the bottom of the converter.
Side-Centered Structure. Term used in describing the internal structure of materials, as determined by X-ray, to designate structures in which the equivalent lattice points are at the corners of the unit cell and at the center of the faces perpendicular to the a-axis and b-axis.
Side Combustion Stove. Blast Furnace (see) stove in which the combustion chamber is located at the circumference.
Sideraphite. Acid-resistant, iron-base alloy, with approximately 23% nickel, 5% aluminum, 5% copper, and 4% tungsten; used for chemical apparatus.
Siderite. Iron ore, chemically essentially ferrous carbonate, $FeCO_3$, relatively rare in this country.
Sideroferrite. Metallic iron, occurring native in petrified wood.
Siemens Direct Process. Process for the manufacture of wrought iron from iron ore, without the intermediate formation of pig iron.
Siemens Furnace. Gas-heated, reverberatory furnace, with the gas preheated by use of a **Regenerator** (see).
Siemens-Halske Process. (a) Process of treating copper sulfide ores with acidified ferric sulfate solutions, followed by electrolysis, the copper being deposited on the cathode, and the ferric sulfate regenerated at the anode. (b) Process of isolating beryllium by electrolysis of beryllium oxy-fluoride in a bath of fused alkaline earth beryllium fluorides. The beryllium is obtained in molten form at the cathode, at the surface of the bath.
Siemens-Martin Process. See **Open-Hearth Process.**
Siemens-Martin Steel. See **Open-Hearth Process.**
Siemens Process. See **Pig and Ore Process.**
Siemens Silesian Furnace. Silesian Furnace (see) for zinc distillation, with heat regenerator for preheating the fuel gas.
Sieurin Process. Process for manufacturing sponge iron, in which layers of high-purity ore, coal, and lime are heated in closed crucibles. After cooling, the porous cake of powdered metal is removed from the crucible and crushed for further processing.
Signal Bronze. High-copper **Phosphor Bronze** (see), with about 1.8% tin; commonly available in wire form.
Silal. Foreign heat-resistant cast iron of about 5% silicon content; used for furnace parts.
Silanca. Silver of at least 92.5% **Silver** (see) content hardened with about 4% antimony, 1–3% cadmium, and up to 2.5% zinc.
Silcaz Alloy. Boron-containing master alloy for increasing the hardenability of steel, with 35–40% silicon, about 10% titanium, 10% calcium, 7% aluminum, 4% zirconium, and 0.5% boron (Electro Metallurgical Co.).
Silcrome. Series of high-chromium steels of excellent heat and corrosion resistance, made by Ludlum Steel Co.,

with up to 0.20% carbon, up to 28% chromium, about 1% silicon, 1% copper and, optionally, up to 11% nickel and 3% molybdenum.

Silesian Furnace. Early type of furnace, usually rectangular, for heating retorts in the manufacture of **Zinc** (see) by distillation.

Silesian Method. (a) See **Silesian Furnace.** (b) Process of lead ore roasting, using a large mass of ore, slowly roasted at a low temperature, the lead being collected outside the furnace as it flows down an inclined hearth.

Sil-Fos. Self-fluxing copper-base alloy for brazing at low temperatures, made by Handy & Harmon, approximating 15% silver and 5% phosphorus, whose presence gives the alloy low melting point (1300°F) and high fluidity. Sil-Fos brazed joints have high strength and conductivity.

Silica. Chemically, silicon dioxide, SiO_2, found principally in nature as sand and quartz; used as strongly acid refractory, and also as an acid flux.

Silicate Cotton. See **Mineral Wool.**

Silicide. Binary compound of silicon with a metal.

Silico-. See also **Silicon-.**

Silico-Manganese. Alloy, in which manganese predominates (about 65–70%), with 12–25% silicon, the balance being iron and carbon. Carbon varies from grade to grade, with 1% carbon as the lowest and 3% as the highest content; used in steel manufacture as a source of silicon and manganese.

Silicon-. See also **Silico-.**

Silicon (chemical symbol Si). Element No. 14 of the periodic system; atomic weight 28.06. Extremely common element, the major metallic component of all rocks and sands; its chemical reactions, however, are those of a metalloid; occurs in at least three forms: A brown powder, a graphite-like form, and as dark gray crystals; the first form may merely be surface-oxidized microcrystals. Silicon is immune to acids and to atmospheric action, but reacts vigorously with alkalies, evolving hydrogen; melting point 2588°F; boiling point 4271°F; specific gravity 2.42. Chemically, it is almost always tetravalent. Pure silicon can be obtained by reaction of potassium fluosilicate with sodium, followed by washing away the soluble alkali salts. Commercially, it is made, in less pure form, by reduction of silica with carbon in the electric furnace. Used in metallurgy as a deoxidizing scavenger; iron-base alloys, particularly those with about 14% silicon, show high acid resistance; copper-base alloys show high strength and corrosion resistance. Silicon is present, to some extent, in all steels, and is deliberately added to the extent of about 4% for electric sheets, extensively used in alternating current magnetic circuits. Silicon cannot be electrodeposited; obtainable from Electro Metallurgical Co.

Silicon-Aluminum. Master alloy of silicon and aluminum, used for the purpose of adding silicon to aluminum-base alloys. Though silicon may range up to 50%, the most typical content is about 25%.

Silicon-Aluminum-Iron. See **Alsifer.**

Silicon Brass. Brass with fractional or low percentages of silicon. See **Silicon-Copper.**

Silicon Bronze. Series of copper-base alloys, the other principal constituent being silicon to the extent of about 4%, showing high strength and good corrosion resistance.

Silicon Carbide. Binary compound of silicon and carbon, SiC, made by electric-furnace interaction of carbon and silica. It is extremely hard, and therefore extensively used in grinding and for similar abrasive purposes. Because of chemical and heat resistance it is also extensively used as a refractory.

Silicon Cast Steel. Cast steel of high strength and corrosion resistance [see **Cast Steel** (a)], with 0.75–1.75% silicon.

Silicon-Chromium Steel. High-alloy steel, extensively used for internal combustion engine valves and similar heat-resistant products, with 9–12% chromium, up to 5% silicon and usually less than 1.0% carbon.

Silicon-Copper. Copper-base alloys with 10–12% silicon, and low percentages of impurities, such as tin, zinc, iron, and aluminum; used as a master alloy for dilution to lower-silicon alloys, such as **Everdur** (see). Other silicon-copper alloys are higher in silicon content (up to 50%).

Silicon Impregnation. See **Siliconizing.**

Silicon Iron. Cast iron of high silicon content, usually between 12 and 15%. Carbon is usually low, about 0.80%; extremely hard and brittle, showing excellent chemical resistance, particularly against oxidizing acids.

Siliconizing. Impregnating iron and steel with silicon; the resultant surface contains about 14% silicon, and is heat- and wear-resistant. The metal is subjected, at temperatures of 1700–1850°F, to the action of chlorine and silicon carbide.

Silicon-Manganese Steel. Series of steels with up to 15% manganese and up to 2% silicon, extensively used for the manufacture of springs because of high fatigue resistance.

Silicon Pig (Iron). **Pig Iron** (see) of relatively high silicon content.

Silicon Punch and Chisel Steel. See **Silicon-Manganese Steel.**

Silicon Red Brass. Red brass (about 17% zinc), with about 1% silicon added, giving the alloy greater adaptability to resistance welding. See **Brass.**

Silicon Spiegel. High-manganese ferrous alloy, with manganese at 25–30%, silicon 5–8%, and carbon approximating 3%.

Silicon Steel. Steel, usually made in the basic open-hearth or electric furnace, with about 0.50–5% silicon, other elements being usually kept as low as possible. Because of high electric resistance and low hysteresis loss, silicon steel sheet and strip are standard in electric magnet manufacture.

Silicon-Zirconium. See **Ferro-Zirconium.**

Silico-Spiegel. Iron-base master alloy for adding manganese and silicon to steel; contains up to 30% manganese, 8–15% silicon, and up to 3% carbon.

Silky Fracture. See **Fracture.**

Sil-Man (Steel). Low-alloy, shock-resistant tool steel, with about 2% silicon, 0.3% vanadium, 0.3% chromium, and 0.60% carbon (Vanadium Alloys Steel Co.).

Silmo (Steel). Series of low-alloy steels (Timken Roller Bearing Co.).

Sil-Ten (Steel). Series of structural steels, with 0.60–0.90% manganese and 0.20–0.30% silicon; copper may be added, up to 0.20%, for corrosion resistance (United States Steel Corp.).

Silumin. German aluminum-base alloy, with up to 14% silicon.

Silundum. See **Silicon Carbide.**

Silvaz Alloy. Boron-containing master alloy for increasing the hardenability of steel, with 35–40% silicon, about 10% titanium, 10% vanadium, 6% aluminum, 6% zirconium, and 0.5% boron (Electro Metallurgical Co.).

Silver (chemical symbol Ag). Element No. 47 of the periodic system; atomic weight 107.880. Very malleable and ductile, soft, lustrous, white metal; the best conductor of heat and electricity among the elements; melting point 1761°F; boiling point 3906°F; specific gravity 10.50. Unaffected by air or water; tends to tarnish only in the presence of sulfur fumes. Chemically, it is always monovalent. Silver is obtained, for the most part, as a by-product of lead and copper metallurgy, or it may be obtained from silver ores by smelting, cyanidation, or amalgamation; used as a secondary monetary medium, for jewelry and tableware and, more recently, as an alloy with lead, for soldering purposes. It is also electrodeposited, usually by means of a double alkali cyanide bath. Standard grades are: (1) fine silver—99.9% pure; (2) sterling silver—92.5% pure; (3) coin silver—90% pure, copper being usually the alloying constituent.

Silver-Bearing Copper. Essentially pure copper, with small quantities of silver added, primarily used for operations involving high temperatures such as would anneal ordinary copper.

Silver Bell Metal. Alloy series, averaging 60% tin and 40% copper, used for bell and chime manufacture.

Silver Bismuth. See **Bismuth Silver** (a).
Silver-Clad Copper. Copper with a cladded layer of silver (see **Cladding**).
Silver-Clad Nickel. Nickel with a cladded layer of silver (see **Cladding**).
Silver-Clad Steel. Steel, usually of relatively low-carbon content, with a cladded layer of silver (see **Cladding**).
Silver Glance. See **Vitreous Silver.**
Silvering. (a) Process of coating glass with a thin layer of reflecting silver, for mirror purposes. After careful cleaning, the glass is usually treated with a silver-salt solution, containing reducing agents; silver, reduced from the salt, is deposited on the glass. (b) See **Silver Plating.**
Silver Lead. Crude lead, before removal of naturally-occurring silver.
Silver Mill. Plant for treatment of silver ores for the recovery of silver metal.
Silver Plate. (a) Household goods, tableware in particular, plated with silver. (b) Product of **Silver Plating** (see).
Silver Plating. Coating metal objects with silver; the object to be coated is made cathode (negative electrode) in an electrolytic bath containing a decomposable silver salt, normally a double alkali cyanide. Silver metal is commonly used as anode (positive electrode).
Silver-Ply (Stainless-Clad) Steel. Clad metal (see **Cladding**) in which a thin layer of stainless steel has been bonded to a thicker base of simple steel (Jessop Steel Co.). Note that silver is completely absent, despite the name.
Silver Solder. Series of low-melting silver alloys with silver usually, but not necessarily, predominant. The alloys are widely used in the soldering of metal sheet, such as copper, stainless steel, etc.
Silver Tin. Tin Plate (see) with a dull surface, obtained by mechanically roughening the base steel.
Silver Top. See **Gray Top.**
Silvery Iron. See **Silvery Pig (Iron).**
Silvery Luster. Sheen of characteristic, bright-gray color.
Silvery Pig (Iron). Ferrosilicon (see), containing 8–14% silicon, and up to 1.50% carbon.
Sil-X. Exothermic Ferrosilicon (see), made by Chromium Mining & Smelting Corp.
Simonds Steel. Series of high chromium-nickel steels, with 15–25% chromium and 20–35% nickel (Simonds Saw & Steel Co.).
Simple Alloy Steel. See **Ternary Steel.**
Simple Blow Test. Test of the resistance of metal to breakage by impact, by determining the weight of the hammer, and the height of the drop required to break the piece at the first blow.
Simple Carbon Steel. See **Carbon Steel.**
Simple Steel. Common steel, without special alloying constituents and with carbon below 1.5%.
Simple Structure. Term used in describing the crystal structure of materials, as determined by X-ray, in which the equivalent lattice points are at the corners of the unit cell.
Singapore Tin. Pig **Tin** (see) of about 99.9% purity, obtained from the Singapore area.
Single-Acting Hammer. Forging hammer which is lifted by steam after the stroke.
Single Crystal. Macroscopic mass of metal, the whole of which represents one single, large crystal, instead of the usual polycrystalline mass. Such crystals are difficult to form, being grown by slow deposition, usually from the liquid state.
Single Draft Drawing. Drawing wire through one die at a time.
Single Draft Wire. See **Single Draft Drawing.**
Single Electrode Potential. See **Electrode Potential.**
Single Furnace. Box **Annealing** (see) furnace with heating chamber adequate for only one annealing box.
Single Iron. Steel rolled singly, as opposed to doubling. See **Doubling** (a).
Single Pickling. Black Pickling (see) of steel, after all operations are completed, as compared with **Full Pickling.**
Single Piece Rolling. See **Single Rolling.**

Single Refined Iron. See **Merchant Bar.**

Single Rolled Iron. See **Merchant Bar.**

Single Rolling. Rolling metal singly, as opposed to **Pack Rolling** (see).

Singles. See **Singling.**

Single Shear. Shear stress, tending to fracture the material in a single plane. See **Shear** (a).

Single-Shear Heat Steel. Steel, with about 0.75% carbon, made by carburizing wrought iron.

Single-Shear Steel. See **Shear Steel.**

Single Treating. Heating medium-carbon steel (approximately 0.35–0.60% carbon) to the 1425–1560°F range, and then quenching in oil.

Singling. Preliminary smelting of **Antimony** (see) ores with iron, to yield an iron- and sulfur-containing crude metal, known as singles.

Sinkhead. See **Riser.**

Sinking Fire. Forge, heated by charcoal, for heating wrought iron scrap, or pig iron, till the metal is partially molten or can be welded together.

Sintered Carbide (Tools). Composite, containing carbides of extremely refractory metals, such as tungsten, tantalum, titanium, etc., cemented together by a relatively low-melting metal, acting as a matrix; cobalt is generally used for the latter purpose.

Sintering. Converting powder into a continuous mass by heating to a temperature considerably below fusion, usually after preliminary compacting by pressure. Union between the particles results from surface cohesion; no fusion is involved.

Sintering Furnace. Furnace designed for sintering or pre-sintering operations, with or without facilities for controlled atmosphere.

Sivyer Steel. Series of low-alloy and high-alloy steels, commonly in casting form, manufactured by Sivyer Steel Castings Corp.

Size Distribution. Analysis of crushed or ground materials on the basis of particle size.

Sizing. Separation of powdered or crushed material into fractions of approximately the same size, by screening, air separation, etc.

Sizing Rolls. Set of rolls for giving material of round cross-section, pipe particularly, a final shaping to correct size.

Skeleton Crystal. Imperfect crystal, formed by rapid crystallization.

Skelp. Plate or strip of hot-rolled steel, used for pipe manufacture. The skelp is rolled longitudinally into circular form and the edges then welded together.

Skelp Mill. Rolling mill for rolling blooms into **Skelp** (see).

Skelp Steel. See **Pipe Steel.**

Sketch Plate. Metal, most commonly steel plate, cut to any shape, except rectangular.

Skew Plate. See **Bloomery.**

Skimmer. Iron plate or bar for deflecting slag into a separate slag runner while pig iron is being tapped from a **Blast Furnace** (see).

Skimming Gate. Channel, in a mold, with a bridge over it to deflect slag from molten metal while being cast.

Skimming Ladle. Ladle possessing a skimming device, such as a guard on its lip, to hold back slag while metal is being poured.

Skimmings. Dross skimmed from the surface of molten metal. See **Dross.**

Skin. Outer, solidified surface of an ingot, formed during the period of cooling of the ingot and while the interior is still molten.

Skin-Dried Mold. Casting **Mold** (see), whose inside surface has been treated with an inflammable solvent, containing rosin and silica, and the solvent then burned away.

Skin Pass. Passage of metal through rolling mill, substantially without reduction, to increase surface stiffness.

Skin-Rolling. See **Temper-Rolling.**

Skleron. See **Scleron.**

Skull. Film of metal and dross, remaining behind in a pouring vessel after pouring of the metal.

Skull Cracker. Apparatus (or zone) in which **Skulls** (see) are broken up after removal from the ladles or similar vessels.

Skull Drop. See **Skull Cracker.**

Skutterudite. Gray-colored, lustrous mineral alloy, cobalt arsenide.

Slab. Semi-finished bar, hot-rolled from an ingot, for use in forging or rolling, with minimum cross section dimensions of 10 and 1½ inches. See also **Billet, Bloom,** and **Sheet Bar.**
Slab-and-Edging Rolling. See **Flat Rolling.**
Slabbing Mill. Rolling mill for the reduction of ingots to slabs.
Slabbing Rolling. See Flat **Rolling.**
Slab Edging Press. Press, in rolling mill train for the manufacture of steel plates, used to square edges and to adjust width of the metal.
Slab Heating Furnace. Reheating Furnace (see) for heating slabs during rolling.
Slab Ingot. Metal **Ingot** (see) intended for rolling directly into metal plate or slab, usually with width more than twice the thickness.
Slab Squeezer. Equipment for pressing sides of hot metal **Slabs** (see) and thus controlling the width and adjusting the edges.
Slab Zinc. Zinc metal, in slab form.
Slacken. Slag, obtained in metal smelting, deliberately added to fresh ore prior to smelting, to prevent fusion during heating.
Slag. Product resulting from the action of a flux on the non-metallic constituents of an ore, or on the oxidized, undesired metallic constituents. Most slags are chemical combinations of basic oxides with acid oxides, with neutral fluxes added to aid fusibility.
Slag Brick. Brick formed of **Slag** (see).
Slag Buggy. Cart for carrying slag away from a furnace, most commonly to a **Slag Dump** (see).
Slag Car. See **Slag Buggy.**
Slag Cement. Cement made by treating blast furnace slag with slaked lime.
Slag Dump. Zone for discarding slag, most commonly while still molten.
Slag Furnace. Furnace for recovering lead from the slags of lead smelting.
Slagging. Fluxing refractory gangue by means of fluxing agents, to form a **Slag** (see).
Slag Hearth. See **Slag Furnace.**
Slag Hole. Hole in side of a furnace, for the withdrawal of fluid slag. See **Slag.**
Slag Inclusion. Inclusion (see) of slag in a metal ingot.
Slag Lead. Lead recovered from the slag of lead smelting.
Slag Notch. See **Slag Hole.**
Slag Pocket. In an open-hearth furnace, a chamber, at the bottom of the uptake and downtake flues, to conduct gas to and from the checkerwork and to catch any fortuitous solid matter.
Slag Pot. Vessel for receiving molten slag when tapped from a furnace. See also **Slag Buggy.**
Slag Squeezer. See **Broadside Mill.**
Slag Wool. See **Mineral Wool.**
Slake Trough. Water tank for cooling metal, after heating in a blacksmith's forge.
Slakin. See **Slacken.**
Slenderness Ratio. Length of a test piece, divided by the square root of the cross-sectional area.
Slick. See **Slime.**
Slickens. See **Slime.**
Slicker. Tool for smoothing **Mold** (see) surfaces, particularly after application of a **Mold Wash** (see).
Slicker Solder. High-quality, soft **Solder** (see) of about two parts tin and one part lead.
Slime. Material of colloidal nature and particle-size, encountered in ore-treatment and in electrolytic refining operations. Since slimes do not settle well but tend to form colloidal muds in water, they present special problems, particularly in ore-treatment.
Slip. (a) Irreversible displacement of one part of a crystal relative to another part along a definite crystallographic plane. (b) Faulty operation of a **Blast Furnace** (see), causing raw material, at first wedged in some part of the furnace, to suddenly drop down, violently releasing gases previously trapped below. (c) In using **Vitreous Enamels** (see), a water suspension of finely-ground enamel ingredients, ready for application to steel. (d) See **Forward Slip** and **Backward Slip.**
Slip Bands. Lines usually parallel, running across a crystalline grain, caused by mechanical working.
Slip Crack. In powder metallurgy, break in the pressed compact, caused

by part of the compact slipping away from the remainder.

Slip Interference Theory. Theory claiming the increased hardness and strength of cold-worked metals as being due to the inability of slip planes in the metal to slip any further.

Slip Lines. See **Slip Bands.**

Slip Planes. Planes of easiest rupture in material, the crystals of which are ductile.

Slip Theory. See **Slip Interference Theory.**

Slit Edge. Type of edge, approximately square, obtained by slitting strip metal, without filing. See also **Number 3 Edge.**

Slitting. Passing strip or sheet metal through rotary knives, to cut the metal to size.

Sliver. Loose metal piece rolled down onto the surface of the metal during rolling operations.

Sloppy Heat. Modification of the **Bessemer Process** (see) yielding a final product in which the manganese content is greater than 0.50%.

Slug. Mass of incompletely roasted ore.

Sm. Chemical symbol for **Samarium** (see).

Small Terne. See **Short Terne.**

Small Tin. Tin obtained from **Slimes** (see).

Smaltine. See **Gray Cobalt.**

Smaltite. See **Gray Cobalt.**

Smelter. Plant for **Smelting** (see).

Smelting. Metallurgical operation in which the metal is separated, in fused form, from non-metallic materials, or other, undesired, metals, with which it is chemically or physically associated in nature.

Smelting Furnace. Furnace, e.g., of the blast or reverberatory type, used in smelting operations.

Smelting House. See **Smelter.**

Smithco Steel. Series of low-alloy machine steels, with 0.80–1.35% manganese, 0.50–1.20% nickel, 0.40–0.75% chromium, and up to 0.75% silicon (George H. Smith Steel Casting Co.).

Smith Process. Process of refining copper, by a series electrolytic operation (see **Series System**), with horizontal electrodes, the top surface being cathodic, the bottom surface anodic, with the crude copper to be refined in between, protected by linen diaphragms to catch the slimes.

Smoke. See **Fume.**

Smoking. Coating of inside of a mold with soot by means of a luminous flame.

S Monel. **Monel Metal** (see), with about 4% silicon as additional constituent.

Smut. See **Smutty Surface.**

Smutty Surface. Carbon deposit, formed on high-carbon steel, if etched too deeply in pickling operations.

SMZ Alloy. Alloy containing 60–78% silicon, 5–8% manganese and 5–8% zirconium; used as ladle addition to cast iron (Electro Metallurgical Co.).

Sn. Chemical symbol for **Tin** (see).

S-N Curve. In fatigue testing, a curve showing the relationship between the maximum stress (S), and the stress cycles (N) necessary to cause failure. The stress, at which the curve becomes asymptotic to the N axis, is accepted as the fatigue limit.

Snort Valve. Valve for reducing the pressure of blast in the cold blast main of a **Blast Furnace** (see) during stopping of the taphole after casting of metal.

Snowflakes. See **Flakes.**

Snyder Furnace. Tilting direct **Arc Furnace** (see), using high voltages and relatively small electrodes.

Soaking. Permitting a metal ingot to stand in a tightly-covered hole or pit, known as soaking pit, generally gas-fired to maintain temperature, so that the hot, usually liquid, interior of the ingot diffuses its heat and the whole ingot will be at an approximately equalized temperature, ready for rolling. More generally, the term covers maintaining a metal at a given temperature, in order to obtain uniformity of temperature throughout.

Soaking Furnace. See **Soaking.**

Soaking Period. Length of time of holding material at a relatively constant temperature, to achieve structural or other metallurgical changes.

Soaking Pit. See **Soaking.**

Soaking Time. See **Soaking Period.**
Soaking Zone. See **Heating Zone.**
Society of Automotive Engineers. See **S.A.E.**
Soda Ash Process. Process of selectively removing sulfur from pig iron, in the fused state, by contacting it with molten sodium carbonate (soda ash).
Soda Ash Stain. Faulty **Tin Plate** (see) coating, caused by improper wet cleaning of the base metal, prior to the tinning operation.
Soderberg Electrode. See **Continuous Electrode.**
Soderberg Shear. Modification of **Rocking Shear** (see).
Sodium (chemical symbol Na). Element No. 11 of the periodic system; atomic weight 22.997. Very soft, malleable, silvery-white metal of the alkali family; reacts vigorously with water, and is therefore rapidly corroded by ordinary atmospheric conditions; melting point 208°F; boiling point 1616°F; specific gravity 0.971, slightly less than that of water. Chemically, it is always monovalent. Since it reacts with water, it cannot be electrodeposited from water solutions. Sodium is manufactured in high tonnages, by electrolytic decomposition of fused sodium chloride. It is extensively used in organic chemistry, notably for the manufacture of tetraethyl lead and of sodium cyanide. A novel utilization is as conductor of electricity, the metal being housed within a steel pipe; by weight, it is a more efficient conductor than copper. Sodium-tin and sodium-lead alloys are also manufactured, and are used principally as scavengers. The metal and its alloys are available from E. I. du Pont de Nemours & Co., Inc.
Sodium Amalgam. Amalgam made by carefully dissolving sodium in mercury. The percentage of sodium may be varied, the amalgam becoming more solid as it is increased. With water, sodium amalgam yields a steady stream of hydrogen.
Sodium Lead (Alloy). Lead-base alloy, containing about 2% sodium, used as a scavenger in lead-base or lead-containing alloys and in chemical syntheses; manufactured by E. I. du Pont de Nemours & Co.
Sodium Zinc (Alloy). Zinc-base alloy, containing about 2% sodium, used as a scavenger in zinc-base or zinc-containing alloys; manufactured by E. I. du Pont de Nemours & Co.
Soft Annealing. Annealing on steel, most commonly steel tubing, which leaves a light, porous oxide coating on the finished metal.
Softening. Removal of hardening metals, such as antimony, arsenic, tin, zinc, etc., from lead. Most frequently, this is accomplished by liquation and oxidation at red heat, at which temperature the alloying metals are more readily oxidized to the corresponding oxides, and are slagged off, together with some lead oxide.
Soft Metal. Molten steel of low (about 0.06%) carbon content, particularly used in **Duplex Processes** (see) for mixing with other steel compositions.
Soft Mill. Roughing mill (see **Roughing Stand**) in pack mill operations.
Softness. See **Hardness.**
Soft Skin-Rolled Temper. In strip or sheet, cold-rolled metal of soft ductile character, suitable for fairly deep drawing.
Soft Solder. See **Solder.**
Soft Spot. Spot in metal, particularly steel, substantially less hard than the remainder.
Soft Spot Test. Method of detecting soft spots in steel, in which the steel, after quenching in water, is lightly etched with dilute nitric acid.
Soft Steel. See **Ingot Iron.**
Soft Temper. See **Soft Skin-Rolled Temper.**
Sol. Colloidal solution. See **Colloid.**
Solar (Steel). Low-alloy tool steel, with about 1% silicon, 0.5% molybdenum, 0.5% manganese, and 0.5% carbon (Carpenter Steel Co.).
Solder. Relatively low-melting alloy, used for uniting metals of higher melting point. The most common solders are those of lead base, and are called soft solders, such as an alloy of approximately equal parts of tin and lead. Bismuth, cadmium, and antimony are

frequently also added to this series. The hard solders are copper-base alloys, usually with tin and/or zinc, and are commonly used for uniting copper-base materials. This process is known as brazing.

Soldering. Process of uniting two metals by the use of a **Solder** (see), usually soft.

Soldering Acid. Hydrochloric acid, also called muriatic acid.

Soldering Flux. See **Flux.**

Solid Diffusion. See **Diffusion.**

Solid Drawing. Drawing or forming metal shapes from ingots of annular cross-section.

Solidification Range. Temperature range through which solidification of a metal takes place.

Solid Solution. Alloy in which the atoms of the components conform to a single lattice arrangement. Usually, atoms of the solute merely replace atoms of the solvent in its lattice, but small atoms occasionally occupy positions inside the parent lattice.

Solid (State). One of the three fundamental states of matter; in the solid state, a material has a specific shape and volume, and resists deformation.

Solidus (Line). Line in any temperature-composition diagram below which the alloy is completely solidified.

Solubility Curve. Curve plotted between solubility and temperature, indicating solubility variation with changes of temperature.

Soluble Anode. Anode (see) in electrodeposition or in electrometallurgy, which goes into solution during the electrolytic process.

Solute. Dissolved substance. See also **Solvent.**

Solution. Homogeneous mixture of substances which is separable into its components by altering the state of one of them.

Solution Potential. See **Electrode Potential.**

Solution Pressure. Pressure of a solute in a solvent, also called osmotic pressure.

Solution Treatment. Heating **Age-Hardening** (see) alloys to cause the hardening constituent to go into solid solution, followed by quenching and age-hardening.

Solvent. Material in which solution takes place. See also **Solute.**

Sonic Test. Determination of certain properties of metal by passage of sound or other sonic effects.

Sonim. Solid non-metallic **Inclusion** (see) found in metal.

Sonorousness. Property of certain alloys of giving bell-like tones when struck. Many copper alloys show this characteristic in marked degree, and bells are usually made of such alloys.

Sorbite. Structure of steel, resulting from the tempering of martensite. In a truly sorbitic structure, the cementite (Fe_3C) is completely dispersed in the matrix.

Sorbitic Pearlite. Structure of steel resulting, on cooling under the proper conditions, from the decomposition of austenite; has a fine, lamellar appearance.

Sorel's Alloy. Zinc-base bearing and casting alloy, with 1–10% copper and 1–10% iron.

Soundness. Freedom, in massive metal, from blowholes, porosity, and similar discontinuities.

Sow. (a) Lateral runner for feeding molten metal from the main runner [see **Runner** (a)] to the pig mold. (b) See **Salamander.**

Space-Centered. See **Body-Centered.**

Space Factor. See **Stacking Factor.**

Space Lattice. Three-dimensional periodic arrangement of atoms, ions, or molecules in space.

Spalling. Tendency of materials to chip away as a result of alternate heating and cooling. The term is frequently used when referring to refractories.

Spang-Chalfant Alloy. Series of low-alloy steels, commonly with carbon below 0.20% (National Supply Co.).

Spangle. Frosted appearance of the surface of zinc coating on galvanized steel. The shape of the spangles is partly determined by the composition of the galvanizing bath, but it also may be controlled by mechanical means. See **Controlled Spangle.**

Spanish Furnace. Reverberatory furnace used in the smelting of lead.

Spanner Bar. Steel bar, connecting, through lever arms, the two **Spanner Wheels** (see) of a rolling mill, so as to adjust both screws equally and thus prevent uneven rolling.

Spanner Wheel. Large wheel on **Rolling Mill** (see), for controlling screws raising and lowering the rolls, and thus the thickness of the rolled metal.

Sparable Tin. Non-metallic mineral tin ore in the form of small grains.

Sparking Alloy. See **Pyrophoric Alloy.**

Sparkle Metal. Crude copper **Matte** (see), of about 74% copper content, obtained from the smelting of sulfide copper ores.

Spark Spectrum. Spectrum of ionized atoms of an element, obtained from spark sources.

Spark Testing. Determining the type of a steel sample by study of the sparks given off when the steel is pressed against a grinding wheel. Both the shape and color of the sparks given off vary with the composition of the steel. An experienced testing operator can determine the carbon content of a steel, semi-quantitatively, very rapidly in this way.

Sparry Iron (Ore). See **Siderite.**

Spathic Iron. Non-metallic mineral, **Siderite** (see), ferrous carbonate.

Spathic Iron Ore. See **Siderite.**

Special Carburizing-Quality Strip Steel. Strip steel ordered or sold on the basis of guaranteed austenitic grain structure. See **Limited Austenitic Grain Structure.**

Special Coated Manufacturing Terne. Short Terne (see) plate of relatively light coating.

Special Cokes. See **Coke Tin Plate.**

Special High-Grade Zinc. Metallic **Zinc** (see), with total impurities not over 0.01%; produced by fractional distillation or by electrolytic means.

Special Killed Steel. Killed steel, usually in sheet form, ordered or sold on the basis of specifications more limiting in chemical and/or physical tolerances than ordinary **Killed Steel** (see).

Special Leaded Phosphor Bronze. See **Leaded Phosphor Bronze.**

Specially Processed Sheet. Steel sheet given any special or non-standard treatment, such as enameling (see **Vitreous Enamel**), phosphate coating (see **Parkerizing**), etc.

Specially Smooth Finish. See **Cylinder Finish.**

Special-Requirement Quality Strip Steel. Strip steel ordered or sold on the basis of specifications including special or additional restrictions above the standard, regarding chemical, physical, or metallurgical properties.

Special-Requirement Wire. Steel wire ordered or sold on the basis of specifications including special or additional restrictions above the standard, as to chemical, physical, or metallurgical properties.

Special Steel. See **Alloy Steel.**

Special-Surface Quality Wire. Steel wire ordered or sold on the basis of specifications directing special surface limitations.

Special Tolerance. Stricter tolerance than the normal (see **Regular Tolerance**).

Specialty Mill. Rolling mill, built for the rolling of a single product.

Special Wire. High-carbon steel wire, of special metallurgical or finish specifications, for a special purpose, such as spoke wire, music wire, etc.

Specific Area. Total surface area of the particles in a gram of a material.

Specification Quality Strip Steel. Strip steel ordered or sold on the basis of any specifications previously agreed to between seller and purchaser.

Specific Density. See **Specific Gravity.**

Specific Gravity. Numerical value representing the comparative weight of a given volume of a substance as compared with that of an equal volume of water, whose specific gravity is taken as 1.000. The specific gravity of metals varies from 0.534 for lithium to 22.47 for osmium. Specific gravity can be affected somewhat by rolling, forging, and similar mechanical treatment. All metals, with the exception of bismuth, have a lower specific gravity when molten than in the solid state.

Specific Heat. Heat capacity of a substance as compared with the heat capacity of an equal weight of water. The value can also be given in terms of the number of calories necessary to raise the temperature of one gram of the substance one degree Centigrade. Obviously, the higher the specific heat of a substance, the more heat it absorbs when its temperature is being raised, and the more heat it will release on cooling.

Specific Heat Pyrometer. See **Pyrometer.**

Specific Resistance. See **Electrical Resistivity.**

Specific Surface. In powder metallurgy, surface area of one gram of powder, expressed in square centimeters.

Specified Discard. Specification, in ordering or selling metal, limiting the proportion of **Discard** (see) in an ingot.

Specimen Test. Physical test conducted on a full-sized completed unit.

Spectacle Furnace. Shaft furnace, with an inclined hearth, the reduced metal flowing out continuously during the smelting operation.

Spectroscope. Apparatus for studying spectra; in metallurgical work, spectroscopes generally have attachments for generating the spectrum of the material under study, as, for example, by arc or flame.

Spectroscopic Test. Chemical analysis, particularly for small amounts of material, by studying the spectrum of the volatilized material.

Spectrum Analysis. Qualitative or semi-quantitative analysis by study of visible and ultra-violet spectra. Because of great sensitivity, the method is frequently used to determine traces of elements.

Specular Iron. Not a metal, but a mineral; a form of **Hematite** (see), iron oxide, Fe_2O_3.

Specular Pig Iron. See **Spiegel.**

Speculum Metal. Copper-base alloy, capable of high polish, and, therefore, usable for reflectors and mirrors; common alloying elements include tin, lead, and zinc.

Speed Case (Steel). Low-carbon open-hearth steel, especially suitable for machining and case-hardening and frequently used in the manufacture of molds (W. J. Holliday & Co.).

Speed Treat (Steel). Open-hearth steel of 0.40–0.50% carbon content (W. J. Holliday & Co.); used for small tools and dies.

Speiss. Metallic arsenides or antimonides, resulting from smelting metal ores, such as those of cobalt, containing arsenic or antimony as contaminants.

Spellerizing. Rolling skelp, process in which the heated bloom is alternately passed through roughened kneading, knobbling, rolls and smooth rolls.

Spellerizing Roll. See **Spellerizing.**

Spelter. Crude zinc obtained on smelting.

Spelter Solder. Series of brazing alloys, melting between 1440 and 1620°F, available in various grain sizes and types.

Spence Furnace. Multiple muffle furnace, used for **Roasting** (see) sulfide ores, the ore being mechanically stirred during roasting.

Sperrylite. Native alloy of platinum and arsenic, chemically approximating $PtAs_2$; frequently contains also antimony and rhodium.

Sp. G. See **Specific Gravity.**

Sphalerite. See **Mock Lead.**

Spheroidal Cementite. Rounded or spheroid cementite, existing in spheroidized iron-base alloys.

Spheroidite. See **Granular Pearlite.**

Spheroidized Cementite. See **Spheroidal Cementite.**

Spheroidized Wire. Steel wire, drawn cold from rods heated slightly below the critical range, yielding a **Divorced Pearlite** (see) structure.

Spheroidizing. Treatment of alloys which causes the coalescence of granules or lamellae of a dispersed phase. In steel technology, the term describes any treatment which will convert the carbide content into rounded or spheroid form.

Spiegel. High-manganese pig iron, containing 15–30% manganese, approximately 5% carbon, and less than 1% silicon, used for the manufacture of steel by the Bessemer or basic open-hearth process.

Spiegeleisen. See **Spiegel.**

Spiegel-Iron. See **Spiegel.**

Spindle. Rod or pipe utilized in forming a mold core. See **Core** (b) and also **Leading Spindle.**

Spindle Temper Steel. Tool Steel (see) with a carbon content of about 1.10%.

Spinning. Procedure of shaping shallow vessels from sheet metal discs, while rotating, by pressure of a tool against the periphery. Frequently, spinning must be carried out in stages, with intermediate annealing to remove the effects of work-hardening.

Spinning Brass. Yellow **Brass** (see), about two parts copper to one zinc, commonly used as sheet for spinning into final form.

Spitting. Phenomenon of forcible ejection of oxygen when silver or platinum, melted in air, is undergoing solidification.

Splasher. Brick-lined metal plate, deflecting the flame from the tap hole of a furnace during casting.

Split Die. See **Segment Die.**

Split Transformation. Simultaneous formation of troostite and martensite, occurring in steel cooled at a rate approaching the critical.

Spoke Wire. Steel wire of high tensile strength and shock resistance, developed by heat treatment, and used in the manufacture of automobile wire wheels; a lower grade, with less heat treatment, is used for baby carriages and toys.

Sponge Gold. See **Cake of Gold.**

Sponge Iron. Finely divided iron, obtained by the direct, low-temperature reduction of iron ores; in this state, iron oxidizes readily.

Sponge (Metal). Metal in porous form. See also **Sponge Iron.**

Sponginess. Porous condition in metal. In powder metallurgy, usually occurs in metal particles which have been obtained by reduction of metal oxides.

Spongy Iron. See **Sponge Iron.**

Spongy Platinum. See **Platinum Sponge.**

Spool Roll. Roll, on roll tables, for carrying away rolled metal, provided with collars or discs, to prevent lateral movement of the metal.

Spool Winding. Winding narrow strip or sheet metal on a central core.

Spoon Test. Rough estimation of the temperature of molten cast iron; the time required for a film to form on the surface of a spoonful of the metal, as taken from the ladle, is used as a rough index of the temperature.

Spot Welding. Resistance **Welding** (see) in which the welding heat is limited to a small area.

Spra Bonderite. Process of forming rust-resisting and paint-adherent coating on steel, by spraying with dilute alkali, and, after washing with water, spraying with a solution of salt and chromic acid.

Spray. See **Mottled Iron.**

Spraying. See **Metal Spraying.**

Spray Tuyere. Box Tuyere (see) of open design, cooling being accomplished by means of a spray of water, which is vaporized by the heat of the furnace.

Spread. See **Lateral Spreading** and **Spread of Metal.**

Spread of Metal. Percentage of widening of metal sheet when passed through a set of rolls.

Sprills. In powder metallurgy, metal particles of cylindrical form, with height not much greater than diameter.

Spring. See **Roll Spring.**

Spring-Back. Characteristic of stiff metals (such as hard-rolled steel sheet), of recovering part of the deformation resulting from a hammer blow or formation in a die. The harder the metal, the greater the spring-back tendency.

Spring Brass. Yellow **Brass** (see) of about 70% copper content, cold-rolled to hard temper.

Spring Bronze. Phosphor Bronze (see), containing about 5% tin, cold-rolled to hard temper.

Spring Heat Steel. Steel of about 0.50% carbon content, made by the carburization of wrought iron.

Spring Plate. See **Bar Steel** (a).

Spring Steel. Steel, normally of the high-carbon or alloy type, used in the manufacture of springs, lending itself to appropriate heat treatment; usually made in the open-hearth or electric furnace.

Spring Wire. Wire, most commonly of steel, used for making simple spiral springs. Steel wire is most commonly of the patented type. See **Patenting.**

Sprouting. See **Spitting.**

Sprue. See **Riser.**

Sprue Hole. See **Gate.**

Square Bar. Steel **Bar** (see) of square cross-section.

Square Edge Strip. Strip metal, with edges produced by hot edge rolling.

Square Mil. Area of a square of one-mil side. A mil is 0.001 inch.

Squaring Shear. Shear for cutting edges and ends of sheet metal, to form rectangular shapes.

Squeezer. See **Puddling** and **Slab Edging Press.**

Squeezer Roll. Cloth-covered roll in terne and tin plate manufacture, for squeezing most of the adhering oil from the surface of the plates.

Squirting Process. See **Extrusion.**

Sr. Chemical symbol for **Strontium** (see).

S Scratch Brush Wire. Steel wire of Bessemer stock, used in the manufacture of fine brushes, usually continuously drawn dry as far as the second annealing and then wet drawn, followed by bright annealing and a final wet drawing.

Stabilized Steel. See **Stabilizer.**

Stabilizer. Metals such as columbium, titanium, vanadium, molybdenum, and tungsten, the presence of which prevents intercrystalline precipitation of chromium carbide in austenitic stainless steels when heated to 930–1300°F. The metals seem to act by formation of stable carbides, insoluble in the austenite and distributed throughout the crystal grains.

Stabilizing Element. See **Stabilizer.**

Stack. (a) See **Blast Furnace.** (b) See **Shaft Furnace.** (c) Pile of lead grids, with organic matter, for the manufacture of **White Lead** (see).

Stack Cutting. Process of cutting stacks of sheet metal, as by sawing, cutting torch, etc.

Stacking Factor. Ratio of weight of laminated metal, most commonly steel, to the same volume of solid metal of the same type.

Staff. Rabble used in **Puddling** (see).

Staff Hole. Hole in **Puddling** (see) furnace for the manipulation of a puddling **Staff** (see).

Sta-Gloss Steel. Series of corrosion-resistant steels (Jessop Steel Co.), with 12–18% chromium and 0.40–0.60% carbon.

Stainex. Series of corrosion-resistant high-chromium (13–17%) steels, with carbon ranging from 0.10 to 0.65% (Columbia Tool Steel Co.).

Stainless Cladding. See **Cladding.**

Stainless-Clad Steel. See **Cladding.**

Stainless Iron. Most **Stainless Steels** (see) are really irons, since the carbon content is normally kept low. The terms are generally used interchangeably.

Stainless Steel. Series of high-chromium steels, frequently with nickel as an additional alloying component, which are resistant to air, oxidizing agents, and other corrosive influences. The best known of the series is the so-called 18–8 stainless, an austenitic steel with approximately 18% chromium and 8% nickel, carbon being kept low.

Stainless Steel Electrolytic Pickling Process. Electrolytic Pickling (see) of stainless and other high-chromium steels; the metal is made anode in a bath of 65% sulfuric and 10% hydrofluoric acid content at about 140° F, for several minutes, with a current density of about 250 amperes per square foot.

Stall. See **Stall Roasting.**

Stall Roasting. Heating of ores, under oxidizing conditions, in small enclosures constructed of brick or cast slag. See **Roasting.**

Stamp Copper. Copper produced from ore which has been crushed in stamps, prior to smelting.

Stamp Mill. Equipment for crushing, using hammers (stamps); frequently used in amalgamation, and on tin and copper ores prior to smelting.

Stand. Operating portion of a set of rolls, consisting of the rolls themselves, the chocks (bearings), supporting the rolls, together with the housings for these chocks, and the screws by which the space between the rolls can be controlled.

Standard Alloy. Series of heat- and corrosion-resistant iron-chromium-nickel alloys (Standard Alloy Co.) with 7–60% nickel and 15–30% chromium; molybdenum is optional up to 2–4%. Carbon is normally about 0.40%.

Standard Beam. See **I-Beam.**

Standard Brass. Yellow **Brass** (see) consisting of 70% copper and 30% zinc.

Standard Copper. Metallic copper, exceeding approximately 96% in purity.

Standard Electrode Potential. See **Electrode Potential.**

Standard Gold. Legally adopted alloy for the coinage of gold. In the USA, standard gold is a 10% copper alloy; in Great Britain, the copper content is 8.33%. In both cases, the remainder is gold.

Standard Oxidation-Reduction Potential. See **Oxidation-Reduction Potential.**

Standard Pipe. Pipe conforming to the American Standard (Briggs) pipe gage.

Standard Rail. See **Standard Railroad Rail.**

Standard Railroad Rail. Rail, for use in railroad service, weighing 60–152 pounds per yard.

Standard Silver. Legally adopted alloy for silver coinage. In Great Britain, where it was originally standardized, it is called sterling silver, and contains 7.5% copper, alloyed with the silver. American standard silver is the 10% copper alloy.

Standard Steel. Steel conforming to chemical and/or physical specifications as dictated by authoritative organizations such as **S.A.E.** (see) and **A.I.S.I.** (see).

Standard Tin. Tin (see) of 99.75% or greater purity.

Standing. See **Doubling Floor.**

Standing Floor. See **Doubling Floor.**

Stand of Rolls. See **Stand.**

Stanley (Steel). Series of corrosion-resistant high-chromium (12–23%) steels, with nickel optional up to 9%, manufactured by Stanley Works.

Stanniferous. Containing tin.

Stannum. Latin name for tin, used in medicine and pharmacy.

Stannum Metal. Tin-base bearing metal (Lumen Bearing Co.).

Star. (a) See **Doubling Process.** (b) Star-shaped unit of hard iron, used in tumbling operations for the cleaning of metal articles.

Star Antimony. Metallic antimony. See **Doubling Process.**

Star Bowl. See **Doubling Process.**

Star Fracture. See **Fracture.**

Star Metal. See **Star Antimony.**

Starring Mixture. See **Doubling Process.**

Starting Sheet. Thin sheet of metal used, in the electrolytic refining of metals, as cathode on which the metal can start its deposition.

Star-Zenith (Tool) Steel. High-carbon, alloy tool steel (Carpenter Steel Co.), with about 18% tungsten, 4% chromium, and 1% vanadium.

Stassano Furnace. Indirect Arc Furnace (see), with electrodes entering the furnace through the sides.

Static Indentation Test. See **Indentation Test.**

Static Stress. Stress which is applied gradually and which remains constant after application.

Statuary Bronze. Bronze (see) capable of good casting; copper is commonly about 90%, with tin, lead, and zinc principal among the alloying elements used.

Staybolt Iron. See **Wrought Iron.**

Staybrite. Series of British **Stainless Steels** (see) of the chromium-nickel type.

Steadite. Eutectic (see) of iron, 6.89% phosphorus, and 1.92% carbon.

Stead's Brittleness. Brittleness which occurs in steel of very low, less than 0.15%, carbon content as a result of long heating in the 930–1380°F range. Enormous crystals result, with consequent loss of strength and ductility.

Stead's Reagent. Metallographic etching solution, consisting of acidified copper chloride, useful in detecting phosphorus segregation in steel; copper tends to deposit on parts of low phosphorus content.

Steam Bronze. Hydraulic Bronze (see), most commonly of about 5% zinc, 5% tin, and 5% lead content.

Steam Metal. Copper-base alloy, with about 7% tin, 4% zinc, and 2% lead, used for valves and other hydraulic fittings.

Steam Pattinsonation. Modification of the **Pattinson Process** (see), using steam to agitate the molten lead and aid in the refining.

Steckel Mill. Four-high **Reversing** rolling **Mill** (see) in which no power is applied to the rolls, the metal being drawn through the rolls by tension reels.

Steel. Iron, malleable in at least one range of temperature below its melting point without special heat treatment, substantially free from slag, and containing carbon more than about 0.05% and less than about 2.00%. Other alloying elements may be present in significant quantities, but all steels contain at least small amounts of manganese and silicon, and, usually, as undesirable constituents, also sulfur and phosphorus. Eutectoid steel contains approximately 0.80% carbon; on moderately slow cooling, eutectoid steel will change entirely to pearlite. Carbon below 0.80% makes a hypoeutectoid steel, carbon above 0.80% a hypereutectoid steel. Special and alloy steels are listed under their commercial names.

Steel Bronze. Modified yellow **Brass** (see), with about 40% zinc, and low percentages of manganese, aluminum, and iron.

Steel Casting. See **Cast Steel** (a).

Steel-Cored Aluminum (Conductor). Electric conductor comprising a core of steel, for mechanical strength, surrounded by one or more layers of aluminum wire for electrical conductivity.

Steel-Cored Copper (Conductor). Electric conductor, comprising a core of steel, for mechanical strength, surrounded by one or more layers of copper wire for electrical conductivity.

Steel Grit. Irregular, jagged pieces of steel, formed by passing molten steel through a jet of steam, removing the globules formed simultaneously and crushing the irregular particles into finer sizes.

Steel Making. Manufacture of steel from pig iron.

Steel Mill. Plant for the manufacture or shaping of steel.

Steel Powder. Steel of any composition, in finely divided form; increasingly used in powder metallurgy.

Steel Press. Apparatus for exerting pressure on molten steel, after casting, so as to produce dense and pore-free ingots.

Steel Refining. See **Restoring.**

Steel Restoring. See **Restoring.**

Steel Roll. Roll (see **Rolling Mill**), made of cast or forged steel.

Steel Spout. Spout for conducting molten steel from an open-hearth furnace to the ladle.

Steel-Through Heat Steel. Steel, with about 1.25% carbon, made by the carburization of wrought iron.

Steel Wire Gage. See **Metal Gage.**

Steel Wool. Fine steel shavings in wool-like masses, used as an abrasive.

Steel Works. See **Steel Mill.**

Stellite. Series of hard, tough, and corrosion-resistant cobalt-base alloys, with 10–40% chromium, 0–25% tungsten, and carbon up to 2.5%, manufactured by Haynes Stellite Co. and used for cutting tools, hard facing work, etc.

Stentor (Steel). Non-Deforming Steel (see), with about 1.5% manganese and approximately 0.90% carbon (Carpenter Steel Co.).

Stereotype Metal. Series of high-alloy **Type Metals** (see), with 3–17% tin and 12–23% antimony, the remainder being lead.

Sterling. See **Standard Silver.**

Sterling Nirosta (Steel). See **Sterling Stainless (Steel).**

Sterling Silver. See **Standard Silver.**

Sterling Spelter. See **Intermediate Zinc.**

Sterling Stainless (Steel). Series of high-chrome and nickel-chrome steels of the corrosion-resistant stainless type (Firth-Sterling Steel Co.).

Sterlite. Copper-base alloy, with 25% nickel, 20% zinc, and fractional percentages of manganese, iron, and silicon (Sterlite Foundry & Mfg. Co.).

Sterro-Metal. Copper-base alloy, with about 40% zinc, 1.5–5.0% iron, and, optionally, up to 1% tin; used principally for hydraulic cylinders.

Stetefeldt Furnace. Furnace for the roasting of silver ores by **Chloridizing** (see), or for roasting low-sulfur copper ores.

Stewart Lumite. Series of aluminum-base casting alloys, of variable composition, with copper, silicon and nickel as the most common alloying elements (Stewart Die Casting Corp.).

Stewart White Brass. Series of zinc-base die-casting alloys, with about 4% aluminum and, optionally, up to 3.5% of copper (Stewart Die Casting Corp.).

Stibium. Latin name for antimony, used in medicine and pharmacy.

Sticker. (a) Faulty open-hearth steel processing, in which carbon removal is accomplished with the bath temperature too low to pour properly. (b) Faulty condition of stacked metal sheet, in which the sheets tend to adhere to each other.

Sticker Hole. Hole in metal sheet, resulting on one sheet when two adhering sheets [see **Sticker** (b)] are torn apart.

Sticker Patch. Extra metal adhering to sheet metal, resulting on one sheet when two adhering sheets [see **Sticker** (b)] are torn apart.

Stiefel Mill. Mill for rolling seamless tubing.

Stiffening Roll. See **Backing-Up Roll.**

Stiffness. Resistance of a material to the action of a bending force.

Still. Apparatus for converting a material from the solid or liquid state to the vapor state, with or without simultaneous chemical decomposition; usually the vapor is condensed in a separate part of the apparatus, becoming liquid or solid again.

Stillbottom Plate. See **Stillbottom Quality Steel Plate.**

Stillbottom Quality Steel Plate. Steel plate ordered or sold on the basis of specifications governing suitability for construction of direct-fired vessels.

Still (Tank) Pickling. Pickling (see) during which both the acid solution and the material being treated are maintained stationary.

Stobie Furnace. Electric direct **Arc Furnace** (see), using one, two, or three-phase current, most commonly each phase having its own return electrode.

Stock. Mixture of ore, reducing agent, and flux, prior to charging into a furnace.

Stock and Goldschmidt Process. Obtaining beryllium, in fused form, by electrolysis of a beryllium oxyfluoride in a fused bath of sodium and barium fluorides. See also **Siemens-Halske Process** (b).

Stock Indicator. Steel rod, which passes through the **Try Hole** (see) of a blast furnace, used to determine the level of the material in the furnace by resting on top of the material, with a steel cable leading to an indicating device.

Stoke Hole. Hole to permit introduction of a **Rabble** (see) into a furnace.

Stone Wire. Soft steel wire, usually in annealed or galvanized form, for general home and farm utilizations.

Stoodite. High-chromium (about 35%) iron alloy, used for hard-facing welding rods (Stoody Co.).

Stoody Welding Rod. Hard-facing welding rod of high alloy content (Stoody Co.).

Stool. Heavy iron or copper plate on which an **Ingot Mold** (see) rests, when ready to receive the molten metal.

Stopper Hole. Hole in a **Puddling** (see) furnace for the introduction of a rabble.

Stopping Off. (a) Protecting desired areas from chemical action, such as carburizing, electroplating, etc., by an appropriate coating of those areas with a material which is unaffected by the chemical action. (b) Filling part of a mold with sand to prevent molten metal from entering.

Storage Battery. Electric battery which can be repeatedly discharged and then recharged by reversing the current flow and passing electricity through it from an outside source, reversing the chemical reactions which created the original electric current. There are two standard types (1) **Lead Storage Battery** (see) (2) **Alkaline Storage Battery** (see).

Stove. See **Blast Furnace.**

Straight Carbon Steel. See **Carbon Steel.**

Straightener. Equipment for removing irregularities of metal in flat or wire form, resulting from rolling.

Straightening Press. Power-actuated apparatus for straightening rolled steel products, such as rails, shafting, etc.

Straightening Rolls. See **Cross Rolls.**

Straight Side Converter. Converter (see **Bessemer Steel**) of essentially cylindrical shape.

Strain. Deformation produced on a body by an outside force. See also **Stress** and **Hooke's Law.**

Strain-Aging. Slow rise of hardness and tensile strength in some metals, at room temperature, after cold working.

Strain Hardening. Hardening which occurs when metals are subjected to deformation below and not too close to their recrystallization temperature.

Straits Tin. Tin (see) metal made from Malay District ores.

Strand. See **Strand Roll.**

Strand Annealed Wire. Steel wire annealed (see **Annealing**) by passage, in single strand form, through a bath of molten lead or through a furnace, followed by cooling in air.

Strand Roll. Rolling mill pass or passes, after the **Leader Pass** (see).

Strauss Copper Sulfate Test. Test to determine susceptibility of **Stainless Steel** (see) to intergranular corrosion, by boiling in a solution of copper sulfate and sulfuric acid. Unsatisfactory metal will show intergranular cracks after a period varying from a few hours to several hundred hours.

Strauss Metal. Cemented tungsten **Carbide** (see), for cutting tools, with cobalt as the binding matrix (Ludlum Steel Co.).

Stream Gold. Native gold, found in streams.

Stream Tin. Tin obtained from ores which have been washed down by streams to lower lands. Bolivian tin is of this type. See also **Lode Tin.**

Strength. Ability of a material to resist external stresses. See **Tensile Strength** and **Compressive Strength.**

Strengthening Rod. Steel rod used for the reinforcing of cores in large molds [see **Core** (b)].

Stress. Deforming force to which a body is subjected, or, the resistance which the body offers to deformation by the force. See also **Strain** and **Hooke's Law.**

Stress Concentrator. Point or line in the external or internal structure of a metal, such as a crack, sonim, etc., which causes a concentration of stress when the metal is subjected to mechanical or quenching stresses; these points are likely to be the start for fatigue cracking.

Stress-Corrosion Test. Determination of susceptibility of steel to combined corrosion and stress, by bending a wire about one-eighth inch thick around a two-inch radius, and then subjecting it to copper sulfate or nitric-hydrofluoric acid solution.

Stress-Cycle Curve. See **S-N Curve.**

Stressite. Series of compacted powdered copper-base and iron-base alloys, for powder metallurgy purposes (Chrysler Corp.).

Stress-Number Curve. See **S-N Curve.**

Stress-Number-of-Cycles Curve. See **S-N Curve.**

Stress Relaxation. See **Relaxation.**

Stress Relief. Low-temperature annealing for removing internal stresses, such as those resulting in a metal from work hardening or quenching.

Stress Relieving. See **Stress Relief.**

Stress-Strain Diagram. Curve showing the relationship between stress and strain on a sample under test. Breaks and deviations on this curve indicate the elastic limit, yield point, proof stress, ultimate strength, elongation, etc.

Stretcher Head. Large clamp for holding sheet metal during **Stretcher Levelling** (see).

Stretcher Levelling. Flattening of metal sheet by stretching.

Stretcher Strains. See **Lüder's Lines.**
Stretching Machine. Machine for **Stretcher Levelling** (see).
Strike. (a) Tool for smoothing sand in a **Mold** (see). (b) Chemical bath into which metal to be plated may be dipped preliminary to electrolytic deposition.
Striking. See **Strike** (b).
Strip. Flat rolled metal, with greatest thickness of about 0.25 inch.
Stripper. (a) See **Ingot Stripper.** (b) See **Strippers.**
Stripper Punch. In powder metallurgy, a punch which forms part of the normal die and, in addition, acts to eject the compact from the die at the end of the forming.
Strippers. Tin Plate (see) with imperfections which allow partial stripping to a lighter coating.
Stripping. (a) Removing coated or electrolytically deposited metal from the base. (b) See **Ingot Stripper.**
Stripping Test. Test of thickness of galvanized coating on steel (see **Galvanizing**), in which the coating on a specimen is dissolved by a chemical, and the quantity of dissolved metal determined by the loss of weight.
Strip Steel. Steel, usually of open-hearth manufacture, after being rolled into strip form. It may have cold finish (cold-rolled strip) or hot finish (hot-rolled strip).
Strip Tester. Apparatus for determining electrical characteristics of **Electrical Sheet** (see).
Strontium (chemical symbol Sr). Element No. 38 of the periodic system; atomic weight 87.63. Moderately hard but ductile, white alkaline earth metal, resembling barium in appearance; melting point 1395°F; boiling point 2491°F; specific gravity 2.6; chemically, nearly always divalent; reacts with water, releasing hydrogen, and therefore corrodes slowly in air. It cannot be electrodeposited from aqueous solution; manufactured by interaction of strontium oxide and aluminum, in vacuo; not much used commercially; obtainable from Kemet Laboratories Co.
Structural Nickel Steel. See **Nickel Steel.**
Structural Quality Hot-Rolled Strip Steel. Hot-rolled strip steel ordered or sold on the basis of specifications governing use in structural and similar fields, such as cars, etc.
Structural Quality Steel. Steel ordered or sold on the basis of specifications governing suitability for the construction of bridges, buildings, locomotives, ships, etc.
Structural Steel. See **Structural Quality Steel.**
Structural Test. Physical test conducted on built-up sections rather than original parts, for determining the strength of the design and joints.
Stubs' Gage. See **Metal Gage.**
Stubs' Iron Wire Gage. See **Metal Gage.**
Stubs' Steel Wire Gage. See **Metal Gage.**
Stupp. Deposit, obtained in the distillation of mercury ores, consisting of mercury metal, mercuric oxide, soot and dust, ore particles, etc.
Sub-Boundary Structure. Network of boundaries which, on etching, appear within the main crystals of the metal structure. The granules within the sub-boundaries seem to have a uniform orientation.
Subcritical Annealing. See **Process Annealing.**
Subcutaneous Blowhole. Blowhole (see) located near the surface of the metal.
Sublimate. Product of **Sublimation** (see).
Sublimation. Direct conversion of a substance from the solid state to the gaseous state, without passing through the liquid state.
Submerged Cathode Cell. Electrolytic cell, used in the isolation of alkali, alkaline earth, and similar easily-fusible metals from fused electrolytes, in which the cathode enters the bath from below, all parts of the cathode being submerged beneath the surface of the fused bath.
Sub-Microstructure. Metal structure with detail too small to be viewed microscopically. Such structure is usually determined by other evidence, such as X-ray or electron analysis.

Subsilicate. Slag in which the basic oxides are chemically equivalent to twice the amount of acid oxides.

Sulfating Roasting. Roasting (see) in which part of the sulfur of the ore is converted to sulfate, and remains with the residual material.

Sulfatizing Roast. See **Sulfating Roasting.**

Sulfide (Sulphide). Any oxygen-free compound with sulfur as one component. When present in metals, sulfur is usually in sulfide form.

Sulfur (chemical symbol S). Element No. 16 of the periodic system; atomic weight 32.06. Non-metal occurring in a number of allotropic modifications, the most common being a pale-yellow brittle solid; melting point of rhombic sulfur 235°F; boiling point 832°F; specific gravity of rhombic sulfur 2.07. Divalent, tetravalent, or hexavalent; can readily be electrodeposited on the *anode* from solutions of soluble sulfides. Sulfur occurs extensively in nature both as the native element and as metal sulfide; obtained commercially from native deposits by the Frasch process which melts the sulfur under ground with superheated water and then subjects it to air-lift action. Extensively used for chemical purposes; metallurgically, it is most commonly encountered as an undesired contaminant, particularly in steel. However, it is frequently deliberately added to cutting stock, to increase machinability.

Sulfurized Steel. Steel containing sulfur as an added constituent, or in excess of the normal fortuitous ranges.

Sulfur Print. Print made from the surface of steel to show the concentration of sulfur in the steel. A sheet of photographic bromide paper, previously soaked in dilute sulfuric acid, is placed on the filed surface of the steel; the acid reacts with the sulfur, being in the form of sulfides, to yield hydrogen sulfide, which turns the paper dark at the appropriate points. The color density of the print is a measure of the sulfur concentration. The usefulness of the test is increased by the fact that combined phosphorus tends to segregate together with sulfur.

Sull. Coating of iron hydroxide produced on steel by controlled oxidation while wet.

Sullage Piece. See **Riser** (b).

Sull-Coated Wire. Steel wire, drawn from rods subjected to water spraying after acid cleaning, in order to form a coating which, together with a lime coating, holds drawing lubricants.

Sulling. Producing **Sull** (see) coating on steel.

Sulphur. See **Sulfur.**

Sumet Brónze. Copper-lead alloy, with up to 30% lead and, optionally, up to 5% tin; used for bearings, etc. (Sumet Corp.).

Sump Solution. See **Barren Solution.**

Sunken Seam. See **Falling Seam.**

Superbronze. Series of yellow brasses, containing low percentages of aluminum and manganese, and, optionally, iron. Harder and stronger than brass, these alloys are, however, not as ductile or machinable. See **Brass.**

Supercut (Steel). High-sulfur, free-cutting steel, manufactured by Republic Steel Corp.

Super Die Steel. High-alloy, wear-resistant die steel (Columbia Tool Steel Co.) of high carbon content (about 2.2%). Other alloying elements are about 11% chromium, 1% tungsten and 1% silicon.

Super-Duralumin. Duralumin (see) with about 1% silicon as an additional constituent.

Superficial Hardness (Test). Determination of the hardness of the surface of a material, usually by scratching or by relatively light penetration loads. See **Rockwell Superficial Hardness (Test)** and **Vickers Hardness (Test).**

Superfines. In powder metallurgy, the portion of a sample having a particle size less than 10 microns, equivalent to 0.01 millimeter.

Super High-Speed Steels. High-carbon (about 0.85%) high-alloy tool steels, with 17–22% tungsten, 8–15% cobalt, 5–6% chromium, 1–2% vanadium, and about 1% molybdenum.

Superior Nickel-Chrome. Series of nickel-chromium alloys, used primarily for electric resistance purposes, chromium approximating 20% (Alloy Metal Wire Co.).

Superior Steel. Series of steels, of diverse composition, manufactured by Superior Steel Corp.

Superloy (Steel). Series of wear- and corrosion-resistant alloy steels, manufactured by Washington Iron Works.

Supermal. Malleable cast iron, made by Jeffrey Mfg. Co.

Super Manganese Bronze. Copper-base alloy, approximating 21% zinc, 7% aluminum, 3% iron, and 3% manganese; has excellent strength and corrosion resistance.

Super-Nickel. Alloy of approximately 70% copper and 30% nickel, manufactured by American Brass Co., used primarily in condenser tubes and similar marine service products in which corrosion-resistance is desired.

Super-Refractory. Refractory capable of withstanding unusually high temperatures and/or other severe conditions of operation.

Supersaturated Solution. Metastable solution in which the dissolved material exceeds the amount which the solvent can hold in normal equilibrium under the temperature and other conditions prevailing.

Superstrength Bronze. Series of yellow brasses (see **Brass**) with zinc ranging from 28% to 40%, with fractional percentages of aluminum, iron, and manganese as hardeners (Baldwin Locomotive Works).

Supertemp (Steel). Low-alloy steel, containing about 1% tungsten and fractional percentages of chromium and manganese, manufactured by Bethlehem Steel Co.

Supremus (Tool) Steel. High-speed alloy tool steel (Jessop Steel Co.), with about 18% tungsten, 4% chromium, 1.5% vanadium, and, optionally, 1% molybdenum.

Surface Absorption. See **Metallic Cementation.**

Surface Conditioning. Removal of surface flaws in ingots, prior to rolling or other working.

Surface Finish. See **Finish.**

Surface Hardening. Treatment of metal, most commonly with chemicals, at elevated temperatures, for hardening its surface.

Surge Pickling. Pickling (see), in which sheet metal is held vertically while acid is agitated around it.

Suttonite. See **Suttonizing.**

Suttonizing. Reclaiming worn or broken tools by adding **Hard-Facing** (see) metal (Welding Equipment & Supply Co.).

Su-Veneer. Steel (Superior Steel Corp.), clad with various metals, on one or both sides. See **Cladding.**

Swaging. Forging, usually to a circular cross-section, by working between dies.

Swamp Ore. See **Limonite.**

Swarf. Oxidized and fritted iron, usually made from thread cuttings, used on the bottom of **Puddling** (see) furnaces.

Sweating. (a) Condensation of moisture on metal surfaces due to humidity and temperature conditions. (b) Slight fusion of brick in normal high-temperature operation of a furnace, usually of the open-hearth type. (c) Formation of **Blisters** (see) on steel when heated to too-high a temperature. (d) See **Sweating Out.**

Sweating Furnace. Furnace for liquating low-melting metals or alloys from dross or oxides.

Sweating Out. Appearance, usually, in small globules, of one constituent of relatively low melting point, of a mixture or alloy on the surface of a solid mass when heated.

Swedesteel. Series of carbon and alloy tool steels, manufactured by Paragon Tool & Steel Co.

Swedging. See **Swaging.**

Swedish Iron. Wrought iron, made in Sweden from relatively pure ores.

Swedish Process. See **Continental Process.**

Sweep. Pattern used for forming molds of round or similar symmetrical cross-section.

Sweepings. Dust swept from the workbenches and floors of jewelers and other workers with precious metals, and heated for recovery of the metals contained in the dust.

Sweeps. See **Sweepings.**

Sweep Washings. Precious metals washed or recovered from **Sweepings** (see).

Sweetaloy. Series of high-chrome and nickel-chrome steels of the corrosion-resistant and stainless type, manufactured by Cooper Alloy Foundry Co. Chromium varies from 10% to 28%, and nickel is optional up to as high as 65%; up to 3% of molybdenum is also optional.

Sweet Roasting. See **Dead Roasting.**

Swinburne-Ashcroft Process. Treating sulfide ores with chlorine gas in a closed shaft furnace at about 1200°F. The metallic values of the ore are converted to fused chlorides, while the sulfur is volatilized.

Swindell Furnace. Electric direct **Arc Furnace,** utilizing three-phase electric current, carried by three electrodes entering from the top; the electrodes are raised to clear the furnace when the latter is tilted for tapping.

Swinging Plate. Copper plate, amalgamated on the surface, hanging in a sluice to recover **Float Gold** (see).

Swording. Separating steel sheets, sticking together, by inserting an instrument, shaped like a broad, long knife, between the sheets.

Sycee Silver. Lumps of silver, frequently warranted, by stamp of a fiscal authority or chemist, formerly used in China for monetary purposes. The larger lumps are often shaped like a boat, and are called shoes.

Sylvan. Native **Tellurium** (see).

Sylvanite. Native gold-silver telluride, approximating $AuAgTe_4$ in chemical composition.

Symington Alloy. Series of ferrous casting alloys of high-alloy composition, manufactured by Symington-Gould Corp.

Synthetic Cast Iron. Cast iron made by melting together steel scrap with carbon, usually in an electric furnace.

System (of Crystallization). See **Crystal System.**

T

Ta. Chemical symbol for **Tantalum** (see).

Tack Plate Quality Strip Steel. Strip steel, ordered or sold on the basis of suitability for the manufacture of tacks.

Tack Welding. Preliminary **Welding** (see) at isolated points, to insure proper alignment of parts during welding.

Taggers. Light-gage (55–65 pounds per base box) **Tin Plate** (see).

Taggers' Tin Plate. See **Taggers.**

Tailing (Tailings). Residue or refuse from wet mining or metallurgical process; used occasionally also to designate furnace **Slags** (see).

Tail Mark. Surface defect on rolled metal, caused by marks on the rolls resulting from cold ends of metal.

Tails. See **Tailing (Tailings).**

Tainton Process. Electrolytic recovery of lead; lead sulfate, formed by roasting of the ore at about 950°F, is leached with hot saturated brine and then subjected to electrolysis with graphite anodes and rotating sheet-iron cathodes.

Take-Up Reel. See **Tension Reel.**

Talbot Furnace. See **Talbot Process.**

Talbot Process. Modification of the basic **Open-Hearth Process** (see) of steel manufacture, in which a tilting furnace is used, retaining a bath of purified metal at all times to speed up oxidation.

Talmi Gold. See **Abyssinian Gold.**

Tam. Titanium alloys manufactured by Titanium Alloy Mfg. Co.

Tambac. See **Tombac.**

Tampico Rolls. Combination of cloth-covered rolls and fiber brushes, for cleaning adhering oil from terne and tin plate by means of bran and sawdust.

Tandem Mill. Arrangement of rolling mills, in direct line, allowing the metal to pass from one set of rolls into the next.

Tank Steel. Steel ordered or sold on the basis of specifications governing suitability for constructions involving no particular stresses.

Tantalum (chemical symbol Ta). Element No. 73 of the periodic system; atomic weight 180.88. A steel-gray, hard metal; very ductile and malleable under special treatment, highly resistant to almost all acids, aqueous alkalies, and atmospheric corrosion; melting point 5480°F; boiling point, over 7400°F; specific gravity 16.6. Chemically, its valence is normally five (tantalic), and occasionally four (tantalous). The electrodeposition of tantalum has not been definitely achieved. The metal can be isolated in pure form by electrolysis of the fused double potassium fluoride. Used for electrodes in electrolytic rectifiers, acid-resisting chemical equipment, and as the carbide, in **Sintered Carbide Tools** (see). The metal is obtainable from Fansteel Metallurgical Corp.

Tantalum Carbide. Extremely hard compound of tantalum and carbon, chemically approximating TaC; used for **Cemented Carbide** (see) tools.

Tantiron. High-silicon (approximately 14%) cast iron, highly resistant to corrosion; manufactured by Bethlehem Foundry & Machine Co.

Tantung. Alloy of cobalt, chromium and tungsten containing tantalum carbide and columbium carbide (Fansteel Metallurgical Corp.).

Tap. Quantity of metal permitted to run out of a furnace or holding vessel at one time.

Tap Bar. Bar, pointed at one end, for opening the **Tapping Hole** (see) of a furnace.

Tap Cinder. Slag tapped from the furnace at the end of the process of wrought iron manufacture. See **Puddling.**

Tap Density. In powder metallurgy, the apparent density of a metal powder determined when the receptacle is tapped or vibrated during the period of loading.

Taper. See **Mold Taper.**

Tap Hole. See **Tapping Hole.**

Tapping. Removing molten metal or slag, by gravity, through an opening at or near the bottom of the apparatus.

Tapping Bar. See **Tap Bar.**

Tapping Clay. Clay used for closing up **Tapping Holes** (see) in a furnace.

Tapping Hole. Hole at or near the bottom of a furnace, for drawing off metal or slag.

Tappings. See **Tap Cinder.**

Tap Steel. Deep-hardening tool steel, used for the manufacture of taps and similar tools.

Tarnish. Surface discoloration on a metal, usually due to a film of oxide.

Tarnowitz Process. Heating lead ores by roasting relatively large charges at low temperatures, followed by remelting, to recover the metal content of the residue.

Tatham Furnace. Crucible furnace for recovering zinc from zinc residues.

Tatmo (Tool) Steel. High-carbon alloy tool steel (Latrobe Electric Steel Co.), with about 8% molybdenum, 4% chromium, 2% tungsten, and 1% vanadium.

Taylor Process. Drawing extremely fine wire, particularly from non-ductile metals. The molten metal is cast into a glass or quartz tube, and the composite drawn down, after heating to a temperature sufficient to plasticize the glass or quartz.

Taylor-White Process. Method of hardening tool steel of the high-speed type (see **Taylor-White Steel**), in which the metal is heated to white heat, cooled first to 1300–1560°F in a molten lead bath, and then to room temperature in oil, followed by reheating to 700–1240°F and air-cooling.

Taylor-White Steel. Tool steel, with about 8% tungsten, 3% chromium and 1% carbon; the first of the high-speed tool steels.

Taylor Wire. See **Taylor Process.**

Tb. Chemical symbol for **Terbium** (see).

Te. Chemical symbol for **Tellurium** (see).

Tea Lead. Lead alloyed with about 2% tin, used in foil form for wrapping tea.

Tedge. See **Gate.**

Teeming. Pouring metal from the bottom of the container, such as a ladle.

Teeming Hole. See **Bessemer Steel.**

Teeming Ladle. Ladle for holding molten metal, prior to casting into molds.

Telegraph Wire. See **Telephone and Telegraph Wire.**

Telephone and Telegraph Wire. Steel wire of relatively low electrical resistance for use in communications work.

Telloy. Finely ground tellurium, used as vulcanizing agent in rubber; manufactured by R. T. Vanderbilt Co.

Telluric Bismuth. Semi-metallic mineral, tetradymite, a sulfo-telluride of bismuth, chemically approximating $Bi_2(TeS)_3$. Occasionally, copper, gold, and iron are also present.

Telluric Silver. Mineral alloy, hessite, chemically approximating silver telluride, Ag_2Te.

Telluride. Compound of tellurium and metal; several metals, gold among them, are found in nature as tellurides.

Telluriferous. Containing tellurium.

Tellurium (chemical symbol Te). Element No. 52 of the periodic system; atomic weight 127.61; a member of the sulfur family. White, brittle metalloid of poor heat and electrical conductance; melting point of 846°F; boiling point 2534°F; specific gravity 6.24. Chemically, it is divalent, tetravalent, and hexavalent. Tellurium is obtained as a by-product of copper refining; when heating the slimes from the electrolytic refinery, tellurium volatilizes together with selenium, though a great deal of the tellurium enters the slag. After solution in acid and removal of the **Selenium** (see) by sulfur dioxide, dilution with water throws down the tellurium as a black powder. It is used to some extent as a strengthening addition to lead, and to copper to improve machinability. Some use of the metal is also made in finely divided form, as a vulcanizing agent in rubber.

Tellurium Bronze. Bronze (see) with about 1% tellurium, added to increase machinability.

Tellurium Copper. Copper to which approximately 1% of tellurium has been added, primarily to achieve easier machinability.

Tellurium Lead. Lead with fractional percentages of tellurium, giving the alloy higher strength and greater chemical resistance, particularly to sulfuric acid.

Tellurnickel. Nickel telluride, occurring native as a white, metallic mineral. Chemically, it approximates Ni_2Te_3.

Tempaloy. Special bronze, akin to **Ambraloy** (see), manufactured by American Brass Co.

Temper. Stiffness and springiness of a metal, developed by cold working. For each metal, a series of standard tempers has been accepted. For tool steels, the term temper refers to carbon content. The tempers are numbered from 7 to 14, each number representing about 0.10% of carbon content.

Temper Brittleness. Brittleness resulting in certain alloy steels from slow cooling following tempering, probably due to carbide separating at the grain boundaries. In chrome-nickel steels, a small percentage of molybdenum is frequently added to prevent this type of brittleness.

Temper Brittleness Range. Temperature range between 425°F and 600°F, in which it is inadvisable to temper steel susceptible to **Temper Brittleness** (see).

Temper Carbon. Free graphitic carbon, in the form of spheroidal aggregates of minute crystals, formed in cast iron by heating below the melting point, and also in high-carbon steels as a result of cementite decomposition.

Temper Colors. Series of colors, ranging from light yellow at about 428°F, to dark blue at about 600° F, resulting from the low-temperature tempering of steel. Can be used as a rough indication of the temperature of tempering, although the presence of alloying elements changes the color materially.

Tempered Lead. Lead hardened by alkaline-earth metals and/or lithium; usually made by electrodepositing the alkaline-earth metal into a fused lead cathode.

Tempered Steel. Steel subjected to tempering. See **Tempering** (a).

Temper Graphite. See **Temper Carbon.**

Temper Hardening. Increasing the hardness of hardened high-alloy steels by tempering the heat-treated steel in the 930–1110°F range, the increase in hardness being known as secondary hardness. See **Tempering** (a).

Tempering. (a) Process of **Annealing** (see) in which the hardened metal is heated, a second time, to a temperature below the critical and then cooled. With steel, tempering involves decomposition of austenite or martensite; if carried far enough, the carbide will be precipitated. (b) In metal casting technique, the addition of water to molding sand mixtures. (c) See also **Temper.**

Tempering Furnace. Furnace used for **Tempering** (see) operations.

Tempering Machine. Equipment for the uniform tempering of large steel plates, in a bath, while pressed between heated metal slabs to maintain shape and flatness. See **Tempering** (a).

Tempering Oil. Quenching Oil (see) usable, without decomposition, up to about 525°F.

Tempering Test. Test on a material, most commonly steel, indicating characteristics on tempering [see **Tempering** (a)] under standard conditions.

Temper Mill. See **Temper Rolling.**

Temper Number. Degree of hardness developed by cold working (see **Temper**).

Temper Number 1. See **Hard Temper.**

Temper Number 2. See **Half-Hard Temper.**

Temper Number 3. See **Quarter-Hard Temper.**

Temper Number 4. See **Soft Skin-Rolled Temper.**

Temper Pass. See **Temper Rolling.**

Temper Rolling. Subjecting metal sheet to **Cold-Rolling** (see), to stiffen the sheet, without significant reduction in thickness.

Temper Test. See **Tempering Test.**

Tempilaq. Suspension, in an organic solvent, of a temperature-sensitive material, changing in color on attaining a specified temperature range.

Tempil Pellet. Mixture of chemicals, in tablet form, having a predetermined melting point; used to indicate attainment of a certain temperature. A series with melting temperatures ranging from 750 to 1600°F is available.

Tempilstick. Crayon for transferring a mark, containing a mixture of chemicals of a predetermined melting point, to indicate temperature. A series ranging in melting point from 125 to 700°F is available.

Temporary Hardness. Hardness in metal caused by heat treatment or by cold working, as opposed to **Permanent Hardness** (see).

Tenacity. See **Tensile Strength.**

Tenor. Metallic content, e.g., of an ore or semi-processed metal.

Tenorite. Non-metallic mineral, chemically approximating cupric oxide, CuO.

Tensile Strength. Breaking strength of a material when subjected to a tensile (stretching) force. Usually measured by placing a standard test piece in the jaws of a tensile machine, gradually separating the jaws, and measuring the stretching force necessary to break the test piece. Tensile strength is commonly expressed as pounds (or tons) per square inch, or, in the metric system, as kilograms per square millimeter.

Tensile Strength Load. See **Tensile Strength.**

Tensile Stress. See **Tensile Strength.**

Tensile Test. See **Tensile Strength.**

Tensilite. Manganese Bronze (see) of approximately 21–29% zinc content, with about 3% manganese, 4% aluminum, and up to 2% iron (American Manganese Bronze Co.).

Tension Reel. Device, in the manipulation of continuous metal, such as strip, etc., for maintaining constant tension on the metal.

Tension Test. See **Tensile Test.**

Terbium (chemical symbol Tb). Element No. 65 of the periodic system; atomic weight 159.2. Rare earth metal isolated by electrolysis of the trichloride in a bath of fused alkali chlorides, but not sufficiently pure for the determination of physical constants. Chemically, it is trivalent. Like all rare earth metals, it cannot be electrodeposited.

Ternary (Alloy). Alloy containing three elements, exclusive of impurities.

Ternary Diagram. Constitution Diagram (see) for a **Ternary Alloy** (see) system.

Ternary Steel. Steel containing one major alloying element in addition to carbon.

Ternary System. See **Ternary Diagram.**

Terne. Lead-tin alloy, used in the manufacture of **Terne Plate** (see).
Terne Coat. See **Terne Plate.**
Terne Plate. Sheet steel, coated with a lead-tin alloy. The percentage of tin is usually kept as low as possible because of its high cost; however, about 15% is normally necessary in order to obtain proper coating of the steel, since pure lead does not alloy with iron and some surface alloying appears necessary for proper adhesion.
Test Coupon. See **Coupon.**
Test to Destruction. Physical test on a material conducted until the point of breakage or destruction is reached.
Testing Machine. Instrument used for determining physical properties, such as tensile strength, fatigue, etc.
Test Lead. Relatively pure silver-free lead, used for assaying, cupelling, etc.
Test Piece. Metal, cut or mechanically shaped to specific dimensions, for use with an instrument for determining physical properties, such as tensile strength, etc.
Test Specimen. Material under test, normally of a standard shape, for subjection to physical test. See **Coupon.**
Tetradymite. See **Telluric Bismuth.**
Tetragonal (Crystal System). Form of crystal arrangement in which the three axes are perpendicular to one another, two being equal in length.
Tetravalent. Having a chemical **Valence** (see) of four.
Tewel. See **Tuyere.**
Th. Chemical symbol for **Thorium** (see).
Thallium (chemical symbol Tl). Element No. 81 of the periodic system; atomic weight 204.39. Very soft, blue-white metal of high malleability but little tensile strength, resembling lead; tarnishes in air and water; melting point about 577°F; boiling point 2655°F; specific gravity 11.85. Chemically, it is both monovalent (thallous) and trivalent (thallic). The metal is obtained from flue dust of sulfuric acid plants and the mud of the chambers, from which it is extracted with water and then precipitated by zinc, or electrodeposited after acidification with sulfuric acid. Not used commercially.
Thallium-Lead. See **Lead-Thallium.**
Therlo. Copper-base, electric-resistance alloy, with 10–13% manganese and 2–6% aluminum.
Therm. 100,000 **B.T.U.** (see).
Thermal Analysis. Determining equilibrium conditions and phase relationships in metals by means of heating and cooling curves.
Thermal Capacity. Heat required to raise the temperature of a body one degree centigrade.
Thermal Coefficient of Linear Expansion. Change of length per unit length for each degree of temperature change.
Thermal Conductivity. Capacity of substances to pass heat mechanically. Metals normally have high thermal conductivity.
Thermal Convection. Transmission of heat through a fluid, liquid or gas, by means of fluid currents.
Thermal Critical Point. See **Critical Point.**
Thermal Diagram. See **Constitution Diagram.**
Thermal Expansion. See **Thermal Coefficient of Linear Expansion.**
Thermal Flaw. See **Internal Crack.**
Thermalloy. Series of nickel-chromium-iron alloys, with 2–65% nickel, 8–30% chromium, about 1% silicon and 1% manganese; iron represents the balance (Electro Alloys Co.).
Thermal Radiation. Transmission of heat by means of ether waves, akin to light waves, but of greater wave length.
Thermal Shock. Sudden change in temperature (usually downward).
Thermal Transformation. Transformation (see) occurring on heating.
Thermal Transmission. Passage of heat through a body, as a result of the combined effects of conductivity, convection, and radiation. See **Thermal Conductivity, Thermal Convection,** and **Thermal Radiation.**
Thermisilid. High-silicon cast iron of foreign manufacture, containing about 15% silicon; resistant to chemical corrosion; used for chemical equipment.
Thermit. Mixture of iron oxide and aluminum; when ignited, reacts vigorously, to yield molten iron. By analogy,

reduction of any metal oxide by aluminum is frequently called thermit reduction.

Thermite. See **Thermit.**

Thermit Process. See **Thermit Welding.**

Thermit Welding. Use of **Thermit** (see) mixture, aluminum and iron oxide, together with any desired alloying agents, as a welding agent, which when ignited, provides both the necessary heat and the molten iron to act as filler metal.

Thermochemical Equation. Equation which shows, in addition to the chemical change, the heat-energy changes involved.

Thermocouple. See **Pyrometer.**

Thermocouple Metal. Metal which, in combination with another, gives a thermocouple. See **Pyrometer.**

Thermodynamics. Science of the relationships of heat and energy. See also **First Law of Thermodynamics, Second Law of Thermodynamics,** and **Third Law of Thermodynamics.**

Thermoelectric Element. See **Thermocouple Metal.**

Thermoelectricity. Voltage or electric potential difference at the junction of two metals kept at a different temperature from that at the point of electric measurement (see **Pyrometer**).

Thermoelectric Metal. See **Thermocouple Metal.**

Thermoelectric Method. Determining the **Critical Point** (see) in iron and steel, based on plotting changes in electrical properties with temperature changes.

Thermoelectric Pyrometer. See **Pyrometer.**

Thermometal. See **Bimetal.**

Thermostatic Metals. See **Bimetal.**

Thies Process. Treating gold-silver ores with a mixture of chloride of lime and sulfuric acid, to obtain a solution of gold chloride, while the silver precipitates as insoluble silver chloride.

Thin Case. Case (see) on metal, with a total penetration less than 0.01 inch.

Third Law of Thermodynamics. Every substance has a positive entropy, which may reach zero at absolute zero temperature in the case of perfect crystalline substances.

Thomas Basic Pig (Iron). See **Basic Pig (Iron).**

Thomas-Gilchrist Process. Process, involving the use of a basic lining and limestone addition in the Bessemer converter, to remove phosphorus from pig iron. See **Bessemer Steel.**

Thomas Process. See **Thomas-Gilchrist Process.**

Thomas Slag. See **Basic Phosphate Slag.**

Thomas Steel. Basic **Bessemer Steel** (see).

Thompson Process. Electric resistance **Welding** (see) process.

Thorium (chemical symbol Th). Element No. 90 of the periodic system; atomic weight 232.12; slightly radioactive, dark gray metal, of moderate hardness and high degree of ductility and malleability; melting point 3348°F; boiling point over 5500°F; specific gravity 12.1; immune to atmospheric corrosion and to alkalies, but slowly attacked by acids. Chemically, it is always tetravalent. Thorium cannot be electrodeposited; isolated, in the form of metal spangles, by electrolysis of anhydrous thorium chloride in a bath of fused sodium chloride. Not commercialized, though alloys with tungsten have been studied for filament use.

Thread Stripping Test. Test on metal, most commonly steel, in which the metal is tapped, or cut, to a standard thread, and the tensile strength of such thread determined.

Three-High Mill. Rolling mill, consisting of three vertically-positioned rolls.

Three-High Train. See **Three-High Mill.**

Three-Pass Stove. Stove in **Blast Furnace** (see) in which the gas from the combustion chamber passes through two regenerative flues.

Through. See **Minus Mesh.**

Throwing Power. Ability of an electroplating solution to deposit metal at points on the cathode more distant from the anode than others. The throwing power of different electroplating solutions varies considerably.

Thulium (chemical symbol Tm). Element No. 69 of the periodic system; atomic weight 169.4. Rare earth metal, which has never been isolated, and, therefore, its physical constants are unknown. Chemically, it is trivalent. Like all rare earth metals, it cannot be electrodeposited.

Thum-Balbach Process. Electrolytic refining of crude silver, using, as electrolyte, a solution of silver nitrate and nitric acid, together with carbon cathodes. The deposited silver does not adhere to the cathodes but falls to the bottom of the electrolytic bath and is periodically scraped out.

Thum Furnace. Gas-fired furnace, used in the smelting of zinc-lead ores.

Thuriting. Heat-treating steel, to attain deep-drawing characteristics; the metal is heated rapidly to about 1790°F, followed by rapid cooling and box annealing at about 1200°F.

Thwing Pyrometer. Radiation **Pyrometer** (see); radiation is permitted to fall on a conical mirror which concentrates the rays on the hot junction of a small thermocouple; the resulting electromotive force, converted to temperature, is read on a millivoltmeter.

Ti. Chemical symbol for **Titanium** (see).

Tiemannite. Naturally-occurring mineral, chemically approximating mercuric selenide, HgSe.

Tiers-Argent. Alloy, used principally for ornamental purposes, consisting of two parts aluminum and one part silver.

Tight Coat. Metal coating, such as galvanizing, relatively free from defects and discontinuities.

Tight Coating. See **Tight Coat.**

Tile Copper. Copper, frequently cast as flat rectangular plates, obtained by re-smelting copper **Bottoms** (see).

Tilt Hammer. Obsolete form of power hammer, in which a metal hammer head was held at the end of a wooden beam, raised by a mechanical cam and falling by gravity.

Tilting. Forging by means of a **Tilt Hammer** (see).

Tilting Converter. Bessemer converter (see **Bessemer Steel**) which can be rotated in a vertical plane on its shorter axis, for easier charging and discharging. All modern converters are of this type.

Tilting Table. Table, for holding metal during and after rolling, which can be tilted so as to permit one or both ends to be raised or lowered, for better manipulation of the metal.

Timang. Steel of high manganese (12–15%) content, with about 3% nickel and about 0.70% carbon (Taylor-Wharton Iron & Steel Co.).

Time Quenching. See **Interrupted Quenching.**

Time-Temperature Cycle. Standard sequence of heat-treatment steps, at varying temperatures and for varying periods of time.

Time Yield Stress. Initial creep stress. See **Creep** (a).

Time Yield Stress Test. See **Initial Creep Stress Test.**

Timken Steel. Series of low-alloy and high-alloy steels, most commonly with carbon below 0.20% (Timken Steel & Tube Co.).

Tin (chemical symbol Sn). Element No. 50 of the periodic system; atomic weight 118.70. Soft, crystalline, silvery-white metal of high malleability and ductility, but low tensile strength; melting point 449° F; boiling point 4384° F, yielding the longest molten-state range for any common metal; specific gravity 7.28; can readily be deposited from stannous sulfate and sodium stannate baths. Tin is both divalent (stannous) and tetravalent (stannic). Because of its existence in allotropic forms, the metal is subject to **Tin Pest** (see). Obtained by smelting cassiterite, stannic oxide; the quality of the metal depends a good deal on the locality from which the ore is obtained. Since it is almost completely immune to the action of air or water, it is used for piping, to convey non-corrosive liquids. It finds much greater outlet as a coating on steel, correctly called tin plate. In alloy form, it is extensively used in solders, terne plate, bronze, white metals for bearings, type-casting alloys, etc. Note that what is popularly called "tin" is really **Tin-Plate** (see).

Tin Bar. Low-carbon steel bar, usually of basic open-hearth or Bessemer process manufacture, about eight inches wide, used for rolling into **Black Plate** (see).

Tin Bath. (a) See **Tin Pot.** (b) See **Tin Plating.**

Tin Coating. See **Tinning.**

Tin Cry. Peculiar crackling sound heard when pure tin is bent. Assumed to be due to the sliding of layers of crystals over one another.

Tin Dross. Dross (see), formed in molten tin baths used in the coating of steel.

Tin Flower. Large spangles on galvanized steel (see **Galvanizing**), resulting from the presence of tin, or tin and cadmium, in the molten zinc bath.

Tin Foil. Tin in **Foil** (see) form.

Ting. See **Sycee-Silver.**

Tin Glass. See **Bismuth.**

Tin House. Building or room for **Tin Plating** (see).

Tinicosil. Copper-zinc hardware and corrosion-resisting alloy consisting of about equal parts of copper and zinc, modified with 10–15% nickel; small percentages of lead are optional (Titan Metal Mfg. Co.).

Tinman's Solder. Soft **Solder** (see) consisting of approximately two parts tin and one part lead.

Tin Mill. Plant for manufacturing **Tin Plate** (see) and **Terne Plate** (see).

Tin-Mill Black Plate. Non-coated sheet steel less than 0.014 inches thick and not more than 32 inches wide.

Tinned Armature Bending Wire. High-carbon steel wire of high tensile and elongation characteristics, tin coated, with good surface finish.

Tinned Rope. Rope made of **Tinned Wire** (see).

Tinned Sheet Iron. See **Tin Plate.**

Tinned Wire. Steel wire, with a thin coating of tin, usually obtained by passing the wire, in strand form, through molten tin.

Tinning. Coating with tin, most commonly by immersion into molten tin; electrodeposition and metal spraying are also used.

Tinning Flux. See **Flux Process.**

Tinning Machine. See **Tinning Stack.**

Tinning Metal. Tin-lead alloy (about equal parts of each), used in preparing electrotyping shells for backing.

Tinning Pot. See **Tin Pot.**

Tinning Stack. Equipment for continuous **Tinning** (see) of steel. Normally consists of a tinning pot to hold the tin, a feeding device for the steel, a tinning machine for carrying the steel through the molten tin, washing devices, and a polishing machine.

Tin Pest. Tendency of tin cooled to low temperatures, particularly below 6.4°F, to crumble into a powder known as gray tin, an allotropic modification also called alpha tin; unaffected white tin metal will also be transformed into gray powder when contacted with alpha tin.

Tin Pickling. Pickling (see) sheet iron or steel prior to tinning (see also **Tin Plate**).

Tin Pickling Machine. See **Pickling Machine.**

Tin Plate. Sheet iron or steel coated with tin, usually by dipping into molten tin, rather than by electroplating, despite the name. Extensively used for tin cans and similar containers. It should be noted that the term "tin," as commonly used in the non-technical sense, refers to tin plate and not to pure tin.

Tin Plate Base Box. See **Base Box.**

Tin Plate Gage. See **Metal Gage.**

Tin Plating. Coating metal objects with tin; the object to be coated is made cathode (negative electrode) in an electrolytic bath containing a decomposable tin salt. Acid tin baths contain stannous chloride or more complex salts, such as the fluoborate and fluosilicate, whereas alkaline baths usually contain the chloride, together with free alkali. Tin metal is normally used as anode (positive electrode).

Tin Pot. Container for holding molten tin in the **Tinning** (see) process.

Tin Pyrites. See **Pyrites.**

Tin Skimming. Material, consisting principally of lead-tin-antimony-oxygen compounds, collecting on the surface of molten lead during the oxidation step of lead **Softening** (see).

Tinstone. See **Cassiterite.**
Tin White Cobalt. See **Gray Cobalt.**
Tin Yellow. Yellowish color in the zinc coating of galvanized steel (see **Galvanizing**), caused by the presence of too high a percentage of tin in the molten zinc bath.
T Iron. Steel or iron, with a cross-section in the shape of a T.
Tisco Steel. Steels of diverse compositions, manufactured by Taylor-Wharton Iron & Steel Co.
Tiska. Series of **Stainless Steels** (see) of up to 22% chromium and 10% nickel content, manufactured by Taylor-Wharton Iron & Steel Co.
Tissier's Metal. High-copper **Brass** (see), containing about 97% copper, 2% zinc, and 1% arsenic.
Titan. See **Titanium.**
Titan Bronze. Series of modified yellow **Brasses** (see) of about 40% zinc and up to 1% tin content (Titan Metal Mfg. Co.).
Titanic Iron Ore. Non-metallic ore of iron and titanium, ilmenite, chemically approximating $FeTiO_3$.
Titaniferous. Containing titanium.
Titanium (chemical symbol Ti). Element No. 22 of the periodic system; atomic weight 47.90; melting point about 3270° F; boiling point over 5430°F; specific gravity 4.5. Bright white metal, very malleable and ductile when exceedingly pure, as obtained by thermal decomposition of the tetraiodide on a tungsten filament. More simply, it is made by interaction of the liquid tetrachloride with sodium in a bomb; resistant to atmospheric corrosion, but is slowly affected by dilute acids. Chemically, it is most commonly tetravalent (titanic), but also trivalent (titanous) and divalent in some compounds; it cannot be electrodeposited. Pure titanium is not used commercially, but as a ferrous alloy it is used as a scavenger, since it reacts vigorously with nitrogen and oxygen, and as a stabilizer in stainless steel. It also enters, in larger percentages, in some alloy steels. In copper-base alloys, the metal acts to give strong bronzes. Titanium can be obtained from Titanium Alloy Mfg. Co.
Titanium Carbide. Compound of approximately 75% titanium and 25% carbon; extremely hard, though not as hard as tungsten carbide, and used for similar purposes. See **Cemented Carbide.**
Titanium-Copper. See **Copper-Titanium.**
Titanium-Manganese. See **Manganese-Titanium.**
Titanium-Nickel. See **Nickel-Titanium.**
Titanium Pig. Pig Iron (see), with fractional percentages of titanium, made from ores containing titanium.
Titanium Steel. Steel, containing substantial quantities of titanium as the principal alloying agent. Up to 1% titanium may be added to toughen and strengthen the steel, but titanium is much more common as a **Stabilizer** (see) in stainless steels.
Tl. Chemical symbol for **Thallium** (see).
Tm. Chemical symbol for **Thulium** (see).
T-Max. Maximum temperature to which steel is heated prior to quenching or normalizing.
Toad's Eye Tin. Massive **Cassiterite** (see).
Tobin Bronze. Modified **Brass** (see), manufactured by American Brass Co., with approximately 60% of copper, 1% of tin, traces of lead and iron, with zinc representing the balance; shows high strength, toughness, and resistance to corrosion, has excellent hot-working and fair cold-working properties; extensively used as a filler rod in the torch welding of cast iron, steel, and copper-base alloys.
Tocco Process. Form of **Induction Hardening** (see), used for the surface hardening of steel; particularly valuable for parts of complicated shape and for brazing operations.
Toledo Alloy (Steel). Series of tough and shock-resistant alloy steels, of varying composition, manufactured by Unicast Corp.
Tolerance. Allowable deviation from nominal measurements or compositions.
Tomassi Process. Electrolytic refining of lead, using crude argentiferous lead

anodes, large discs of copper, about half immersed in the electrolyte, as cathodes, and an electrolyte consisting of a solution of a double alkali lead acetate. Lead crystals are periodically scraped off the copper cathodes.

Tombac. Series of copper-base alloys, copper being 70–92%, with the remainder essentially zinc. Tin, up to 3%, is optional. See **Brass.**

Tombak. See **Tombac.**

Tombasil. Brass (see), containing about 14% zinc and 5% silicon; used for wear-resisting products (Ajax Metal Co.).

Toncan Iron. High grade iron, with carbon at about 0.03%, containing about 0.5% copper and 0.07% molybdenum; manufactured by Republic Steel Corp. Because of its high purity and the presence of the alloying elements, the metal has a relatively high atmospheric corrosion resistance.

Tongue. See **Former.**

Tongue and Groove Pass. Closed Box Pass (see) in which a groove is cut in one roll, enclosing one side and the edges of the metal being rolled, with a tongue on the opposite roll closing the pass from the other side.

Tool Endurance. See **Cutting Test.**

Tool Steel. Steel intended for manufacturing cutting tools and similar products. Medium and high-carbon steels are used for the cheaper tools; higher types are served by the alloy steels. The better tool steels are made by the electric furnace or crucible process.

Top. Top sheet in pile, in pack mill operations. See **Pack Rolling.**

Top Casting. Casting metal into a mold through the top of the mold. See **Casting** (a).

Top End. End of steel plate nearest the upper part of the original ingot.

Tophet. Series of chromium-nickel alloys, optionally containing iron, manufactured by W. B. Driver Co. Carbon and silicon are present in small or fractional percentages; because of high electrical and high-temperature corrosion resistance, used for electric heating wire and similar products.

Topping. Breaking off pieces from the top of an ingot of **Crucible Steel** (see) until the fracture shows complete soundness in the metal.

Top Pouring. See **Top Casting.**

Torch. Device for creating sharply-localized flame, by bringing jets of combustible gas and of air or oxygen to a common point.

Torch Brazing. See **Gas Brazing.**

Torch Hardening. See **Flame Hardening.**

Torch Welding. See **Gas Welding.**

Torque. See **Torsion.**

Torsion. Strain created in a material by twisting action; correspondingly, the stress within the material resisting the twisting.

Torsional Strength. Maximum shear strength which a material will withstand in torsion.

Torsional Stress. Any stress due to twisting.

Torsion Test. Determination of the resistance of a material in shear, or to torsional stress. See **Torsion.**

Tossing. Purifying tin by pouring the metal from a ladle into a kettle, so as to expose thin sheets of the liquid metal to air, forming oxide dross. See **Dross.**

Total Calorific Value. See **Gross Calorific Value.**

Total Carbon. Total content of carbon in cast iron, including free carbon, present as graphite, and combined carbon, chemically combined with the iron or alloying constituents.

Total Discard. Proportion, by weight, of metal discarded because of flaws (pipe, etc.) (see **Crop**) to the total weight of the original ingot.

Total Heating Value. See **Gross Calorific Value.**

Touch Needle. Needle (or narrow strip) of gold or gold alloy of specific composition, for use as a basis of comparison in judging other gold alloys, by means of comparative marks made on a **Touchstone.** (see)

Touchstone. Hard stone on which gold or gold alloys are tested by comparing streaks left on the stone with streaks made by **Touch Needles** (see) of known composition.

Tough. (a) See **Tough Cake.** (b) See **Toughness.**

Tough Cake. Refined, commercial **Copper** (see).

Toughening. (a) Procedure of heating medium-carbon steels of about 0.35–0.60% carbon content to the 1425–1560°F range, quenching in oil or water, and then drawing to a relatively high drawing temperature—approximately 840–1200° F. This treatment adds to the strength and toughness of the steel, but little quenching hardness remains after the drastic drawing. (b) Final refining step in copper smelting, most commonly **Poling** (see).

Toughness. Property of resisting fracture or distortion. Usually measured by impact test, high impact values indicating high toughness.

Tough Pitch Copper. Copper, containtaining fractional percentages of oxide (0.03–0.05% oxygen), remelted after electrolytic refining, or resulting from **Poling** (see).

Tozing. See **Tossing.**

T. P. G. See **Metal Gage.**

Trabuk. Tin-base alloy with about 5% antimony, 5% nickel, and 2% bismuth.

Trace. Extremely small quantity of an element; the term is frequently used in chemical analysis and in assaying to indicate the presence of elements in amounts too small to determine quantitatively.

Tracer Rolls. See **Dancer Rolls.**

Traction Test. See **Tensile Test.**

Trade Heat. First heats (about five or six in all) in breaking in a new basic open-hearth furnace, in which the charge consists of approximately equal parts of molten metal and of scrap steel.

Tramp Iron. Stray metallic iron found, as a contaminant and usually in relatively large masses, in non-metallic materials such as ores, foods, etc.

Trancor. High-Silicon Steel (see), used for magnetic and transformer purposes (American Rolling Mill Co.).

Transcrystalline Failure. See **Transcrystalline Fracture.**

Transcrystalline Fracture. Fracture in metal, not following crystal boundaries.

Transformation. Internal change in solid metal.

Transformation Range. Temperature range over which a chemical or metallurgical change takes place. See also **Critical Range.**

Transformation Temperature. See **Transformation Range.**

Transformer Steel. See **Electric Sheet.**

Transition Point. Temperature of transformation of one solid crystalline form of a substance to another. More broadly, the point where different **Phases** (see) can exist in equilibrium.

Transition Structure. Metallographic structure developed in a metal as a result of control of time and temperature while the metal is in its **Transformation Range** (see).

Transition Type System. See **Peritectic System.**

Transmission. See **Thermal Transmission.**

Transmutation. Conversion of one element into another, either by natural radioactivity or by artificial action. Alchemy was dedicated primarily to efforts at transmuting base metals into gold and silver.

Transuranic Metals. Hypothetical elements with atomic weight greater than that of uranium (238.07). Traces of such elements, in highly radioactive form, can be synthesized by bombarding heavy elements with neutrons.

Transverse Bend Test. Longitudinal bend made on a transverse test specimen, cut perpendicularly to the direction of rolling of the metal.

Transverse Fissure. See **Progressive Fracture.**

Transverse Specimen. Specimen, for test purposes, with length of the specimen perpendicular to the direction of rolling.

Transverse Strength. Breaking strength, usually of brittle materials, as determined from a concentrated central stress in the material held as a simply supported beam.

Transverse Test. Test, commonly used on steel specimens, to indicate that the longitudinal axis is perpendicular to the direction of the greatest extension of the material from which the test specimen has been taken. See also **Longitudinal Test.**

Traveling Rack. Continuous belt conveyor, used for conveying terne plates, horizontally, from the coating bath, while the plates are air-cooled.

Traveling Table. Roll Table (see) which moves on tracks alongside the rolls.

Tray Annealing. See **Box Annealing.**

Tree. See **Lead Tree.**

Treeing. See **Lead Tree.**

Triclinic (Crystal System). Crystal arrangement in which all three axes are unequal, and none is perpendicular to the others.

Trip Hammer. See **Tilt Hammer.**

Triple Point. Temperature and pressure at which three phases of any substance exist in equilibrium.

Triplexing. See **Triplex Process.**

Triplex Process. Preparing steel by first treating pig iron in a Bessemer converter, followed by removal of phosphorus in an open-hearth furnace and, finally, by treatment of the steel in an electric furnace, under a reducing slag.

Triturating. Grinding to a fine powder under substantially dry conditions.

Trivalent. Having a chemical **Valence** (see) of three.

Troostite. Former name for fine **Pearlite** (see), resulting from decomposition of **Austenite** (see).

Tropenas Converter. Converter (see) in which the air enters from tuyeres on the side; the air stream is deflected onto the top of the bath when the converter is tilted, and the carbon is oxidized to carbon dioxide instead of carbon monoxide, as in the standard Bessemer process.

Troy Weight. System of weights, in contradistinction to the more common avoirdupois system, used in weighing precious metals.

True Annealing. See **Full Annealing.**

Truform (Steel). Non-Deforming Steel (see), with about 1.2% manganese, 0.50% chromium, 0.50% tungsten and 0.90% carbon (Jessop Steel Co.).

Trumpet Brass. Special **Brass** (see) with about 18% zinc and 1% tin, used in the manufacture of brass tubing.

Try Hole. Opening in the top of a **Blast Furnace** (see) through which the **Stock Indicator** (see) is passed in order to determine the height of the material in the furnace.

Tube Annealing. Continuous **Annealing** (see) used on steel wire, in which the metal passes continuously through externally-heated tubes.

Tubing. Pipe, most commonly of the seamless type made from a solid billet by **Piercing** (see).

Tuc-Tur. Copper-base alloy with 21–29% zinc and, optionally, up to 15% nickel and up to 18% tin.

Tuf-Stuf. Copper-base alloy, with 8–14% aluminum, and 2–4% iron (Mueller Brass Co.).

Tula. Silver, containing small amounts of copper and lead.

Tumbling. Cleaning articles by rotating in a cylinder with cleaning materials, usually also with steel pieces to increase the cleaning action.

Tumbling Barrel. See **Tumbling.**

Tumbling Box. See **Tumbling.**

Tungo Steel. Low-alloy steel (Colonial Steel Co.) containing approximately 2% tungsten, 1.5% chromium, 0.25% vanadium and 0.50% carbon; used for tools and as hot-die steel.

Tungsten (chemical symbol W). Element No. 74 of the periodic system; atomic weight 183.92. Gray metal of high tensile strength, ductile and malleable when specially handled; melting point 6098°F; boiling point 8540°F; specific gravity 19.3. It is immune to atmospheric influences and most acids, but not to strong alkalies. Chemically it shows valences from two to six. It can be obtained pure by the reduction of the trioxide with hydrogen and the resulting powder is used for making filaments and other products by powder metallurgy technique. Electrodeposition is quite difficult and requires codeposition of another metal, like iron; not much used commercially. The metal is used as filament and in thin sheet form in incandescent bulbs and radio tubes. Tungsten carbide is the most important cemented carbide. Important alloying constituent in many high-speed steels. Tungsten is obtainable, in fabricated form, from Fansteel Metallurgical Corp. and from General Electric Co.

Tungsten Carbide. Compound of tungsten and carbon, of composition varying between WC and W_2C; imbedded in a matrix of soft metal, such as cobalt, extensively used for **Sintered Carbide Tools** (see).

Tungsten Cast Iron. Cast iron containing up to 2% of tungsten.

Tungsten Chisel Steel. See **Chisel Steel.**

Tungsten-Chromium Steel. See **Chromium-Tungsten Steel.**

Tungsten Magnet (Steel). Steel, for use in the manufacture of magnets and electrical equipment, with tungsten approximating 5%. About 0.5% chromium and manganese are usually present, and carbon ranges normally from about 0.70% to about 1%.

Tungsten Steel. Steel, usually made by the electric furnace or crucible process, containing up to 20% tungsten as alloying constituent, generally with chromium as an additional element in percentages up to 5%, and optionally, with vanadium, cobalt, and molybdenum. Tungsten steels are extensively used for tools and dies; one of the most common high-speed tool steels is the 18% tungsten, 4% chromium, 1% vanadium type.

Tunnel Furnace. Continuous Furnace (see) in which the material moves on cars pushed or pulled slowly through the furnace.

Tunnel Kiln Annealing Furnace. Furnace for continuous box annealing of metal products, particularly steel. Consists of a long tunnel through which the metal moves on small cars; as a loaded car is pushed into the tunnel at the entrance, a completed car comes out at the other end.

Tup. See **Drop Forging.**

Turbadium Bronze. Copper-base alloy, with 41–46% zinc, about 2% manganese, 2% nickel, and optional fractional percentages of aluminum, lead, and tin.

Turbine Blade Steel. Low-carbon (up to 0.10%) **Stainless Steel** (see), most commonly of the 12% chromium type.

Turbiston's Bronze. See **Turbadium Bronze.**

Turk's Head. Frame with four rolls so arranged in the same plane as to form a shaped hole through which tubing can be drawn to special section shapes.

Turner (Hardness) Test. Early hardness test, in which a diamond point, under varying load, was used to create a visible scratch, and the required load served as a measure of the hardness.

Turner Impact Test. Method of measuring resistance of a material to impact, in which a falling hammer is used, with heights of drop varied, to cause bending or fracture.

Tutania. Tin-base alloy, used for bearings and household ware, with up to 16% antimony, about 3% copper, 8% lead, and 1% zinc.

Tutenag. See **Spelter.**

Tutty. Crude zinc oxide, obtained from the flues of zinc smelting furnaces.

Tuyere. Pipe, or similar apparatus, passing through a furnace wall to deliver air blast into the furnace. See **Blast Furnace** and **Cupola.**

Tuyere Arch. Arch, in a furnace, such as a blast furnace, for passage of a **Tuyere** (see).

Tuyere Cap. See **Wicket.**

Tuyere Pipe. See **Tuyere.**

Tuyere Plate. Metal plate in the side of a furnace or forge, to admit a **Tuyere** (see). See also **Bloomery.**

Tuyere Stock. Pipe, on the outside of a **Blast Furnace** (see), for carrying hot blast to the blowpipe, and, through the latter, to the tuyeres. Also known as leg pipe, boot leg, and pen stock.

Tweer. See **Tuyere.**

Twere. See **Tuyere.**

Twin Crystals. Structure in which the crystal forms appear as mirror images of each other across the interface, called the twinning plane.

Twin Heddle Wire. Tinned steel wire composite, consisting of two parallel strands of wire, soldered together by a tin coating covering the wires.

Twinned Crystals. See **Twin Crystals.**

Twinning. See **Twin Crystals.**

Twinning Axis. Axis around which one part of a twin crystal may be viewed as having been rotated, 180°, relative to the other part.

Twinning Plane. See **Twin Crystals.**

Twins. See **Twin Crystals.**

Twist Drill Gage. See **Metal Gage.**

Twisting Moment. See **Torque.**

Two-High Mill. Two-roll rolling mill in which one roll is above the other. Such mills can either be continuous, i.e., always operating in one direction, or reversing.

Two-Minute Wire. Galvanized wire which will withstand, for two minutes, immersion in a neutral copper sulfate solution.

Two-Pass Stove. Most common type of **Blast Furnace** (see) stove, in which the combustion gases pass through one regenerative flue.

Twyer. See **Tuyere.**

Tymp. Hollow water-cooled casting, usually cast iron, protecting the arch over the dam, called the tym-arch, of a blast furnace having a fore-hearth.

Tymp-Arch. See **Tymp.**

Type. Referring to steel, type indicates the general group of composition and metallurgical characteristics to which the metal belongs.

Type Metal. Series of lead-tin-antimony alloys, with tin at 3–17% and antimony 4–23%. Copper may also be present. For detailed composition, see also **Electrotype Metal, Linotype Metal, Stereotype Metal,** and **Intertype Metal.**

U

U. Chemical symbol for **Uranium** (see).

Udylite. Process for cadmium plating, controlled by Udylite Co.

Ultimate Strength. See **Tensile Strength.**

Ultimate Tensile Strength. See **Tensile Strength.**

Ultracut (Steel). High-sulfur, free-cutting steel, manufactured by Bliss & Laughlin, Inc.

Ultradie (Steel). Series of high-speed, alloy steels, with about 13% chromium, and fractional percentages of vanadium or molybdenum (Cyclops Steel Co.).

Umangite. Native mineral copper selenide, chemically approximating Cu_3Se_2.

U.M.A. Steel. Series of low-alloy steels (Republic Steel Corp.), with 0.50–1.10% chromium, about 0.70% manganese, and carbon up to 0.50%; used for axles, connecting rods, and products requiring hardening.

Umpire Analysis. Chemical or metallographic analysis by a neutral party, usually an outside expert, to resolve differences between differing analyses.

U. M. Plate. See **Universal Plate.**

Unassorted. Specification, in the ordering or selling of metal sheet or plate, permitting the inclusion of **Seconds** (see) with **Prime** (see) products.

Uncoiler. Device for holding coiled metal and aiding in feeding it into a processing or working apparatus.

Underfill. Metal being rolled, but inadequate in one or more directions to fill out the pass, is known as underfilled.

Underpoled Copper. See **Underpoling.**

Underpoling. Treatment of copper by **Poling** (see) for too short a period of time, thereby leaving too high a content of oxygen in the metal.

Undersize. Portion of crushed or ground material passing through a particular screen or sieve. See also **Minus Mesh.**

Unequal Draft. Defect in metal, rolled lighter on one edge than on the other, due to non-parallel rolls.

Uniformity. Consistency of grain structure and size in metal, and absence of abnormal segregation.

Uniloy Steel. Series of high-chromium steels (Cyclops Steel Co.), with 5–30% chromium, optionally nickel up to 12%, and fractional percentages of other elements. Carbon ranges from 0.07 to 0.60%. Many of the series are of the **Stainless Steel** (see) type.

Union Freecut. See **Freecut.**

Union Supercut. See **Supercut.**

Unit. Term used in economic evaluation of ores, equivalent to one per cent of a ton (or twenty pounds) of the constituent under consideration.

Unit Cell. Smallest unit showing the full symmetry of the crystal.

Unit Cubic. Unit cell in cubic crystals.

Unit Deformation. See **Unit Strain.**

United States Standard Gage. See **Metal Gage.**

United States Steel Wire Gage. See **Metal Gage.**

Unit Strain. Change in dimension per unit length.

Unit Stress. See **Fiber Stress.**

Univalent. See **Monovalent.**

Univan Steel. Low-alloy cast steel (Union Steel Casting Co.), with about 1.5% nickel, 1% manganese, and 0.10% vanadium; used for cast parts requiring high strength.

Universal Impact Testing Machine. Machine for making a variety of impact tests, usually including the **Charpy (Impact) Test** (see) and **Izod (Impact) Test** (see).

Universal Mill. Two-high or three-high **Rolling Mill** (see), with both horizontal and vertical rolls; frequently, the vertical rolls are doubled, with one pair on each side of the horizontal rolls.

Universal Mill Beam. Steel beam [see **Beam** (b)] rolled between horizontal and vertical rolls.

Universal Mill Plate. See **Universal Plate.**

Universal Plate. Steel **Plate** [see **Plate** (a)] rolled between straight horizontal and vertical rolls, followed by trimming of the ends only.

Universal Train. See **Universal Mill.**

Unshielded Arc Welding. Arc welding in which the arc is open, i.e., not shielded. See **Shielded Arc Welding.**

Unsoundness. Presence of **Blowholes** (see) or similar imperfections in ingot metal.

Unstable Compound System. See **Peritectic System.**

Unstable Equilibrium. See **Metastability.**

U Pipe Stove. Recuperative furnace (see **Recuperator**) having a series of air or fuel passageways or spent gas outlets in U form.

Upper Inwall. See **Blast Furnace.**

Upper Punch. In powder metallurgy, the upper unit in a die assembly, which moves into the die and transmits pressure to the powder in the die cavity.

Upset Butt Welding. Resistance Butt Welding (see) in which electricity is applied after contact of the materials being welded, so that heat develops mainly through resistance, as contrasted with **Flash Butt Welding** (see).

Upset Forging. Deforming steel while hot, in an upsetting forging machine, by holding the metal firmly and subjecting it to pressure in the direction of the axis of the bar, so as to enlarge the end subjected to pressure.

Upset Test. Test used on steel, wherein a piece of the metal is heated and then upset under a hammer, in order to detect seams, laps, etc.

Upsetting. Process of forging in which the length of a bar is reduced by hammering in the direction of the length, thereby simultaneously increasing the diameter. Rivets, valve heads, etc., are generally manufactured by upsetting technique.

Uptake. See **Bulkhead** and **Offtake.**

Upton-Lewis Endurance Machine. Machine for testing endurance of flat metal specimens which are held between upper and lower halves of a lever, pivoted near the top, with the lower end vibrating at a preselected rate.

Uranium (chemical symbol U). Element No. 92 (the highest known) of the periodic system; atomic weight 238.07; grayish-white, slightly radioactive and malleable metal; softer than steel; melting point about 3075°F; boiling point about 6330°F; specific gravity 18.68; immune to atmospheric influences, readily reacts with acids but not with alkalies. Cannot be electrodeposited. Chemically, it exhibits valences from three to six, hexavalence and tetravalence being most typical. The metal can be isolated by electrolysis of the anhydrous tetrachloride in a bath of fused mixed alkali chlorides. The metal is not used commercially; formerly, it found some utilization as a substitute for tungsten in high-speed steels.

Uranium Steel. Steel containing uranium as a substitute for vanadium and tungsten; not commercialized.

Ure's Process. Recovering mercury from mercury ores by roasting with lime in iron retorts.
U. S. S. G. See **Metal Gage.**
U.S.S. (Steel). Trade-mark designating steels manufactured by United States Steel Corp. and their subsidiaries.
Utica Steel. Low-alloy steel (Ludlum Steel Co.), used for small tools, such as taps, drills, etc. In addition to about 1.3% carbon, it contains about 1.5% tungsten and fractional percentages of chromium and vanadium.

V

V. Chemical symbol for **Vanadium** (see).
Vac-Melt. Series of nickel-base, electric-resistance alloys, with 15–20% chromium, up to 7% molybdenum, up to 3% manganese, and up to 16% iron (George W. Prentiss & Co.).
Vacuum Fusion Furnace. See **Vacuum Refining.**
Vacuum Melting. See **Vacuum Refining.**
Vacuum Refining. Removal of gaseous contaminants from metal by melting under vacuum. The heating is normally accomplished by electrical induction.
Valence. Chemical combining power of atoms or radicals (groups of atoms), measured by the number of hydrogen atoms (or those of another monovalent element) with which the given atom or radical can combine.
Vall Alta Furnace. Modification of the **Hähner Furnace** (see).
Valley Brown Ore. See **Limonite.**
Values. Quantity or proportion of metals, or other valuable constituents, present in raw ores and during smelting operations.
Valve Bronze. Copper-base alloy series, with 2–10% tin, up to 9% zinc, and up to 6% lead; used for casting valves. See **Bronze.**
Valve Copper. High-copper alloy for casting valves and fittings. A typical composition comprises 4% tin, 3% lead, 3% nickel, and 3% zinc.
Valve Effect. Phenomenon shown by certain metals, e.g., aluminum, silicon, tantalum, etc., immersed in electrolytic solutions; electric current is conducted in one direction only, due to oxide formation on the metal.
Valve Spring Wire. Steel wire, of 0.60–0.80% carbon content, quenched in oil, especially selected for resistance to bending and fatigue, and suitability for internal combustion engine valve springs.
Valve Steel. Heat-Resistant Steel (see) for the manufacture of internal combustion engine valves.
Vanadium (chemical symbol V). Element No. 23 of the periodic system; atomic weight 50.95. Gray-white, hard metal, unaffected by atmospheric influences or alkalies but soluble in most strong acids; melting point 3119°F; boiling point about 6150°F; specific gravity 5.87. Chemically, it shows valences varying from two to five. Vanadium cannot be electrodeposited; it can be isolated by calcium reduction of the trioxide. It is seldom used in this form, but ferro-vanadium, made either by alumino-thermic or electric furnace methods, finds its way into steel manufacture as a master alloy for vanadium steels. In steel, vanadium acts as a scavenger; when present in quantities in excess of scavenging needs, it increases toughness and tensile strength. It is frequently used with chromium and manganese for the latter purpose. The metal can be obtained from Vanadium Corp. of America.
Vanadium Cast Steel. Cast steel of high impact resistance [see **Cast Steel** (a)], containing about 0.2% vanadium.
Vanadium Steel. Steel, usually made by the basic open-hearth or electric process, containing small quantities of vanadium (from fractional percentages to 4%, about 0.20% being most common). The use of vanadium in steel is based on its scavenging action, as well as the hardening and toughening effect. Most commonly, vanadium is used with other alloying elements, such as chromium, manganese, and nickel, particularly in high-alloy steels.
Vandex. Finely ground selenium, used as a special vulcanizing agent in rubber (R. T. Vanderbilt Co.).

Vapor. Gas below the critical temperature (the limit at which liquefaction can be accomplished by pressure).

Vapor Density. Relative weight of a gas or vapor, as compared with an equal volume of hydrogen.

Vapor Galvanizing. Process of coating steel with zinc (see **Galvanizing**), using zinc metal vapors instead of the more common method of dipping into molten zinc.

Vaporization. Conversion of a substance from the liquid to the gaseous state.

Vapor Phase Degreasing. See **Degreasing.**

Vapor Pressure. Pressure exerted by the vapor of a substance, whether in the gaseous, liquid, or solid state. At any given temperature, every substance has a definite vapor pressure; when the vapor pressure of a liquid substance reaches that of the surrounding atmosphere, boiling results.

Varnished Wire. Steel wire with a final protective coating of a resinous varnish.

Vascoloy-Ramet. Series of tungsten and tantalum cemented carbides, with nickel or cobalt as binding matrix; used for cutting tools and drawing dies (Fansteel Metallurgical Corp.). See **Sintered Carbide (Tools).**

Vasco (Steel). Series of carbon and low-alloy chromium-vanadium tool steels; the alloy steels contain 0.15–0.25% vanadium, with chromium up to 1.5% (Vanadium Alloys Steel Co.).

V-2-A Steel. See **KA-2 Steel.**

Veining. See **Sub-Boundary Structure.**

Vein-Tin. See **Lode Tin.**

Vent. See **Vent Wire.**

Vent Hole. See **Vent Wire.**

Vent Wire. Metal wire used to puncture molding sand in a metal casting mold in order to form vent holes (or vents) which allow gas to escape when the metal is poured.

Verdigris. Basic copper carbonate, green in color, formed as a coating on copper by oxidation in moist atmosphere.

Veriloy. High-nickel (about 21%), high-chromium (about 11%) steels for chemical and heat-corrosion resistance (Driver-Harris Co.).

Vertical Mill. (a) **Rolling Mill** (see) in which the material is held in a vertical position during rolling. (b) Rolling mill in which the axes of the rolls are vertically placed.

Very-High-Carbon Steel. Steel with carbon in excess of 0.90%.

Vi. Chemical symbol for **Virginium** (see).

Vibratory Stress. Stress on a body, regularly and usually rapidly repeated, in opposing directions.

Vickers Hardness (Test). Standard method for measuring the hardness of metals, particularly those with extremely hard surfaces; the surface is subjected to a standard pressure for a standard length of time by means of a pyramid-shaped diamond. The diagonal of the resulting indentation is measured under a microscope and the Vickers Hardness value read from a conversion table.

Victor Metal. Copper-base alloy of **Nickel Silver** (see) type, with approximately 35% zinc and 15% nickel.

Vielle-Montagne Furnace. Furnace for the roasting of ores, mechanically rabbled.

Virginium (chemical symbol Vi). Element No. 87 of the periodic system; estimated atomic weight 224. Neither the element nor any of its compounds has been isolated, though there is magneto-optic evidence of its existence. Theoretically, it should be a member of the alkali metal family, chemically extremely reactive, like its congener, cesium.

Virgin Metal. Metal obtained directly from the ore, and not previously used, in contradistinction to metal remelted from scrap or previously fabricated material.

Viscosity. Resistance of fluid substances to flowing, quantitatively characteristic for each individual substance under given temperature and other external conditions.

Vitreous Antimony. Non-metallic mineral, antimony sulfide, Sb_2S_3.

Vitreous Cement. See **Amorphous Cement.**

Vitreous Coating. See **Vitreous Enamel.**
Vitreous Copper. Non-metallic mineral, chalcocite, approximating cuprous sulfide, Cu_2S.
Vitreous Enamel. Coating of adherent fused glass on a steel surface. The glass must have a temperature coefficient of expansion akin to that of the steel. See also **Ground Coat** and **Cover Coat.**
Vitreous Silver. Non-metallic mineral, argentite (also known as silver glance); approximates silver sulfide, Ag_2S.
Vitrification. Conversion to glass or glass-like material.
V-Mang (Steel). High-manganese steel (12–14%) with low percentages of molybdenum (American Manganese Steel Div.).
Void. (a) In powder metallurgy, undesirable empty space in a compact, resulting from faulty technique. (b) See also **Pipe.**
Volatile Flux. See **Gasflux.**
Volatility. Tendency of a substance to vaporize at the temperature under consideration. At ordinary temperatures, mercury is the only metal showing significant volatility. At somewhat higher temperatures, the metals arsenic, cesium, rubidium, potassium, sodium, zinc, and cadmium also become appreciably volatile.
Volatilization. See **Vaporization.**
Volta Furnace. Three-phase, direct-arc electric furnace generally analogous to the **Heroult Furnace** (see).
Voltage. Difference of electrical potential, measured in volts, causing a flow of current.
Voltaic Cell. See **Cell.**
Volta's List. See **Electromotive Series.**
Volumetric Analysis. Quantitative chemical **Analysis** (see) in which standardized solutions are used for determining the desired constituent.
Vom Baur Furnace. Electric direct **Arc Furnace** (see), with three electrodes in a straight line, using either two- or three-phase current (generally the former, in which event the central electrode acts as a common return).
V-Segregation. Differences in composition, lying deep in ingots, below the pipe cavity.
Vug. See **Pipe.**
Vulcan (Steel). Series of carbon and low-alloy tool and structural steels, manufactured by Vulcan Crucible Steel Co.
Vul-Mo (Tool) Steel. High-carbon alloy tool steel (Vulcan Crucible Steel Co.), with about 8% molybdenum, 4% chromium, 2% tungsten, and 1% vanadium.

W

W. Chemical symbol for **Tungsten** (see).
Wad. (a) See **Bog Manganese.** (b) See **Graphite.**
Waffle Ingot. Small flat ingot of **Aluminum** (see) metal.
Walking. Inspecting (term used in connection with the inspection of the manufacture of certain types of steel products).
Walking Beam Furnace. Continuous Furnace (see) in which the material moves through the furnace by means of beams in longitudinal slots in the bottom, attached to cams raising the beams and moving them forward.
Walloon Process. Early indirect process for producing wrought iron, in which pig iron was held in the fire with tongs, so as to permit the end to melt gradually and to trickle down through the air blast, burning out the carbon and silicon content; the final metal collected as a pasty mass at the hearth bottom.
Wando Steel. Non-Deforming Steel (see), with about 1.00% manganese, 0.50% tungsten, 0.50% chromium and 1.00% carbon (Cyclops Steel Co.).
War Bronze. Zinc-base alloy (not a true bronze) with about 5% copper, 2% aluminum, and, optionally, fractional percentages of lead and tin.
Warman Steel. Series of low-alloy and high-alloy casting steels, manufactured by Warman Steel Casting Co.
Warne's Metal. Jewelry alloy of about 37% tin, 26% each of bismuth and nickel, and 11% cobalt.
Warrington Wire Gage. See **Stubs' Iron Wire Gage.**
Wash. See **Mold Wash.**

Washburn & Moen Gage. See **Metal Gage.**
Washed Metal. Semi-purified pig iron, with substantial portions of the manganese, phosphorus, and sulfur removed.
Wash Heat. See **Washing** (a).
Washing. (a) Final step in the preparation of a basic open-hearth furnace, in which basic cinder is melted in the furnace, and the molten cinder then packed up against the sides of the hearth to make it solid. (b) See **Mold Wash.** (c) See **Wash Metal Process.** (d) Coating metal with a thin layer of another metal.
Wash Metal Process. Treatment of high-phosphorus pig iron, in a Bessemer converter, with a highly siliceous slag, followed by pouring through a very basic slag in an open-hearth furnace. A slag containing about 12% of "citrate-soluble" phosphorus pentoxide results, which is usable as a fertilizer.
Washoe Process. See **Pan-Amalgamation Process.**
Wash Pot. Pot for holding tin in the manufacture of **Charcoal Tin Plates** (see).
Waste. Broken or otherwise-defective castings, available for remelting and recasting.
Waste Plate. Defective **Tin Plate** (see) not suitable for use, either as is or in retinned form.
Waster. (a) Second quality of coke tin plate, with minor imperfections, though still suitable for commercial use. (b) Defective or rejected casting.
Waster Waste. See **Waste Plate.**
Waste Waste. See **Waste Plate.**
Water Annealing. Cooling steel of low carbon content by immersion in water after air cooling till blackness in color has been reached.
Water-Back Hearth. Hearth (see **Ore Hearth**) with water-cooled walls and sides.
Water-Back Ore Hearth. Ore Hearth (see), with the back cooled by a stream or spray of water, to prevent burning out.
Water Block. See **Water Box.**
Water Box. Hollow iron box or case, cooled by a constant stream of water, for cooling any equipment or portion of a furnace.
Water Core. Hollow, water-cooled core, used in a mold to enable rapid cooling of the inner portions of the casting.
Water Crack. See **Quenching Crack.**
Water-Hardening. Hardening steels by **Quenching** (see) in water, after heating.
Water-Hardening Steel. Steel adaptable to hardening by heat treatment and quenching in water.
Water Jacket. See **Water Box.**
Water Pyrometer. See **Pyrometer.**
Water-Quenching. In the heat treatment of metals, immersion of the hot metal in water, for rapid cooling.
Water Tuyere. Tuyere (see) protected by means of a **Water Box** (see).
Wattage. Electrical power, measured in watts.
Watt Loss. See **Core Loss.**
Wazau (Cup) Test. Form of **Cupping Test** (see).
Wear-Resistant Steel. High-alloy tool steel, usually of the high-carbon-content type, used for the manufacture of dies and products requiring resistance to abrasion and wear.
Weather Roasting. Oxidation of sulfide ores by exposure to the air, at ordinary temperatures, for periods of many years.
Wedge Furnace. Modern type of vertical, cylindrical roasting **MacDougall Furnace** (see), mechanically rabbled.
Wehralloy. Series of high-alloy, nickel-chromium steels, including **Stainless Steel** (see), and of nickel-chromium-iron alloys, of diverse composition, manufactured by Wehr Steel Co..
Weighting. Placing weights, or other means of pressure, on the upper part of a mold, to prevent separation under pressure of the molten metal during casting.
Weld. Zone which has been subjected to **Welding** (see).
Weld Cladding. Process in which the **Cladding** (see) metal is built up on the base metal by welding.
Weld Decay. Intercrystalline cracking and lowering of corrosion resistance, in austenitic stainless steel, at the side

of a weld, due to precipitation of chromium carbide at the grain boundaries in the zone which reached the 930–1300°F range during welding. Weld decay can be prevented by use of a **Stabilizer** (see) in the stainless steel.

Welded Pipe. Pipe made from **Skelp** (see), by shaping into tubular form and then welding the edges.

Welder. Any equipment for **Welding** (see), especially automatic electric arc or resistance welding.

Welding. Union of metals, usually at the edges, effected through the application of heat, with or without pressure. In fusion welding, the common edges are melted and contact is made by molten metal obtained from the edges themselves or from added metal, with welding accomplished either by electric arc or by gas torch technique. In forge welding, the metal units are heated until plastic, and then hammered or pressed till united. In resistance welding, electric resistance creates the necessary heat which, together with pressure, effects union.

Welding Bell. Die used to bend **Skelp** (see) into pipe shape, in the manufacture of butt-welded pipe.

Welding Flux. See **Flux.**

Welding Heat. Temperature at which union of metals is effected in welding operation.

Welding Powder. See **Welding Flux.**

Welding Rod. Metal or alloy rod or wire, usually of steel, used in gas or electric **Welding** (see), for joining metals, building up surfaces, etc.

Welding Scale. Scale consisting of a mixture of metal oxide and flux; results from welding operations.

Welding Steel. See **Pipe Steel.**

Welding Stress. Residual Stress (see) resulting from welding operations.

Welding Test. Test on steel to determine suitability for welding; usually, a test-weld on the metal.

Welding Transformer. Alternating current transformer, used as a source of power in resistance **Welding** (see).

Welding Wire. Wire, most commonly of steel, used in torch or electric arc welding. See **Welding Rod.**

Weld Iron. See **Wrought Iron.**

Weld Metal. Metal, either base or filler or both, melted during **Welding** (see).

Weldrawn. Tubing of stainless steel, etc., made by Superior Tube Co.

Well. (a) See **Uptake.** (b) See **Crucible** (b).

Welshman. Small steel ring used in removing a steel bar stuck in a **Skull** (see).

Welsh Process. See **English Copper Process.**

Wenstrom Mill. Special type of **Universal** rolling **Mill** (see) used in rolling metal bars.

Wessell's Silver. Copper-base alloy series of the **Nickel Silver** (see) type, with 20–32% nickel and 12–17% zinc.

West (Steel). Series of simple and low-alloy machinery and tool steels of diverse composition, manufactured by West Steel Casting Co.

Wet Assay. Determination of metal values in an ore by chemical methods involving solution of the ore. See also **Assay.**

Wet Cleaning. (a) Removing nongaseous materials from blast furnace or similar gas by washing with water. (b) Removing oil from tin, terne, or similar coated steel, after leaving the coating bath, by washing with alkaline solutions.

Wet Drawing. Procedure of drawing wire, in which cleaned wire, usually after having been partially drawn down by dry drawing, is immersed in a solution of a copper or tin compound, to coat the surface with these metals, and then drawn wet while immersed in a liquid lubricant bath.

Wet Dross. Hard Zinc (see) removed from galvanizing pots. Part of the excess zinc can be shaken out, leaving a granular dry dross of about 5% iron content.

Wet Galvanizing. See **Electrogalvanizing.**

Wetherill Furnace. See **Wetherill Process.**

Wetherill Process. Recovering zinc from ores by heating the oxidized ore with coal on an open iron grate, with the resulting zinc vapor immediately reoxidized to zinc oxide, which is caught in dust chambers.

Wethey Furnace. Horizontal furnace of multiple-hearth construction, used for **Roasting** (see) sulfide ores.

Wet Metallurgy. See **Hydrometallurgy.**

Wet Method. See **Hydrometallurgy.**

Wet Process. See **Hydrometallurgy.**

Wet Puddling. Common **Puddling** (see) process, carried out in a furnace lined with iron oxide or ore rich in iron.

Wheel Blank. Blank [see **Blank** (a)] for forming forged wheel.

Wheel Steel. Steel used in the manufacture of railroad wheels; usually contains about 0.70% carbon and 0.70% manganese.

Wheel Swarf. Broken pieces of grindstone, used for covering iron bars in the manufacture of steel by the **Cementation Process** (see).

Whelco (Steel). Series of high-alloy (particularly nickel-chromium) steels, including **Stainless Steel** (see), manufactured by Wheelock, Lovejoy & Co.

Whirl. Device for straightening wire, in which a series of staggered dies, in succession, bend the wire slightly in opposite directions.

White Annealing. Heating cold-rolled sheets to about 1100°F, to restore ductility.

White Antimony. Non-metallic mineral, valentinite, chemically approximating antimonious oxide or antimony trioxide, Sb_2O_3.

White Arsenic. Arsenic oxide, As_2O_3; non-metal.

White Bloom. Formation of white oxidation or corrosion products on zinc-base metals; presumably due to temperature-humidity conditions.

White Brass. Either of two series of alloys: (1) Copper-base alloy, with approximately one-third zinc. (2) Tin-base alloy, with approximately 32% zinc and, optionally, about 3% copper.

White Bronze. Series of copper-base alloys, with approximately 40% zinc and 4% nickel.

White Cast Iron. See **Cast Iron** and **Pig Iron.**

White Cobalt. (a) See **Gray Cobalt.** (b) Non-metallic mineral, cobaltite or cobalt glance, a cobalt-arsenic sulfide, chemically approximating CoAsS.

White Copper. Series of corrosion-resistant, copper-base alloys of the **Nickel Silver** (see) type.

White-Finished Hot-Rolled Pickled and Annealed Sheet. Steel sheet, hot-rolled, annealed, and finally pickled to remove annealing oxide. Alternatively, pickling may be followed by box annealing.

White-Finished Hot-Rolled Pickled and Annealed Strip. See White-Finished Hot-Rolled Pickled and Annealed Sheet.

White-Finished Pickled Black Plate. Black Plate (see) subjected to pickling and, optionally, to annealing in a reducing atmosphere.

White-Finished Sheet. Sheet steel, hot-rolled, pickled, annealed, and then given a final pickling or box annealing to remove the oxide coating resulting from the annealing step.

White-Finished Strip. See **White-Finished Sheet.**

White Flux. Mixtures of sodium nitrate and sodium nitrite, occasionally used as welding fluxes.

White Furnace. See **Howell Furnace.**

White Gold. Gold-base alloy, with sufficient alloying constituents to give a white color. Palladium to the extent of 10% in the alloy gives an excellent white gold; a cheaper alloy contains about 10% nickel and 5% zinc. As the gold content is lowered, other alloying constituents may be used; thus, in 14 and 18 carat alloys, nickel, copper, zinc, and silver are among the elements to be found.

White Gold Solder. White gold-base alloy of relatively low melting point, capable of soldering gold alloys; usually high in silver, and zinc or nickel; other metals, often present, are copper and cadmium.

White Heart (Malleable) Iron. Malleable Iron (see), with a coarse grain due to a high degree of decarburization, resulting from long annealing at high temperatures.

White Heat. Temperature above about 2200°F.

White Hot. See **White Heat.**

White-Howell Furnace. See **Howell Furnace.**
White Iron. See **Cast Iron** and **Pig Iron.**
White Lead. Basic carbonate of lead, a non-metal, corresponding essentially to the formula $2PbCO_3.Pb(OH)_2$.
White Metal. (a) General term covering alloys of tin-base, lead-base, or antimony-base type, such as bearing metals, type metals, babbitt, etc. (b) Copper matte, of about 77% copper content, obtained from the smelting of sulfide copper ores. (c) Colloquial name for **Silver** (see).
White Nickel. (a) See **Rammelsbergite.** (b) See **Chloantite.**
White Nickel Alloy. Copper-nickel alloy (about 32% nickel typical), with about 3% aluminum.
White Nickel Brass. See **Nickel Silver.**
White Pickling. Cleaning the surface of **Black Plate** (see), after annealing or normalizing and prior to tinning, by immersion in dilute sulfuric acid.
White Pig Iron. See **Cast Iron** and **Pig Iron.**
White Rust. Thick white film, chemically a mixture of zinc carbonate and hydroxide, forming on zinc and galvanized surfaces as a result of exposure to moisture.
White Slag. Slag, white in color, used in basic electric steel processes, with the carbon content converted to calcium carbide.
White-Souther Endurance Machine. Machine for testing the endurance characteristics of metal, in which the specimen is held on one end while rotated, with weights at the other end.
White Tellurium. See **Sylvanite.**
White Tin. Metallic tin, white as compared with **Black Tin** (see).
White X-rays. X-rays of many wave lengths, as opposed to monochromatic rays.
Whitneyite. Native alloy of copper and arsenic, chemically approximating Cu_9As.
Whitwell Stove. Hot Blast Stove (see) of fire brick, operating on a regenerative basis.
Whole Sheet Tester. See **Strip Tester.**
Wicket. Cap for closing the opening in the tuyere stock through which a rod can be inserted for cleaning out the **Tuyere** (see).
Wide Cold-Rolled Strip. See **Cold-Rolled Strip.**
Wide Hot-Rolled Strip. See **Hot-Rolled Strip.**
Wide Strip. See **Cold-Rolled Strip** and **Hot-Rolled Strip.**
Widia. Composite, consisting of **Sintered** tungsten **Carbide** (see), cemented together by about 13% metallic cobalt. Because of extreme hardness and strength, the material is used extensively in tool bits and other machine tools.
Widmanstatten Structure. Metallographic structure in steel, resulting from overheating, in which a new phase forms along definite planes in a solid solution during cooling.
Wilcoloy. Cemented tungsten carbide, used for relay contact points (H. A. Wilson Co.).
Wilco Thermometal. Series of **Bimetals** (see), manufactured by H. A. Wilson Co.
Wild Gas. Faulty gas from a blast furnace, which fails to burn properly.
Wild Heat. Heat (see) of steel resulting in **Wild Steel** (see).
Wild Lead. See **Mock Lead.**
Wild Steel. Steel poured into molds without **Killing** [see **Killing** (a)], so that the reaction between the ferrous oxide and carbon during solidification causes evolution of gaseous carbon monoxide and blowholes throughout the ingot.
Wind. Air blast in a **Blast Furnace** (see).
Wind Furnace. Furnace relying on natural draft, without mechanical air blast.
Window. Space in the housing of roll mills, holding the roll bearings.
Winning. Extracting metal from ores.
Winnowing Gold. Separating gold from gangue, by **Air Separation** (see).
Wipe. Device for removing excess zinc from galvanized steel wire, consisting of two asbestos balls pressing against the wire as it comes from the **Galvanizing** (see) pot.

Wiper. See **Brush.**
Wiping Solder. Soft **Solder** (see) consisting of approximately 60–70% lead and 30–40% tin.
Wire Bar. Electrolytically refined copper, in bar form, ready for drawing into wire.
Wirebar Copper. See **Electrolytic Copper.**
Wire Copper. See **Wire Bar.**
Wire Drawing. Reducing the diameter of metal by drawing it through progressively smaller dies or openings in a hardened steel plate.
Wire-Drawing Bench. See **Wire-Drawing Frame.**
Wire-Drawing Block. See **Wire-Drawing Frame.**
Wire-Drawing Frame. Equipment for drawing wire; consists of a bench supporting the equipment, a block for drawing the wire through the die held in a die holder, and corollary units for rotating the blocks.
Wire Fence. Galvanized steel wire in woven or barbed form, used for the manufacture of fencing.
Wire Gage. See **Gage.**
Wire Rod. Metal, in the form of rod or thin bars, from which wire is drawn. Such rod is made in many sizes and shapes; round is most common.
Wire Rope. See **Rope Wire.**
Wire Shape. Any shape, other than the standard circular cross-section, into which wire is drawn.
Wire Silver. Native silver, occurring in thread or wire form.
Wire Spring. See **Spring Wire.**
Wissco (Steel). Series of high-alloy chromium and nickel-chromium steels, including **Stainless Steel** (see), manufactured by Wickwire Spencer Steel Co.
Wobbler. Extension of neck of **Roll** (see) with notch, to which a coupling box is fitted, connected to the source of power.
Wohwill Process. Process for refining gold, commonly used on bullion of over 85% gold content, in which the impure metal acts as anode in an electrolytic cell, with pure gold sheet as cathode, and a solution of acid gold chloride as the electrolytic bath.
Wolfram. (a) See **Tungsten.** (b) Non-metallic mineral, wolframite, an iron-manganese tungstate, chemically approximating $(Fe, Mn)WO_4$.
Wolframinium. Aluminum-base alloy, with about 1.5% antimony, 1% tin, and fractional percentages of copper and tungsten.
Wolframite. See **Wolfram** (b).
Wollaston Process. Process of drawing extremely fine wires; a partially drawn wire is closely fitted into a tube of another, ductile metal, followed by rolling and drawing the composite down to the desired size; the jacketing metal is dissolved off with acid by the ultimate user. Gold can be drawn down to as small a diameter as 0.00001 inch.
Wollaston Wire. See **Wollaston Process.**
Wolverine—. Series of copper-base alloys, available in bar and tube form, of the **Brass** (see) and **Bronze** (see) types (Wolverine Tube Co.).
Wompco. Cast iron of relatively high strength, with about 2.90% carbon, 2% silicon, and 0.75% manganese (Worthington Pump & Machinery Corp.).
Wood Copper. Non-metallic, fibrous mineral, olivenite, basic copper arsenate, chemically approximating $Cu_4As_2O_9.H_2O$.
Wood Iron. Fibrous variety of **Siderite** (see).
Wood's Alloy. See **Fusible Alloys.**
Wood's Fusible Metal. See **Fusible Alloys.**
Wood's Metal. See **Fusible Alloys.**
Wood Tin. Fibrous variety of **Cassiterite** (see).
Woody Fracture. See **Fracture.**
Wootz. Steel made in India by the **Cementation Process** (see) by reduction of ore in a hearth resembling a **Catalan Forge** (see).
Work-Hardening. See **Strain Hardening.**
Working. (a) Mechanical treatment of metal to change dimensions and other physical characteristics. (b) Black droplets rolling on the surface of a molten lead-base metal sample before completion of **Softening** (see).

Working Period. In the manufacture of basic open-hearth steel, the period between the **Lime Boil** (see) and the tapping of the metal, during which the slag and carbon are adjusted and the temperature raised to the optimum for tapping.

Working Roll. See **Four-High Mill.**

Working Strength. See **Safe Load.**

Working The Heat. See **Heat Working.**

Work Lead. Lead before subjecting to desilverization.

Work Roll. See **Working Roll.**

Works Annealing. See **Annealing.**

Work Stove. Grooved plate, in a lead-reducing furnace or hearth, for carrying molten lead from the furnace to an outside collecting vessel.

Worms. See **Lüder's Lines.**

Worthite. Series of corrosion-resistant, high-alloy steels (chromium about 20%, and nickel about 24%), with low percentages of molybdenum, copper, and silicon (Worthington Pump & Machinery Co.).

Wortle (Steel). Series of low-alloy tool and magnet steels, most commonly, with several per cent of tungsten.

Wrought Iron. Commercially pure iron, obtained in pasty condition in masses composed of adherent particles intermingled with slag. As a consequence of mechanical operations on these masses, finished wrought iron always shows a fibrous structure, with some of the slag remaining as strings in the iron; it is ductile and malleable, but for most purposes it has been supplanted by the simple steels, which are not only stronger and more uniform, but less costly to manufacture. See also **Puddling.**

Wrought Pipe. Forged iron or steel pipe, as distinguished from cast iron pipe.

Wrought Steel. Object made by hot-working or hot-rolling of steel.

W.W.G. See **Stubs' Iron Wire Gage.**

X

X-Ite. High-nickel (about 38%), high-chromium (about 18%), iron-base alloy series, resistant to heat and corrosion (General Alloys Co.).

X-ray Analysis. Method of determining internal atomic structure of crystals by subjecting the material to X-rays.

X-rays. Extremely short waves propagating with the velocity of light but lying beyond the visible spectrum. The wavelength ranges from about 0.01 to 50 Ångstrom units.

X-ray Test. Examination of a material by X-rays, to detect sub-surface flaws.

Y

Y. Chemical symbol for **Yttrium** (see).

Y Alloy. Aluminum alloy containing low percentages of copper (about 4%), nickel (about 2%), and magnesium (about 1.5%); used for aircraft and piston castings, the latter usually with heat treatment, since it does not appreciably affect dimension stability.

Yankee. Curved lifting tool used in operating on molds in metal founding.

Yb. Chemical symbol for **Ytterbium** (see).

Yellow Antimony. Allotropic modification of **Antimony** (see), yellow in color, obtained by oxidizing antimony hydride at very low temperatures.

Yellow Arsenic. (a) Allotropic modification of **Arsenic** (see), yellow in color, obtained by condensing arsenic vapor under regulated conditions. (b) Orpiment, non-metallic mineral, chemically approximating arsenic trisulfide, As_2S_3.

Yellow Brass. See **Brass.**

Yellow Casting Brass. Yellow **Brass** (see), of good casting quality, commonly with up to 5% lead added to improve machinability.

Yellow Copper. See **Copper Pyrites.**

Yellow Gold. Gold-base jewelry alloys, with about 25% silver and 25% copper.

Yellow Metal. (a) See **Muntz Metal.** (b) Colloquial name for **Gold** (see).

Yield. See **Recovery** (b).

Yield Point. Point on any stress-strain diagram at which there is marked increase in strain, without material increase in stress; characteristic of soft irons and steels, i.e., annealed ferrite.

Yield Point Load. Load equivalent to the **Yield Point** (see), determined during a tensile test.

Yield Strength. Stress corresponding to a definite strain during tensile testing. The strain may be 0.1, 0.2, or 0.5 per cent of the elongation.

Yoloy. Low-alloy steel (Youngstown Sheet & Tube Co.), containing about 2% nickel, 1% copper and up to 0.20% carbon; shows considerable resistance to atmospheric corrosion and relatively high strength.

Young Blow. Stopping the "blow" in a Bessemer converter (see **Bessemer Steel**) as soon as the flame begins to drop off.

Young's Modulus. See **Modulus of Elasticity.**

Young Turndown. See **Young Blow.**

Ytterbium (chemical symbol Yb). Element No. 70 of the periodic system; atomic weight 173.04. Rare earth metal, isolated in relatively impure form by electrolysis of the anhydrous trichloride in a bath of fused alkali chlorides; melting point about 3270°F; boiling point unknown. Chemically, it is trivalent. Like all rare earth metals, it cannot be electrodeposited.

Yttrium (chemical symbol Y). Element No. 39 of the periodic system; atomic weight 88.92. One of the most common rare earth metals, steel gray in color, melting point 2714° F, boiling point about 4530°F, specific gravity 5.51. It reacts with hot water, releasing hydrogen, and oxidizes slowly in air. Chemically, it is always trivalent. Like all rare earth metals, it cannot be electrodeposited. The metal can be isolated in the form of small spangles by electrolysis of the anhydrous trichloride in a fused bath of sodium chloride. The metal and its alloys are not used commercially.

Z

Z. A. M. (Metal). Zinc-base alloy, used as anode in the electrodeposition of zinc; low percentages of aluminum and mercury are present (Hanson-Van Winkle-Munning Co.).

Zamak. Series of zinc-base alloys, with up to 4% aluminum, about 1% copper, and fractional percentages of magnesium optional (New Jersey Zinc Co.).

Z Beam. Metal beam of Z cross-sectional shape, rolled to standard dimensions.

Z-D Process. Treatment of aluminum-base alloys, for protection from corrosion, with sodium silicate, followed by heating.

Zellweger Furnace. Reverberatory furnace with a long hearth.

Zeppelin Alloy. Series of foreign high-strength aluminum alloys, with low and fractional optional percentages of copper, silicon, manganese, and zinc.

Ziervogel Process. Extracting silver from sulfide ores by oxidizing roasting, forming soluble silver sulfate, which is dissolved in hot water and then cemented out by copper.

Zilloy. Zinc-base alloy, containing 1% copper and fractional percentages of magnesium (New Jersey Zinc Co.).

Zimmerman-Daniels Process. See **Z-D Process.**

Zinc (chemical symbol Zn). Element No. 30 of the periodic system; atomic weight 65.38. Blue-white metal; when pure, malleable and ductile even at ordinary temperatures; melting point 787°F; boiling point 1665°F; specific gravity 7.14. Chemically, it is always divalent; can readily be electrodeposited; commercially, the process, also known as electrogalvanizing, is accomplished by use of baths containing the sulfate or the double alkali cyanide. Zinc is obtained from its ores, most commonly the sulfide, sphalerite, by roasting and then reduction by carbon in retorts, the zinc distilling over, or the roasted ores can be leached and the metal recovered electrolytically. Metal of very high purity can be obtained by redistillation. It is extensively used as a coating for steel (see **Galvanizing**), and sheet zinc finds many outlets, such as dry batteries, roofing, etc. Its most important alloy is **Brass** (see), and it also enters into many commercial alloys with lead and the other white metals (see **White Metals**). Zinc-base alloys are of great importance in die castings.

Zinc Ashes. Oxidized zinc floating on molten zinc in galvanizing baths.

Zinc Blende. See **Mock Lead.**

Zinc Box. Box or compartment holding zinc dust or shavings, for contact with gold or silver cyanide solutions in the cyanide extraction of these metals, to precipitate the precious metals.

Zinc Coating. See **Galvanizing.**

Zinc-Copper Couple. Zinc superficially coated with copper by immersion in a solution of copper sulfate; frequently used in organic chemistry as a reducing agent, the copper acting to form an electric couple with the zinc, which reacts (usually with acid) to yield hydrogen and itself go into solution.

Zinc Dust. Finely divided zinc in powder form, used principally for paint formulation.

Zinc Gage. See **Metal Gage.**

Zinc Fume. Gray powder collecting at the end of zinc furnace condensers; consists of fine zinc powder and a small amount of zinc oxide.

Zinc Gray. Paint made with **Zinc Dust** (see).

Zincing. See **Galvanizing.**

Zincite. Non-metallic mineral, chemically approximating ZnO.

Zincote. Process of rust-proofing steel and forming a paint-bonding surface, in which the metal is sprayed with a solution of zinc nitrate, sodium bisulfate, and sodium acetate.

Zincoting. See **Zincote.**

Zinc Plating. Coating metal objects with zinc or high-zinc alloy; the object is made cathode (negative electrode) in an electrolytic bath containing a decomposable zinc salt. Acid zinc baths usually involve the use of zinc chloride or sulfate; alkaline baths are normally of the double cyanide type. Zinc metal or zinc-base alloys are normally used as anode (positive electrode).

Zinc Powder. See **Zinc Dust.**

Zinc Scum. Alloy of zinc and silver removed from lead in the **Parkes Process** (see) of desilverization.

Zinc Skimmings. Mixture of zinc metal and oxide, rising to the surface of the molten zinc bath in the **Hot Galvanizing** (see) process.

Zinkite. See **Zincite.**

Zin-O-Lyte. Process for the electrodeposition of bright zinc, controlled and licensed by E. I. du Pont de Nemours & Co. Special Zin-O-Lyte anodes and Zin-O-Lyte zinc brightener are part of the process.

Zircofrax. Series of zircon-base refractories of exceptionally high heat and chemical resistance (Titanium Alloy Mfg. Co.).

Zircon. Zirconium silicate, $ZrSiO_4$, used as a refractory because of its exceptional resistance to heat.

Zirconia. Zirconium dioxide, a refractory material of exceptional heat resistance and a very low coefficient of expansion.

Zirconium (chemical symbol Zr). Element No. 40 of the periodic system, atomic weight 91.22. Fairly hard, steel-gray metal, obtained in spangled form by electrolysis of the tetrachloride in a fused bath of mixed alkali chlorides, or as a brownish-black powder by magnesium reduction of the oxide; melting point believed to be about 3375°F; boiling point, even more uncertain, probably over 5250°F; specific gravity 6.4. Chemically, it is normally tetravalent, but may be also divalent and trivalent. It cannot be electrodeposited. Elementary zirconium has been used, to some extent, in flash bulbs; in finely divided form, it has a very low ignition temperature. In alloy form, it is used as ferro-zirconium and as nickel-zirconium; the element acts as a scavenger, being particularly effective on sulfur, oxygen, and nitrogen. Zirconium steels are also known, but are not of great importance. Zirconium-copper is used to scavenge copper-base alloys and increase their hardness. The metal is obtainable from Foote Mineral Co.; alloys, from Electro Metallurgical Co.

Zirconium-Copper. Master alloy, containing 2–35% zirconium, for addition to copper-base alloys as a scavenger and to confer increased strength on the metal.

Zirconium Ferro-Alloys. See **Ferro-Zirconium.**
Zirconium-Ferrosilicon. See **Ferro-Zirconium.**
Ziv's Steel. Series of carbon and alloy tool steels of varying composition, manufactured by Ziv Steel & Wire Co.
Z Metal. Malleable cast iron, with about 1.7% tempered carbon and 0.7% combined carbon (Acme Steel & Malleable Iron Works).
Zn. Chemical symbol for **Zinc** (see).
Zonal Structure. Metallic structure with composition gradients due to differential solidification of the components of the metal.
Zone. Distinctive portion of a processing apparatus, usually designed to accomplish a specific purpose, such as fusing, reducing, smelting, etc.
Zorgite. Native copper-lead selenide of variable composition.
Zorite. Electric-resistance iron-base heating alloy containing approximately 37% nickel, 16% chromium, 2% manganese, 1% silicon, and 0.5% carbon (Michiana Products Corp.).
Zr. Chemical symbol for **Zirconium** (see).
Zurco (Steel). Series of simple carbon and high-chromium (up to 30%) steels, manufactured by Zurbach Steel Co.
Zyglo Process. Modification of the Magnaflux process. See **Magnaflux Testing.**

www.ingramcontent.com/pod-product-compliance
Ingram Content Group UK Ltd.
Pitfield, Milton Keynes, MK11 3LW, UK
UKHW020126290926
15471UKWH00003B/305